Introduction to
Horticulture
Seventh Edition

N Kumar
BSc (Hort), MSc (Ag), PhD
Dean (Horticulture)
Tamil Nadu Agricultural University, Coimbatore

Oxford & IBH Publishing Co. Pvt. Ltd.
New Delhi
(*A Unit of* CBS Publishers & Distributors Pvt Ltd)

CBS Publishers & Distributors Pvt Ltd

New Delhi • Bengaluru • Chennai • Kochi • Kolkata • Lucknow • Mumbai
Hyderabad • Jharkhand • Nagpur • Patna • Pune • Uttarakhand

Introduction to
Horticulture
Seventh Edition

ISBN-13: 978-81-204-1754-0
ISBN-10: 81-204-1754-2

© 2010, Dr N Kumar
CBS Reprint: 2016, 2019 , *2022*

OXFORD & IBH
New Delhi
(*A Unit of* CBS Publishers & Distributors Pvt Ltd)

Published by Satish Kumar Jain and produced by Varun Jain for
CBS Publishers & Distributors Pvt Ltd
4819/XI Prahlad Street, 24 Ansari Road, Daryaganj, New Delhi 110 002, India.
Ph: 23289259, 23266861, 23266867   Fax: 011-23243014   Website: www.cbspd.com
e-mail: delhi@cbspd.com; cbspubs@airtelmail.in.
Corporate Office: 204 FIE, Industrial Area, Patparganj, Delhi 110 092
Ph: 4934 4934   Fax: 4934 4935   e-mail:publishing@cbspd.com;
                                                                              publicity@cbspd.com

Branches

- **Bengaluru:** Seema House 2975, 17th Cross, K.R. Road, Banasankari 2nd Stage, Bengaluru 560 070, Karnataka
  Ph: +91-80-26771678/79   Fax: +91-80-26771680   e-mail: bangalore@cbspd.com
- **Chennai:** 7, Subbaraya Street, Shenoy Nagar, Chennai 600 030, Tamil Nadu
  Ph: +91-44-26260666, 26208620   Fax: +91-44-42032115   e-mail: chennai@cbspd.com
- **Kochi:** 42/1325, 1326, Power House Road, Opp KSEB Power House, Ernakulam 682 018, Kochi, Kerala
  Ph: +91-484-4059061-65   Fax: +91-484-4059065   e-mail: kochi@cbspd.com
- **Kolkata:** No. 6/B, Ground Floor, Rameswar Shaw Road, Kolkata-700014 (West Bengal), India
  Ph: +91-33-2289-1126, 2289-1127, 2289-1128   e-mail: kolkata@cbspd.com
- **Mumbai:** 83-C, Dr E Moses Road, Worli, Mumbai-400018, Maharashtra
  Ph: +91-22-24902340/41   Fax: +91-22-24902342   e-mail: mumbai@cbspd.com

Representatives

- Bhopal   0-8319310552
- Jharkhand   0-9811541605
- Pune   0-9623451994
- Dhaka (Bangladesh)   01912-003485
- Bhubaneswar   0-9911037372
- Nagpur   0-9021734563
- Uttarakhand   0-9716462459
- Hyderabad   0-9885175004
- Patna   0-9334159340

Printed at Mudrak, Noida, UP, India

Foreword

India is endowed with a diverse climate ranging from tropical to temperate that it is possible to grow an array of horticultural crops in one or other parts of our country. These crops assume great significance as they provide nutritional security, increased employment opportunities, supply raw materials for the expanding agro-processing industries and enhance foreign exchange earnings through export. Further, the horticultural crops particularly plantation crops, trees and ornamental plants provide environmental security for our immediate living surroundings.

Horticultural crop production has been expanding worldwide. It has doubled during the past two and a half decades both due to increase in area particularly in developing countries, and improved productivity in developed countries. Asian countries namely China and India have emerged as the global leaders in fruit and vegetable production with a share of 34 per cent and 10 per cent, respectively, yet their productivity is far below as compared to many developed countries. However, India occupies the first position in production of many fruit crops like mango, banana, pomegranate, sapota and aonla and the second position in production of vegetables like brinjal, cabbage, cauliflower, peas, onion and tomato etc. On the other hand, India is the largest producer, consumer and exporter of spices and spice products in the world and produces more than 50 spices. In plantation crops also, coconut, tea, coffee and cashew assume greater importance. In Floriculture, India has been making rapid strides in the last two decades.

Realising the importance of Horticulture, Indian Government encourages a strong institutional support for R & D programmes in Horticulture with research infrastructure spread over in 10 Central Institutes with 24 regional stations, 12 crop specific National Research Centres, 9 Multi-disciplinary Institutes, 13 All India Coordinated Research Projects with 215 centres, 36 State Agricultural Universities and 15 central general/deemed to be universities with Horticulture discipline dealing with research on different horticultural crops. Besides research, some of these institutes particularly State Agricultural Universities have to fulfill its another mandatory function *viz.,* generating qualified human resources in Horticulture as they offer B.Sc. (Ag.)/(Hort.) degree programmes. These students need a basic text book on Horticulture for their ready reference. I am happy to note that *Dr. N. Kumar, Prof. (Hort.),* the author of this text book entitled *"Introduction to Horticulture"* took up the initiative of bringing out this for the benefit of the students as early as 1986 and he has been periodically revising it; duly incorporating the latest findings as well in this book.

I am sure that this text book will be quite useful to the students of Agriculture/Horticulture, the extension functionaries in the development departments and the enthusiastic horticulturist besides the biological students in conventional universities opting for Horticulture as a special course.

I congratulate *Prof. Dr. N. Kumar,* Tamil Nadu Agricultural University, Coimbatore for his concerted efforts in bringing out this revised seventh edition.

H.P. SINGH
Deputy Director of Hort.,
I.C.A.R,
New Delhi-12

Preface to the Seventh Edition

Horticultural crop production has doubled during the past two and a half decades both due to increase in area and productivity. Besides, horticulture has emerged as a viable diversification option in the Agricultural sector. Realizing its importance and scope, the Government of India including many State Governments have been allocating sizeable funds for development in Horticulture. To meet the growing demand for human resources in Horticulture and allied fields, many State Agricultural Universities offer now specialized undergraduate programme in Horticulture. Similarly, many conventional universities have Botany programme with Horticulture as their optional field of Applied Botany. These students need a basic book in Horticulture for ready reference. I am happy to note that my textbook meet this end. The constant encouragement offered by the staff and students of these universities prompted me to revise this present edition, duly incorporating the latest findings and technologies in many new chapters so as to keep pace with the development in Horticultural science. The present enlarged edition covers four aspects in Horticulture *viz.*, Fundamentals of Horticulture, Plant Propagation, Pomology and Ornamental Horticulture. I do sincerely hope that this edition will certainly meet the requirement of the above group of students besides the Horticultural enthusiasts.

I am extremely thankful to Dr. H.P. Singh, Deputy Director General (Horticulture), I.C.A.R., New Delhi to have adorned this book with his foreword.

Personally, I am deeply indebted to my wife and daughter for their constant encouragement to bring out this revised edition.

Coimbatore   N. Kumar

31.03.09

Preface to the First Edition

Horticulture contributes to the welfare of mankind in many ways. Fruits and vegetables form an indispensable part of our diet providing the essential vitamins and minerals from the nutritive and health point of view, besides their economic considerations. Spices and condiments add flavour and taste to our dishes. The aesthetic value provided by the ornamentals through annuals, shrubs, creepers and trees is of immense value and provides a happy and pleasant environment to live in.

Horticultural science deals with highly diversified crops, which differs extremely in their growth habits and production. As the crops vary, the propagation techniques and crop production aspects including their management practices also vary considerably and thus require an intimate knowledge of these aspects. The wide range of soil and climatic conditions that exists in the states of South India offer immense scope for growing many tropical, subtropical and temperate crops of fruits, vegetables and ornamentals.

This book has been written with the hope that it may provide a satisfactory guidance for students of agriculture or horticulture or students of botany as a text of applied botany and for the horticultural enthusiasts. In this book, more emphasis has been given to explain the principles and practices of horticulture including gardening. Cultivation aspects of only selected fruits and hints on the cultivation of vegetables are alone dealt with as it is difficult to cover all the horticultural crops in a single volume.

In this humble attempt, I am under immense obligation to Dr. V. Rajagopalan, Vice-Chancellor of Tamil Nadu Agricultural University,

Coimbatore who was kind enough to provide the foreword to this text. I am also highly indebted to Dr. M. Kulasekaran and Dr. K.M.B. Nambisan, Professors of Horticulture for their sincere efforts in going through the entire manuscript and having offered valuable suggestions in the preparation of the same.

My sincere thanks are also due to Mrs. A. Susheela Thirumaran, Associate Professor (Food Technology); Dr. R. Baskaran, Associate Professor (Pathology) and Dr. A. Dhakshina Moorthy, Scientist-2 (ICAR) who have offered very good comments while preparing the manuscript in their fields of interest.

I gratefully acknowledge the encouragement and enthusiastic cooperation extended by all my colleagues in the Faculty of Horticulture, more particularly by Prof. S. Sundararajan, Prof. Dr. J.B.M. Md. Abdul Khader, Dr. I. Irulappan, Dr. S. Thamburaj, Dr. P. Rengasamy, Dr. E.Vadivel, Dr. D. Veeraragavathatham and Mr. R. Venkatachalam in the production of this book.

I owe a lot to my brother Mr. N.T. Krishnan, Department of Zoology, S.T. Hindu College, Nagercoil for his untiring help and constant encouragement in making this venture a concrete reality.

Finally, I am indebted to M/s. Bharathi Press, Nagercoil for the timely publication of this book. Permission granted by Tamil Nadu Agricultural University to publish this book is also gratefully acknowledged. This being my maiden, attempt, it may contain some omissions. I will be grateful to the readers if they help me with their valuable suggestions and constructive criticism.

March, 1986   N. Kumar

Contents

PART 1-FUNDAMENTALS OF HORTICULTURE

Part 1

FUNDAMENTALS OF
HORTICULTURE

1

Introduction

The term "Horticulture" is probably of recent origin and it first appeared in writings of 17th century. This word is derived from the Latin word **'Hortus'** meaning **'garden'** and **'cultura'** meaning 'cultivation'. Today horticulture encompasses more than garden cultivation. The modern 'horticulture' may be defined as the crop science which deals with the production, utilization and improvement of fruits, vegetables, ornamental plants, spices and plantation crops including medicinal and aromatic plants.

The term 'agriculture' refers broadly to the technology of raising plants and animals. On the other hand 'horticulture' which is a part of 'agriculture' in the present context is concerned with raising of so called 'garden crops'. Horticulture deals with enormous number of crops. Garden crops traditionally include fruits, vegetables, spices and plantation crops, ornamental crops including medicinal and aromatic plants.

Horticultural science can be distinguished from agricultural or forestry science in one or more ways.

1. Horticultural produces are utilized in the fresh state and are highly perishable. In contrast, field and forestry crops are often utilized in the dried state and they are usually high in dry matter content.
2. Horticultural crops generally require intensive cultivation warranting a large input, capital, labour and technology per unit area of land, whereas agricultural crops require only extensive cultivation.

3. Cultural operations such as propagation, fertilizer application, training, pruning, harvesting, post harvest handling and marketing are skilled operations and are specific to horticultural crops.
4. Horticultural crops are the richest sources of vitamins and minerals whereas agricultural crops are generally rich in carbohydrates or proteins.
5. Aesthetic sense or gratification is an exclusive phenomenon of horticultural science.

DIVISIONS OF HORTICULTURE

Scientific horticulture consists of the following divisions:

A. Pomology

Pomology refers to the study of fruit crops. A fruit in horticultural science represents a plant, the product of which is edible on ripening. Fruit crops are classified as 'woody plants' and 'herbaceous plants'. The woody plants may be trees, shrubs or vines. Fruits borne on herbaceous perennial plants may be further classified as those with prostrate growth and upright growth. The woody plants may be either **deciduous** or **evergreen**. Deciduous plants are those which have a distinct 'rest period' during which all growth activities stop. The plants normally shed their leaves before entering into the rest period. This period of inactivity coincides with 'winter'. After the 'rest' (at the end of winter) they produce new growth and flowers. There is a definite cyclic growth in such plants coinciding with a particular season as shown below:

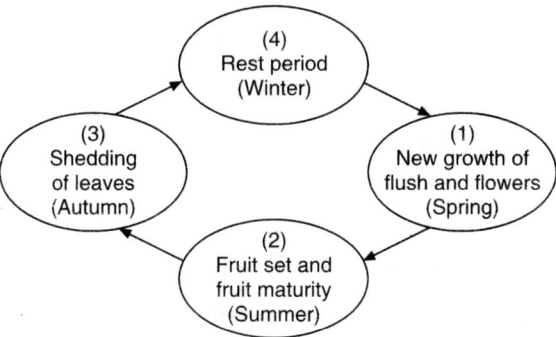

Under South Indian conditions, the plant produces new growth, flowers and fruits in February to June. After July, there is a gradual stoppage of growth and the leaves are shed by October, November and the plants enter into rest.

Evergreen plants do not have any distinct rest period. They retain leaves always. But in such plants there may be a period of comparative inactivity. The plants will not shed their leaves during this period. This period of comparative inactivity generally occurs just before flowering. Based on temperature requirements also, the fruit crops can be classified as temperate, sub-tropical and tropical fruits. Generally the temperate fruits are deciduous and the subtropical and tropical fruits are 'evergreen'. Based on the tolerance to the relative humidity of the atmosphere, the fruit crops can be also classified as arid, semi-arid and humid zone crops. The arid/semi-arid regions have peculiar eco-climatological features and they can exist in tropics, sub-tropics as well as temperate zones also. In these zones, a deficiency of moisture restricts but not necessarily inhibits plant growth. The rainfall in this zone is very low (0-250 mm) and is confined to 2 to 4½ months and the remaining are dry months. Besides, in these regions, higher solar radiation incidence (450-500 cal/cm2/day) and high wind velocity (20 km/hr) results in a high potential evapotranspiration (6 mm/day). The soils are of poor quality, with low fertility level and poor water holding capacity (20 to 25%) and with high infiltration rate (9 cm/hr). Ber, annona, date palm, phalsa are typical examples of arid or semi-arid crops. Mangosteen, litchi and mandarin oranges are examples for humid zone fruits.

An outline of fruit crops with some typical examples are given below.

I. Fruits borne on woody plants:

   (i) *Tree fruits*

      1. Deciduous (temperate)

         (a) Pome-Apple, Pear, Quince
         (b) Drupe (stone fruits): Peach, Plum, Apricot

      2. Evergreen

         (a) Sub-tropical: Mangosteen, Litchi, Sweet Oranges
         (b) Tropical: Mango, Sapota, Guava

   (ii) *Small fruits*

      1. Deciduous: Raspberry, Blackberry
      2. Evergreen: West Indian Cherry

II. Fruits borne on herbaceous perennial plants:

         (a) Prostrate growth: Strawberry
         (b) Upright growth: Banana, Pineapple

B. Olericulture

It refers to the study of vegetables. Vegetable in horticultural science is an edible herbaceous plant or part their of which is commonly used for culinary purposes or as salads.

C. Floriculture

It is the art of growing, selling, designing and arranging flowers and foliage plants. It includes the following sub-divisions.

1. **Commercial floriculture:** It deals with the cultivation of economic flowers like jasmine, roses, chrysanthemum, production of cut flowers and gift plants.
2. **Arboriculture:** It refers to the growing of trees for aesthetic or scientific or education purpose.
3. **Landscape gardening:** It consists of planning and arrangement of home grounds, public area and business establishments. It involves not only the use and placement of horticultural plants, but also the placement of buildings and other accessories in an aesthetic manner.
4. **Ornamental floriculture or gardening:** It refers to the study of various groups of ornamental plants which are used to decorate indoor and outdoor gardens.

D. Spices, Plantation Crops, Aromatic and Medicinal Plants

Spices are those plants, the products of which are made use of as food adjuncts to add aroma and flavour (e.g. pepper, cardamom, clove and nutmeg etc.). Condiments are also plants, products of which are used as food adjuncts to add 'taste' only (e.g. coriander, cumin etc.). Both spices and condiments contain essential oils which provide the flavour and taste. They are of little nutritive value.

The crops like coconut, arecanut, tea, coffee and rubber are known as plantation crops which are grown in extensive scale. Aromatic plants are those plants, the part of which yields aromatic essential oils on steam distillation or solvent extraction (e.g. Geranium, Patchouli and lemon grasses). The plants yielding alkaloid and steroid principles which have got preventive and curative properties are known as medicinal plants (e.g. *Catharanthus*, Periwinkle, pyrethrum and fox-glove).

Other branches of horticulture include fruit nurseries, seed production in vegetables and flowers, processing in fruits and vegetables, and marketing.

Importance of Horticulture

1. Horticultural crops contribute to **national income**. This can be well judged from the total value of the produce available annually from these crops. Horticultural crops occupy only 13 per cent of the cropped area but their contribution to the national income is roughly 28 per cent of the total value of the agricultural produce. The value of the Horticultural produces has been also increasing steadily. For instance, during a period of 15 years (i.e. from 1970–71 to 1984–85), the average increase in the value has been 336 per cent, while the value of the total agricultural produce has gone up only by 268 per cent during the same period.

   The contribution of horticultural crops to the national income is also perceptible by the annual foreign exchange earnings achieved through their export. The agriculture sector accounts for about 25 per cent of our export basket. Of this, horticultural crops alone account for over 56 per cent. These crops fetch 20–30 times more foreign exchange per unit area than cereals due to higher yields and higher prices available in the international market. In India, horticultural products including fruits, vegetables, their processed products, spices and plantation crops, cut flowers and live ornamental plants and the planting materials such as seeds of vegetable crops and flowering plants are being exported. Since these horticultural crops bring money/gold to farmers, the development in horticulture is often called as 'Golden Revolution'.

S. No.	Particulars	Exports (Rs. In cross)
1.	Fruits	143.70
2.	Vegetables	195.00
3.	Spices	240.00
4.	Plantation crops	1,600.00
5.	Flower crops (cut flower, cut foliage, life plants, seeds, tubers, corms)	33.00
6.	Medicinal and Aromatic plants	171.00

2. Fruits and vegetables are regarded as 'protective foods' since they supply minerals such a Calcium, Iron and Phosphorus and Vitamins like A, B Complexes and C in adequate quantities. They also contain a variety of polyphenols which are increasingly regarded as protective agents against chronic diseases. These are required to keep our health in good condition. Iron from fruits are better absorbed and become easily available than the iron from meat. Fruit and vegetables also serve as good laxatives.

For the predominantly vegetarian Indians, the Nutrition Expert Group prescribes daily a minimum of 2400–3900 calories of energy, 55 g proteins, 0.4–0.5 g calcium, 20 mg Iron, 3000 mg of β Carotene for the supply of Vitamin A, 1.2–2.0 mg thiamine, 1.3–2.2 mg riboflavins, 16–26 mg nicotinic acid, 50 mg ascorbic acid, 100 mg folic acid and 1 mg vitamin B12. To obtain this, Dietitians recommend 300 g of vegetables (i.e. 125 g of green leafy vegetables, 100 g of roots and tubers and 75 g of other vegetables) and 90 g of fruits per day. But the per capita availability works to about 30 g of fruits and 92 g of vegetables. This gap has to be bridged by increasing the production and productivity of fruits and vegetables. Recent recommendation from FAO/WHO shows that men and women should respectively consume atleast 800 and 400 g of fruits and vegetables daily to reduce the burden of diseases like diabetics, cardiovascular diseases, obesity and even cancer.

Nutritive value of common fruits and vegetables are furnished in table 1.

Horticultural crops are also good sources of carbohydrates, proteins and fats as indicated below:

(a) Protein rich sources	Protein content (g/100 g)
Agathi (pods and tender leaves)	8.4
Cluster beans (pods)	3.2
Peas (green seeds)	7.2
Curry leaves	6.1
(b) Fat rich sources	Fat content (g/100 g)
Cashewnut	46.9
Almond	58.9
Coconut (fresh)	41.6
Walnut	64.5
(c) Carbohydrate rich sources	Carbohydrate (g/100 g)
Potato	22.6
Sweet Potato	28.2

3. Horticultural crops yield more produces per unit area compared to cereals

Crops	Yield (t ha)
Fruits (average)	8.4
Banana	40–60
Pineapple	45
Grapes	40
Vegetables (average)	12.2

Table 1. Nutritive value of important fruits and vegetables (per 100 g of pulp).

S. No.	Name	Moisture (g)	Protein (g)	Fat (g)	Minerals (g)	Fibre (g)	Carbohydrate (g)	Energy (K. cal)	Calcium (mg)	Phosphorus (mg)	Iron (mg)	Carotene (mg)	Thiamine (mg)	Riboflavin (mg)	Niacin (mg)	Vita min C (mg)
FRUITS																
1.	Apple	84.6	0.2	0.5	0.3	1.0	13.4	59	10	14	1.0	0	–	–	–	1
2.	Avocado	73.6	1.7	22.8	1.1	–	0.8	215	10	80	0.7	–	–	–	–	–
3.	Banana	70.1	1.2	0.3	0.8	0.4	27.2	116	17	36	0.9	78	0.05	0.08	0.5	7
4.	Ber	81.6	0.8	0.3	0.3	–	17.0	74	4	9	1.8	21	0.02	0.05	0.7	76
5.	Dates (fresh)	59.2	1.2	0.4	1.7	3.7	33.8	144	22	38	–	–	–	–	–	1
6.	Grapes	82.1	0.6	0.4	0.9	2.8	13.1	58	20	23	0.5	3	0.04	0.03	0.2	–
7.	Guava	81.7	0.9	0.3	0.7	5.2	11.2	51	10	28	1.4	–	0.03	0.03	0.4	212
8.	Jack	76.2	1.9	0.1	0.9	1.1	19.8	88	20	41	0.5	175	0.03	0.13	0.4	7
9.	Mango	81.0	0.6	0.4	0.4	0.7	16.9	74	14	16	1.3	2143	0.08	0.09	0.9	16
10.	Orange	87.6	0.7	0.2	0.3	0.3	10.9	48	26	20	0.3	1104	–	–	–	30
11.	Papaya	90.8	0.6	0.1	0.5	0.8	7.2	32	17	13	0.5	666	0.04	0.25	0.2	57
VEGETABLES																
12.	Agathi	73.1	8.4	1.4	3.1	2.2	11.8	93	1130	80	3.9	5400	0.21	0.09	1.2	169
13.	Arakeerai	87.0	2.8	0.4	2.4	–	7.4	44	364	52	38.5	–	–	0.01	0.4	–
14.	Ashgourd	96.5	0.4	0.1	0.3	0.8	1.9	10	30	20	0.8	0	0.06	0.01	0.4	1
15.	Cabbage	91.9	1.8	2.1	0.6	1.0	4.6	27	39	44	0.8	1200	0.06	0.09	0.4	124
16.	Carrot	86.0	0.9	0.2	1.1	1.2	10.6	48	80	530	2.2	1890	0.04	0.02	0.6	3
17.	Drumstick leaves	75.9	6.7	1.7	2.3	0.9	12.5	92	440	70	7.0	6780	0.06	0.05	0.8	220
18.	Chillies (green)	85.7	2.9	0.6	1.0	6.8	3.0	29	30	80	1.2	175	0.19	0.39	0.9	111
19.	Field beans	86.1	3.8	0.7	0.9	1.8	6.7	48	210	68	1.7	187	0.10	0.06	0.7	9
20.	Onion	88.6	1.2	0.1	0.4	0.6	11.1	50	47	50	0.7	0	0.08	0.01	0.4	11
21.	Potato	74.7	1.6	0.1	0.6	0.4	22.6	97	10	40	0.7	24	0.10	0.01	1.2	17
22.	Sweet potato	68.5	1.2	0.3	1.0	0.8	28.2	120	46	50	0.8	6	0.08	0.04	0.7	24

Crops	Yield (t/ha)
Potato	20–25
Tapioca	30–40
Tomato	10–14
Spices (average)	10–00
` Paddy	5–6
Wheat	4–5

4. Horticultural crops generate more employment opportunities, especially in the rural sector, uplifting the rural economy (Table 2).

5. Horticultural crops are also highly remunerative and profitable than cereals. Case studies conducted in many parts of India showed than the net income per ha and the cost benefit ratio for horticultural crops are always higher than the cereals. Globally, fruits and vegetables occupy 7.0 per cent of total arable area while in India, they occupy 13.0 per cent of the total arable area. Similarly, the plantation crops in our country occupy only 2.0 per cent of the total cultivatable land, but they account for about 3/4th of total export earnings from all the agriculture commodities.

Table 2. Labour requirement in certain horticultural crops.

Crops	Labour requirement (Man days/ha)
(a) Fruits	860
Grapes	2500
Banana	1000
(b) Vegetables	200
(c) Plantation crops	
Tea	875
Coffee	250–390
Cardamom	190
Cereals	143

6. Fruits yield more calorific value. An adult normally requires 11,00,000 calories per year through the food he consumes. This can be obtained from 0.44 hectare of wheat. But the same quantity can be obtained from 0.03 hectare of fruits like banana or 0.06 hectare of mango. Thus, fruits can go a long way to reduce the pressure on food grains such as wheat and rice.

7. Many horticultural produces and their by-products are the important raw materials for many industries, thus providing more employment opportunities e.g. rubber, coir and sago industries. Major developments are taking place in this direction since last few years involving several multinational companies seeking collaboration for setting up fruit and vegetable processing units in India.

8. Social importance: It is a part of civilization, wherever civilization is highly advanced, horticulture is widely developed. Flowers are being used for worshipping Gods in temples. Every woman in our country considers her adornment complete only when her hair is decorated with aromatic and attractive colorful flowers. Flower offering is a symbol of affection in other countries. Often in hotels and business establishment, keeping the cut flowers in vases is becoming very common in India.

9. Horticultural Therapy: In some parts of the U.S.A., people who are unhappy and do not have mental power and balance are given horticultural therapy, a treatment by means of which their attention is diverted to ornamental gardening, flower decoration etc. and thus they are made free from their unhappy mood. Also, by making them to be with flowers of particular colour, the mental stress or depression can be removed.

   Flower therapy or aroma therapy is the art and science of using essential oils from plant sources for keeping us healthy. Although this is well mentioned in early civilization, nowadays, it is gaining scientific, medical and popular recognization. It helps to eliminate stresses and improve overall health without using chemical drugs. The essential oils, found in petals, leaves, roots and heartwood of plants act as anti-bacterial, anti-viral and anti-fungal agents besides help to cure many ailments.

10. Orchard tourism: This new concept is slowly developing in many countries. Visiting orchards for recreation has become an important tourist activity for urban citizens where the tourists are encouraged to hands-on involvement in activities like planting, pruning, grafting and harvesting fruits in order to freely engaged in agricultural production activities, which is now a different experience from their urban life.

11. Government of India is also attaching much importance for development of horticulture by allocating more funds now as evident below:

Plan	Rs. (in millions)
Fifth (1973–78)	76.1 (actual)
Sixth (1980–85)	91.3 (actual)
Seventh (1985–90)	241.9 (actual)
Eighth (1992–97)	10,000
Ninth (97–02)	14,000

Recognizing the importance of horticulture sector in the growth of Indian agriculture, Government of India has launched National

Horticulture Mission with the objective of doubling the horticulture production, i.e. to achieve a production of 300 million tonnes by 2011–12, establishing convergence and synergy among various on-going and planned programmes in the field of horticulture development and to promote the development and dissemination of technologies by blending traditional wisdom and frontier knowledge. To meet this objective, a sum of Rs. 6500 crores has been allotted during X Plan.

Agro-climatic Zones for Horticultural Crops

The Horticultural crops show repetitive wider adaptation and therefore exact delineation of the agro-climatic areas specific to each crops is not possible, except for the typical temperate and tropical species. However, our country can be broadly divided into seven zones for horticultural crops as indicated below:

1. **Temperate Northern region:** Comprising of states of Jammu and Kashmir, Himachal Pradesh, hilly areas of U.P., West Bengal. Temperate fruits, cool season vegetables are the important Horticultural crops in this zone.

2. **North Western arid region:** Comprising of the entire Rajasthan states, Gujarat and parts of Punjab and Haryana. Arid fruit crops like Ber, Pomegranate, Aonla, Date palm, wood apple and seed spices like coriander, cumin, fennel occupy major area in this zone.

3. **North Eastern sub-tropical humid regions:** Comprising of states and union territories like Arunachal Pradesh, Assam, Manipur, Meghalaya, Mizoram, Nagaland and Tripura. Fruits like Banana, Pineapple, Citrus, Jack besides plantation crops like tea and large cardamom are also grown here.

4. **North Central sub-tropical region:** Includes parts of U.P. Bihar, entire M.P. and parts of Maharashtra. Mango, Sapota, Sweet Oranges, Guava are grown here.

5. **South Central tropical region:** Includes the central area bordered by the Western and Eastern ghats in the states of Tamil Nadu, Andhra Pradesh, Karnataka and parts of Maharashtra. Mango, Guava, Sapota, Pineapple, Turmeric etc. are chiefly grown here.

6. **Coastal tropical humid region:** The entire coastal region stretching along the Bay of Bengal in the east and Arabian sea in the west comprises of this region. Bananas, Mango, Cashew, Coconut are the major crops.

7. **Southern hilly zone:** Comprising of Western ghat and Eastern ghat hills above 800 m MSL, where Coffee, Tea, Cardamom, Pepper, Oranges, Pineapple etc., are extensively grown.

2

Soil and Climate–
Limiting Factors

SOIL

The soil acts as a substratum and provides nutrients and water necessary for plant growth. The kind of soil, its fertility status, soil reaction and drainage influence the productivity of the horticultural crops.

Kinds of Soil

Soils are classified on the basis of particle size, the relative amount of various sizes and the content of organic and inorganic matter. Inorganic soils generally have less than 10% of organic mater in the surface layer. The inorganic portion consists of varying amounts of sand, silt and clay.

Sandy soils are coarse textured and have large pore-spaces. They have poor water-holding capacity and low nutrient retaining ability. This type of soil is suitable for vegetables to obtain early crops and rapid root growth. It is the most common medium for propagating cuttings. Horticultural crops when grown on sand needs to be supplied with considerable quantities of nutrients for best production.

Loamy soils have significant amount of sand, silt and clay. Soil is described as sandy loam or silty loam or clay loam depending upon the predominating component of the soil. Sandy loams are the soils to get early crops than silt or clay loams. Loams of all types are highly suitable for horticultural production.

Clay soils are fine textured and have very small pore spaces. They are not suitable for raising of most of the horticultural crops, more so with root and tuber crops, unless they are 'opened up' and improved by incorporation of large quantities of organic matter. But this clay soil attracts, holds and releases nutrients for plant growth in a better way than other types of soils. Mangoes can be practically grown in rich clayey soil; it produces more of vegetative growth with little fruit production.

Organic Matter

The organic matter of the soil is derived not only from the decomposed plant and animal tissues but also from the microorganisms themselves. The decomposition of plant and animal material is accomplished by enzymatic digestion carried out by soil microorganisms. During decomposition, certain substances such as lignins, waxes, fats and some proteinaceous materials resist decomposition, but through complex bio-chemical processes, form a dark non-crystalline colloidal substance called humus. Humus has more absorbing properties for nutrients and moisture that are even higher than those of clay.

These organic matters are reservoirs of plant nutrients and on decomposition they release the nutrients required for plant growth. A rapid method of increasing the organic matter content in the soil may be achieved by growing two or three green manure crops or by the addition of compost. Organic soils are high in organic matter, usually 20% or more. They are found in swamps, bogs, shallow lake bottom and river beds. The organic soils are classified as peats and mucks. The former type contains more than 50% organic matter and is sometimes as high as 95% while the latter contains 20 to 50% organic matter. The decomposition of organic matter is advanced in peat soils and has got high water holding capacity. Onion, potato, carrot, cabbage, cauliflower and lettuce prefer peat soils. Muck soils do not have a high water-holding capacity as that of peat soils and its period of productivity is usually shorter.

Soil Fertility

The capacity of soil to nourish and sustain plant growth is known as soil fertility. The mineral particles of the soil become available to the plants by their interaction with soil, air, soil moisture, soil microbes and humus. Fertility of the soil is low, if its mineral particles predominate the other constituents as seen from the fact that the top layer of earth, 15 to 60 cm in thickness, is more fertile since it contains the largest quantity of humus and soil microbial population. The fertility status of the soil declines as it goes deeper and deeper. Plantation crops like coffee, cardamom and spices like pepper, ginger, clove, vanilla prefer fertile soils rich in organic matter and humus content.

Depletion of soil fertility occurs as a result of four specific factors. These are crop removal, erosion, leaching and volatilization. The crops remove nutrients from the soil and hence continuous cropping for generations will deplete the soil nutrients. Erosion, the washing off or blowing away of soil results in loss of large quantities of nutrients from the top soil. Leaching of nutrients down through the soil profile by percolating water on its way to the water table will reduce the fertility level. Volatilization of soil nutrients is the most common to occur. Volatilization of nitrogen, in the ammonia form and also by denitrification process is evident. The soil fertility can be maintained in a good state by following judicious soil management practices. Most of the horticultural crops prefer a good to moderate fertile soil for higher productivity.

Chemical Composition of Soil

The acid, alkaline or neutral reaction of a soil is expressed on the pH scale. A pH of 7 is neutral and points below 7 are acidic and those above 7 are alkaline. The soil reaction influences the nutrient availability. For instance, boron is deficient in alkaline soils, but is unavailable in very acidic soils. The soil reaction also influences the activity of soil bacteria, which in turn affects the plant nutrients.

Table 3. Optimum pH Range For Certain Horticultural Crops

Crops	Optimum pH range
Mango	6.0–7.5
Citrus	5.5–7.5
Pineapple	5.5–7.0
Avocado	5.0–7.0
Grape	6.0–7.5
Banana	5.5–8.0
Onion	6.2–7.4
Beet root	6.0–7.5
Cauliflower	6.0–7.3
Cabbage	5.8–7.3
Carrot	5.8–7.0
Sweet Potato	5.2–6.0
Chrysanthemum	6.0–6.5
Rose	6.0–6.5
Tea	4.5–5.0

Some diseases are promoted in acid or alkaline soils. Club root disease of crops is more prevalent in acidic soils. Most horticultural crops are adapted to a pH range just below neutral, that is, of a slightly acidic nature.

The soils can be made more acidic (or less alkaline) by applying acid producing chemicals such as aluminium sulphate, sulphur and gypsum (calcium sulphate). Similarly, soils can be made more alkaline (or less acidic) by applying hydrated lime or lime stone or epsum (magnesium sulphate). In areas of heavy rainfall, leaching will remove many soluble carbonates resulting in more acid soil.

In alkaline soils concentration of sodium salts (above 0.1%) is dangerous. Information on soil salinity is necessary to select tolerant varieties and to adopt proper soil management practices. Relative tolerance of fruit trees to salinity is as below:

Tolerant	(8 m mhos/cm)	Date, Guava, fig and grapes
Moderately tolerant	(3 to 6 m mhos/cm)	Pomegranate, Grapefruit, Apple, Pear and Plums
Sensitive	(1.5 to 3 m mhos/cm)	Orange, Peach, Avocado, Strawberry.

Soil Depth

Soil depth should be 2.0 m for most of the fruit crops for normal growth and development. Shallow soils with hard and compact subsoil layers like kankar, rock, and heavy clay should be avoided for fruit crops as in such soils, the roots grow in the top layer of soil parallel to the surface. Such trees are easily-uprooted by winds or premature death of trees may also occur.

Drainage

Apart from the kind of soil, fertility and its composition, drainage of soil should also be considered. Drainage depends upon the nature of sub soil. In a good sub-soil, trees can stand drought better because of deeper root penetration. On wet, heavy soils with impermeable sub-soil, poor performance is noticed due to poor aeration and inadequate drainage. A well drained soil is therefore essential. Water table should be below 2.0 m at all times of the year. Higher water table leads to poor aeration of sub-soils and water saturation. Rotting of roots occurs due to prolonged submergence of roots in moisture. Disease incidence will be more in such cases. The failure of sweet oranges in several places in Uttar Pradesh and Punjab is due to this factor only. The decline of citrus orchards is favoured by poor aeration of the soil. The success of citrus industry in Florida is due to the soil factor only where it is grown in coarse sandy, well drained soils with a water table never exceeding above 1.2 m.

Biological Properties of Soil

Soil is very thickly populated with life of a microscopic order. More than 2000 millions of microbes are estimated to be contained in a teaspoonful of soil. The rich population is of a very varied character. Some of the micro-organisms are harmful and many are useful. The net result of their presence and activities has an important bearing on the fertility of the soil. The main groups of soil microbes are 1. Algae 2. Fungi and 3. Bacteria. Algae are simple microscopic chlorophyll plants living in the soil and trap the energy of sunlight and leave their bodies when dead to accumulate as humus in the soil. Fungi are minute plants without chlorophyll either living as parasites on other plants or thriving on dead matter saprophytically and thus helping the further break down of organic material by bacteria. Bacteria are unicellular organisms and less than 1/10,000 of a centimeter in diameter. They multiply rapidly in prodigious numbers. They are the most useful soil organisms. There are hundreds of species of bacteria. The useful ones are generally classified under the following groups:

(a) Those that cause organic matter to generate heat, decay and release the nitrogen contained in them chiefly as ammonia.

(b) Those that convert ammonia into nitrous acid and nitrites by uniting it with the basic elements in the soil.

(c) Those that convert nitrites to nitrates, in which form the plants get their nitrogen from the soil through their roots in a state of solution in water.

(d) Those particular kinds which live in the nodules of leguminous plant and take nitrogen directly from the air and 'fix' it in those nodules to be built into nitrogen compounds useful to themselves and to plant by other sets of organisms.

The activity of soil organisms should be taken into consideration when cultural and management practices are considered. For instance, the amount of humus added to the soil through decomposition processes will be greater if the supply of nitrogen is relatively high. Thus, it is advisable to apply fertilizer to cover crops before it is ploughed into the soil. The activity is again dependent on soil moisture, temperature, pH and oxygen supply.

CLIMATE

Climate is one of the important complex factors which influence the production of horticultural crops. Climate includes several basic environmental components such as temperature, rainfall, atmospheric humidity, wind, hail storms etc.

Temperature

Every plant requires particular temperature range at which all fundamental processes like photosynthesis, respiration, water absorption and transpiration, cell differentiation proceed favourably leading to the highest marketable yield. This optimum temperature range varies with crops and based on the optimum temperature requirement, horticultural plants are broadly classified into (a) temperate (b) sub-tropical (c) tropical plants. If a tropical plant is grown in sub-tropical zone or *vice versa,* their growth and performance may not be good. However, there are certain exemptions, for instance, grape is a temperate crop but can be grown in subtropical and tropical regions.

In a particular crop itself, the temperature requirement may vary with the stage of crop e.g. tomato at early stages prefer higher night temperature ranges (18° to 27°C) while the optimum night temperature during fruit set ranges from 13° to 17°C. Temperature affects flowering in certain horticultural crops. The optimum temperature for flowering in banana is 10° to 40°C and if low temperature (10°C) prevails, it inhibits proper emergence of the bunch from the pseudostem, a disorder commonly described as 'choking' in banana. Temperature affects the quality of fruits. Low temperature during ripening period favours high acid content in grapes while high temperature favours sweetness.

Extremes in temperature often cause certain injury in plants. Temperature close to or below the freezing point of water may cause permanent damage resulting in death. This is called 'winter kill' and some plants are found to be sensitive to temperature slightly above freezing point and is called 'chilling injury', plants that are resistant to cold injury are known as 'hardy plants' (e.g. asparagus) as compared to those susceptible to cold injury are called 'tender plants' (e.g. cucumber). Cold injury in most of the herbaceous plant is due to the formation of ice crystals which cause mechanical injury to the cells. Similarly, high temperature injury is often related to dessication, as a result of excessive water loss in transpiration compared to water uptake.

Light

The intensity of light, quality and its duration affect the growth and development of horticultural plants. Light influences many development processes of the plant such as germination, tuber and bulb formation, flowering and sex expression and is mainly related to the length of light and dark period (Photo period). For example, in cucurbits, long day conditions favour the initiation of male flowers while short day conditions favour production of female flowers. Certain horticultural crops like coffee, cardamom, cocoa if grown under full sunlight conditions experience reduced vegetative growth, die back and lower yields and for their normal growth filtered shade is required. Fruits like apple and mango when exposed to adequate light, obtain good colour and quality when compared to shaded fruits.

Atmospheric Humidity

Humidity affects the growth and development of horticultural crops. Based on the tolerance to relative humidity of the atmosphere, the horticultural crops can be classified as humid, semi-arid and arid zone fruits. If a humid zone fruit crop is grown in an arid zone, its growth and development is significantly affected. High humidity during flowering and fruiting promotes the incidence of pests and diseases especially in mango, grapes, potato and tea etc. Propagation of many horticultural crops through cuttings, layers and grafting are more successful if performed during the period in which the atmospheric humidity is relatively more than during the dry months.

Rainfall

Total rainfall and its distribution decide the growth and development of horticultural crops especially the rainfed crops. Horticultural crops require varying quantum of rainfall for their successful growth and development (Table 4). However, continuous heavy rain for a period of 2-3 days during the peak flowering phase washes off pollen, hinder insect pollination, injure pollen and dilute the stigmatic fluid. Timely receipt of summer showers (Blossom showers) in February-March decides the flowering and fruiting in coffee during the current as well as the following year. Similarly, summer shower during February-April is essential for panicle initiation in cardamom; otherwise, it will affect the yield. Fruits generally harvested during rainy months are inferior in quality than those harvested during other months e.g., grapes.

Table. 4. Rainfall requirement of certain horticultural crops.

S.No.	Crop	Quantity of rainfall required per year
1.	Mango	025 to 250 cm
2.	Pepper	125 to 200 cm
3.	Cardamom	200 to 250 cm
4.	Rubber	200 to 250 cm
5.	Dates	015 to 025 cm

Wind

Heavy winds cause physical damages by shedding of flowers and fruits, breaking of branches and uproot the trees. Frequent dry and high winds cause rapid loss of moisture and necessitate frequent irrigation.

Air Pollutants

The air pollutants like ozone (O_3) and photo chemical oxidants such as sulphur-di-oxide and nitrogen reduce the assimilation rate, growth and development of plants leading to reduction in yield and quality of the produces. Mangoes grown in Punjab, Uttar Pradesh, Bihar and West Bengal suffer from physiological disorder called 'Black tip' wherever the orchards are situated within a distance of 1.5 KM form the brick kilns. The gases like carbon-di-oxide, sulphur dioxide and acetylene which constitute the brick-kiln fumes are reported to be responsible for this disorder. Black tipped fruits fetch very low price.

Frost

Frost is a thin layer of ice crystals deposited on soil and plant surfaces as a result of freezing temperatures. It occurs commonly in the hill stations during winter months especially on clear sunny days above 2000 m MSL, more particularly in orchards located either at the bottom of the slope or in the valley areas near water sources or damp places. In Nilgiris and Kodaikanal, frost normally damages crops like tea, potato, cole crops during winter months in places above 2000 m.

Hail Storms

Occurrence of hails during pre-blooming or blooming period of apple, plum, peaches generally affect the fruit set. In peaches prevalence of frequent hail storms can be destructive to the fruits which impair its quality or completely destroy them if the stones are large.

Altitude

It also plays a critical role in affecting the various components of climate, thereby limiting the production of certain horticultural crops. Generally, altitude has a profound effect on the temperature. For every increase in 100 m in elevation above mean sea level, a decrease of about 1 to 2°C is normally observed. Under South Indian hills, warm temperate fruits are successfully grown above 1800 m MSL, humid zone fruits and plantation crops like coffee, tea are grown from 1000 m to 1800 m MSL and below that, mostly tropical fruits are grown. Altitude plays a critical role in the flowering and yielding performance of certain horticultural crops. For example, coconut, a tropical plant, if planted in the hills at an elevation of 1000 to 1200 m MSL, it takes nearly 10–12 years for flowering and after flowering also, their yield performance is relatively poor. Similarly hill grown papaya fruits are relatively poor in taste than from those grown in plains. In the case of tea, altitude significantly affects productivity and quality of the tea. High grown teas are superior in quality than those grown in middle elevations but the productivity per unit area is more in tea grown at middle elevations.

3

Growth and Development

The term 'growth'f refers to an irreversible increase in size and weight of the plant. It normally reflects an increase in protoplasm, which may occur through increase in cell size and number of cells. Hence, the process of cell elongation and cell division provides the basis for growth. The division and enlargement of cells are a complicated process involving synthesis of many organic compounds such as protein, cellulose and nucleic acid and the required physical forces that cause cell enlargement. The growth of the whole plant is compartmentalized into areas called 'meristem' where cell division and cell elongation are localized. The meristem cells undergo repeated division *via.*, mitosis, wherein DNA and nuclear materials replicate and cell mass increases. These are the shoot apex, root apex, cambium and the intercalary zones. Adjacent to this area of meristematic activity is the region of cell elongation, which phases into the region of differentiation and maturation.

The increase in protoplasm (occurring in cell division and elongation) is brought about through a series of events in which water, carbon-di-oxide and inorganic salts are transformed into living materials. These events include (a) photosynthesis (production of carbohydrates) (b) nutrient absorption and translocation (uptake of water, nitrogen and other mineral elements) and (3) anabolism (synthesis of various materials produced in plant from the carbon fragments produced by photosynthesis and from the inorganic nutrients and water absorbed from the soil). The energy required for all these processes is provided by a catabolic process *viz.*, respiration (degradation of sugars and fats to release energy). The net balance of anabolism and catabolism results in growth of the plant (Fig. 3.1).

Development

In a plant, the development cycle begins with germination of seed followed by juvenility, maturity, flowering and fruiting phase in general. The cycle of plant growth is completed with fruiting phase and in annuals, it enters into final phase of plant growth *viz.*, senescence and death. In perennials, the plant is ready to recycle after a period of quiescence after fruiting. The growth and development of plants consist of two distinct phases *viz.*, 1. vegetative and 2. reproductive phase. The vegetative phase essentially consists of germination and subsequent development of plants and reproductive phase consist of the formation and development of flower buds, flowers, fruits and seeds or the enlargement and maturation of storage organs–fleshy stems and fleshy roots.

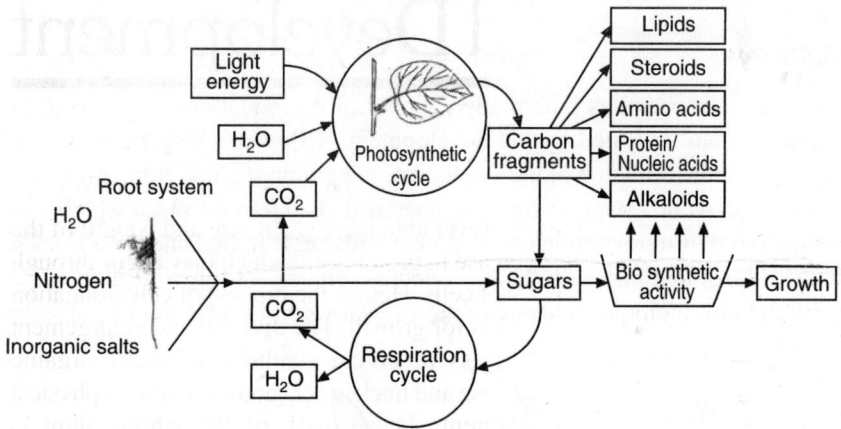

Figure 3.1   Metabolic activity in plant system.

Vegetative Phase

Germination includes all the sequential steps from the time the seed imbibes water till the seedling is self sustaining. During germination, the complex reserve substances are converted enzymatically to simple soluble substances, which are translocated to the embryonic plant. From the stage of germination until the first flower primordium is initiated, the plant is said to be in a vegetative phase of growth. During this phase, if a plant cannot produce flower, regardless of the environmental conditions imposed on it, then it is said to be in 'Juvenile phase'.

The Juvenile phase is characterized by

  (a) Relatively most rapid rate of growth
  (b) Distinct morphological and physiological features in some plants.

Ex: In English Ivy (*Hedera helix*), the leaves are 'lobed' in the juvenile phase while the leaves are 'entire' at the maturity stage. Similarly, the juvenile leaves of Eucalyptus are large and broad at the base and sessile, while matured leaves are elongated and have distinct petioles.

The growth of the bud is not a continuous process but is often associated with periods of arrested development. It is of two types, one is brought out by the unfavourable environment and another one is the bud dormancy in which the growth is temporarily stopped, eventhough all the external conditions required for normal growth are provided. The later one is more associated with internal factors, often due to the formation of growth inhibiting substances. It is broken naturally with cold temperature endured during the period of winter months.

Maturity

When a plant becomes potentially capable of reproduction, it is said to be 'mature' often indicated by the development of flowers. Just proceeding to this stage, transformation of the vegetative stem primordia into floral primordia takes place. At this stage, distinct biochemical changes take place which alter the pattern of differentiation of vegetative bud (i.e. leaf, bud and stem tissues) to the tissues that make up the reproductive organs. At this transition stage, a distinct morphological change in the vegetative bud is also microscopically evident (Fig. 3.2).

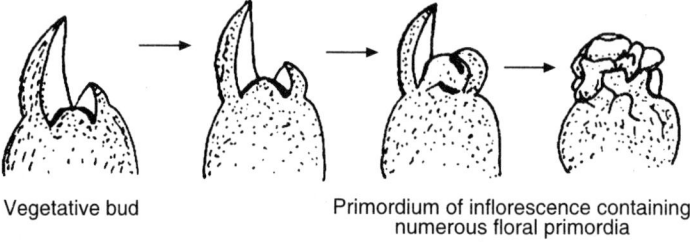

Vegetative bud

Primordium of inflorescence containing numerous floral primordia

Figure 3.2   The transition of a vegetative bud to a floral primordia.

At this transition stage, growth of the central portion is reduced or inhibited and the meristem becomes flattened in contrast to the conical shape characteristic of the vegetative bud. Thus, small protuberances develop in a spiral of whorl arrangements around the meristem which subsequently develop into different floral parts. This stage of transition of vegetative buds to floral primordia is called as 'fruit bud initiation and differentiation'. The time of fruit bud initiation is very important to the horticulturists to adopt certain important horticultural practices such as pruning, application of fertilizers and irrigation etc. Many perennial plants initiate the floral primordia from few to many months before flowering as shown below:

Coconut   Primordia of inflorescences begin to form in the leaf axils about 32 months before the opening of spathe.

Mango   Flower bud differentiation takes place between August to October while flowering takes place between December to February, i.e. 5–6 months before flowering.

Banana   Fruit bud initiation takes place between 3–5 months from planting i.e. 3–6 months before shooting (flowering) depending upon the varieties.

Sweet oranges Flower bud differentiation takes place 3–4 months prior to flowering.

Coffee   Flower bud initiation takes place between August–September but flowering takes place in February–March after the receipt of the summer shower.

This transition stage is influenced by the following factors:

1. **Carbohydrate: Nitrogen ratio (C:N ratio):**   It has been established in many plants that if C:N ratio is optimum, it favours the flower induction, on the other hand, if nitrogen level is more, it will favour only vegetative growth and reduce or delay flowering.

2. **Photoperiodic effect:**   The growth response of a plant to the length of a day i.e. length of the light and dark period is called 'Photoperiodism'. Based on the response to the light, plants can be divided into three groups *viz.*, short-day, long-day and day neutral plants. Short-day plants initiate flowers only when the day length is below 12 hours e.g. Chrysanthemum, coffee and Poinsettia. Long day plants initiate flowers only if day length exceeds 12 hours (e.g. Beetroot, Radish, Spinach). Day neutral plants can initiate flowers under any day length (e.g. Tomato, Broad bean).

3. **Moisture effects:**   Moisture may influence flower initiation in certain horticultural crops.

Fruit Development

Fruit development may be divided into following phases:

1. Initiation of the fruit tissues
2. Pre pollination development
3. Post pollination growth and
4. Ripening, maturation and senescence

The first two phases develop concomitantly with the flowers as a result of cell division. After pollination, cell enlargement is responsible for the major portion of size increase in most of the fruits. However, in some large fruited plants like apple, pear, watermelon and squash, cell division continues even after pollination and hence, both increase in cell number and size contribute for final size.

Pollination plays two functions in the growth and development of a fruit.

(a) It helps in the inhibition of flower or fruit abscission i.e. it helps in the setting of fruits.

(b) It provides male gametes for fertilization.

It has been established in many cases, that pollen contains an auxin or an enzyme that converts auxin precursors present in the stigma to auxin or provides a 'synergist' that helps in the fruitset, but it need not result in fertilization in all case. This is due to the fact that pollen does not germinate or it does, the pollen tube may burst in the style. This results in the development of fruits without seed or small seed-like structures (e.g. grapes, certain varieties of banana).

During pollination, which normally results in fruitset and fertilization of the pollen in rare cases exerts a direct influence on the matured tissues of the fruit. This phenomenon is called 'metaxenia'. This kind of pollen affects the size, quality of the fruit and time of ripening in date palms. On contrary, the direct effect of pollen on embryonic or endosperm tissue is known as 'xenia'.

After fertilization, the developing fruit grows rapidly as a result of the growth stimuli (presumably auxins) received from developing seed within the fruit with the exception of parthenocarpic fruit. This has been demonstrated in many species wherein fruit size increases progressively as seed number increases from one to three or more. Seedless grapes are quite small in most cases. If an apple fruit contains only one seed on one side instead of two, that side tends to be larger. Asymmetrical fruits are often seen in stone fruits, if only one ovule develops instead of two ovules present in the ovary.

Although the control of fruit growth is located in the seed, the raw materials for development of fruits are supplied by the plant. Hence, the nutrition and moisture availability of the plant directly affect the fruit size. It has been estimated that to support the growth of one apple fruit, 40 mature leaves are required and if this ratio is altered, due to abnormally higher fruit set or excessive defoliation due to some foliar diseases, the quality and size of the individual fruits is greatly reduced.

Ripening

A matured fruit becomes ripe and senesce subsequently. Maturation may refer to the attainment of full size or the desired size for commercial use, eventhough the fruit may not be ripe. Ripening, a stage immediately after maturation includes the qualitative changes which occur after the fruit has reached full size, such as changes in pigmentation, firmness and chemical constituents. 'Senescence' is closely related to ripening but consists of the catabolic phase of fruit development during which the tissues deteriorate and organ dies.

Several processes characterize ripening.

(1). In some fruits, a marked rise in the rate of respiration occurs during ripening. Such fruits are called as 'climacteric' e.g. Apple, Banana, Tomato, Bread fruit, Papaya, Peach, Pear, Persimmon etc.

In other fruits e.g. Cucumber, Fig, Grapes, Grapefruit, Lemon, Orange, Pineapple, Strawberry such sudden rise does not occur and are called non-climacteric fruits. (Fig. 3.3).

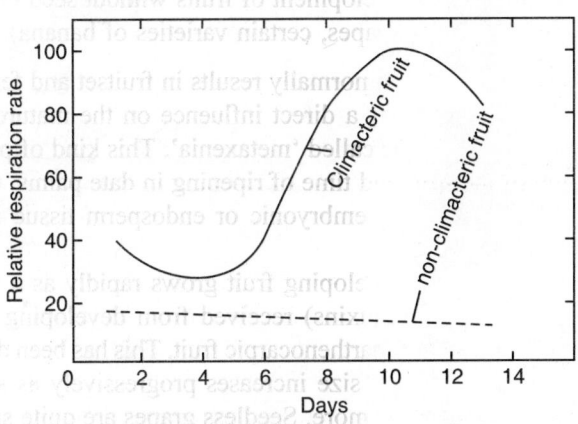

Figure 3.3 Respiration rate in climacteric and non-climacteric fruits.

(2) Colour changes often accompany in ripening, this is either due to the degradation of chlorophyll or production of new pigments or due to both.

(3) Fruits soften as they ripen as a result of enzymatic break down of pectins which cement the cell wall together.

(4) Starch content gets decreased and converted into sugar as ripening proceeds.

(5) Ethylene gas rises prior to the onset of the climacteric to the threshold level which is a prerequisite for ripening. Fruits must have reached a certain stage of maturity before ripening to produce ethylene, as immature fruits, which are unable to ripen lack the capacity to produce ethylene.

(6) Chemicals giving fruits their characteristic flavour, accumulate as fruits ripen. Such chemicals are volatile in nature and are present in minute quantities.

Senescence

Senescence is the erosive process that accompanies ageing prior to death. After ripening of the fruit, its entire organ except the seed senesces. In the case of plant as a whole, the senescence may be partial or complete. Deterioration and death of plant organs such as leaves, stem, flowers and fruits, as in perennial plants, are known as 'partial senescence' while 'complete senescence' is the aging and death of the entire plant as in annual and biennials as a result of termination of the life cycle.

4

Establishment of Orchard

Establishment of an orchard is a long term investment and deserves a very critical planning. The selection of proper location and site, planting system and planting distance, choosing the varieties and the nursery plants have to be considered carefully to ensure maximum production.

Location and Site

Proper selection of site is important. Selection may be made based on the following criteria:

1. The location should be in a well established fruit growing region because one could get the benefit of experience of other growers and also get the benefit of selling the produce through co-operative organizations with other fruit growers.
2. There should be a market close to the area.
3. The climate should be suitable to grow the chosen fruit crops.
4. Adequate water supply should be available round the year.

Before a grower selects a site for establishing a new orchard, he must have assessed the following factors:

1. Suitability of soil, its fertility, the nature of subsoil and soil depth.
2. Site must have proper drainage and no water stagnation during rainy season.
3. Irrigation water must be of good quality.

4. There must be proper transport facilities either by road or rail within the reach.

5. Whether the climatic conditions are suitable for the fruits to be grown and are whether site is free form the limiting factors such as cyclones, frost, hailstorms and strong hot winds.

6. Whether there are seasonal gluts or over production in any particular period of the year.

7. Whether there is assured demand in the market for the fruits to be grown.

8. Whether his orchard is a new venture or whether there are already other growers.

9. Availability of labour.

Preliminary Operations

After selecting the suitable location and site, some preliminary operations have to be done. Trees are felled without leaving stumps or roots. The shrubs and other weedy growth are also cleared. Deep ploughing is essential to remove big roots. The lands should be thoroughly ploughed, leveled and manured. Leveling is important for economy of irrigation and preventing soil wash. In the hills, the land should be divided into terraces depending upon the topography of the land and the leveling is done within the terraces. Terracing protects the land from erosion.

If the soil is poor, it would be advisable to grow green manure crops and plough it *in-situ* so as to improve its physical and chemical conditions before planting operations are taken up.

Planning of an Orchard

A careful plan of the orchard is necessary for the most efficient and economic management. The following points should be borne in mind in preparing the plan:

1. Optimum spacing to accommodate maximum number of trees per unit area.

2. Stores and office building in the orchard should be constructed at the centre for proper supervision.

3. Wells should be located at convenient places in different parts at the rate of one well for 2 to 4 hectares.

4. Each kind of fruit should be assigned in a separate block.

5. Fruits ripening at the same time should be grouped together.

6. Pollinators should be provided in deciduous fruits. In deciduous fruit trees, there are some varieties which require pollen from another variety

to set fruits in them; otherwise, they will be barren. Such pollen donors are known as pollinators. Every third tree in every third row should be planted with a pollinator.

7. Irrigation channels should be laid along the gradients for most economical conduct of water. For every 30 m length of channel, 7.5 cm slope should be given.

8. Roads should occupy minimum space for the economy of transport. The clearance between wind break and first row of trees is advantageous for the road.

9. Short growing trees should be allotted at the front and tall at the back for easy watch and improve the appearance.

10. Evergreen trees should be in the front and deciduous ones behind.

11. Fruits attracting birds and animals should be close to the watchman shed.

12. A good fence is essential. 'Live fencing' is economic and cheap to other kind of fences. The plants suitable for live fencing should be drought resistant, easy to propagate from seed, quick growing, have dense foliage, should stand severe pruning and should be thorny. *Agave, Prosopis juliflora, Pithecolobium dulce* and *Thevetia* if closely planted in 3 rows would serve as a good live fencing. Recently, in many horticultural estates, solar fencing is erected to protect the crops against wild and domestic animals. Though it is relatively costlier, it is very effective and is long lasting without much maintenance cost than barbed wire fencing.

13. 'Wind breaks', rows of tall trees planted close together around the orchard, are essential to resist velocity of wind which cause serve ill-effects particularly moisture evaporation from the soil. Since the wind breaks are very effective in reducing the wind velocity and minimizing the damage to the fruit trees and to other crops, their presence in regions where strong winds prevail is of paramount importance. A wind break ordinarily has its maximum effectiveness for a distance about four times its height but its effect diminishes with distance.

The most effective wind break is a double row of tall trees alternately placed. There should be at least as much as space between the windbreak and the first row of the fruit trees as between fruit trees. It is preferable to dig a trench of 90 cm deep at a distance of 3 m from the wind break trees and prune and cut all the roots exposed and again fill up the trenches. This may be repeated for every 3 or 4 years in order to avoid the competition between the wind breaks and fruit trees for moisture and nutrition.

Trees suitable for wind break should be erect, tall and quick growing, hardy and drought resistant and mechanically strong and dense to offer maximum resistance to wind. The trees which are suitable for growing as wind breaks are *Casuarina equisetifolia, Pterospermum acerifolium, Polyalthia longifolia, Eucalyptus globulus, Grevillea robusta, Azadirachta indica etc.*

Laying Out of Orchards

Any method of layout should aim at providing maximum number of trees per hectare, adequate space for proper development of the trees and ensuring convenience in orchard cultural practices. The system of layout can be grouped under two broad categories *viz.* (a) vertical row planting pattern and (b) alternate row planting pattern. In the former planting pattern (e.g. square system, rectangular system), the trees set in a row is exactly perpendicular to those trees set in their adjacent rows. In the latter planting pattern (i.e. Hexagonal, Quincunx and Triangular), the trees in the adjacent rows are not exactly vertical instead the trees in the even rows are midway between those in the odd rows.

The various layout systems (Fig. 4.1) commonly followed are:

(a) Vertical row planting pattern

1. **Square system:** In this system, trees are planted on each corner of a square whatever may be the planting distance. This is the most commonly followed system and is very easy to layout. The centre place between four trees may be advantageously used to raise short lived filler trees. This system permits inter cropping and cultivation in two directions (Fig. 4.1. A).

2. **Rectangular system:** In this system, trees are planted on each corner of a rectangle. As the distance between any two rows is more than the distance between any two trees in a row, there is no equal distribution of space per tree. The wider alley spaces available between rows of trees permit easy intercultural operations and even the use of mechanical operations (Fig. 4.1. B).

3. **Cluster system:** In this system, trees are planted on each corner of a square forming a cluster and each cluster is set apart at double the distance of trees planted in a cluster. Although there is no equal distribution of space per tree, the wider alley space around each cluster permits easy cultural operation. It accommodates nearly twice the population of square system (Fig. 4.1. F).

(b) Alternate row planting pattern

4. **Hexagonal system:** In this method, the trees are planted in each corner of an equilateral triangle. This way six trees form a hexagon with the seventh tree in the centre. Therefore this system is also called

as 'septule'. This system provides equal spacing but it is difficult to layout. The perpendicular distance between any two adjacent rows is equal to the product of 0.866 x the distance between any two trees. As the perpendicular distance between any two rows is less than unity this system accommodates 15% more trees than the square system. The limitations of this system are, that it is difficult to layout and the intercultivation is not so easily done as in the square system (Fig. 4.1. E).

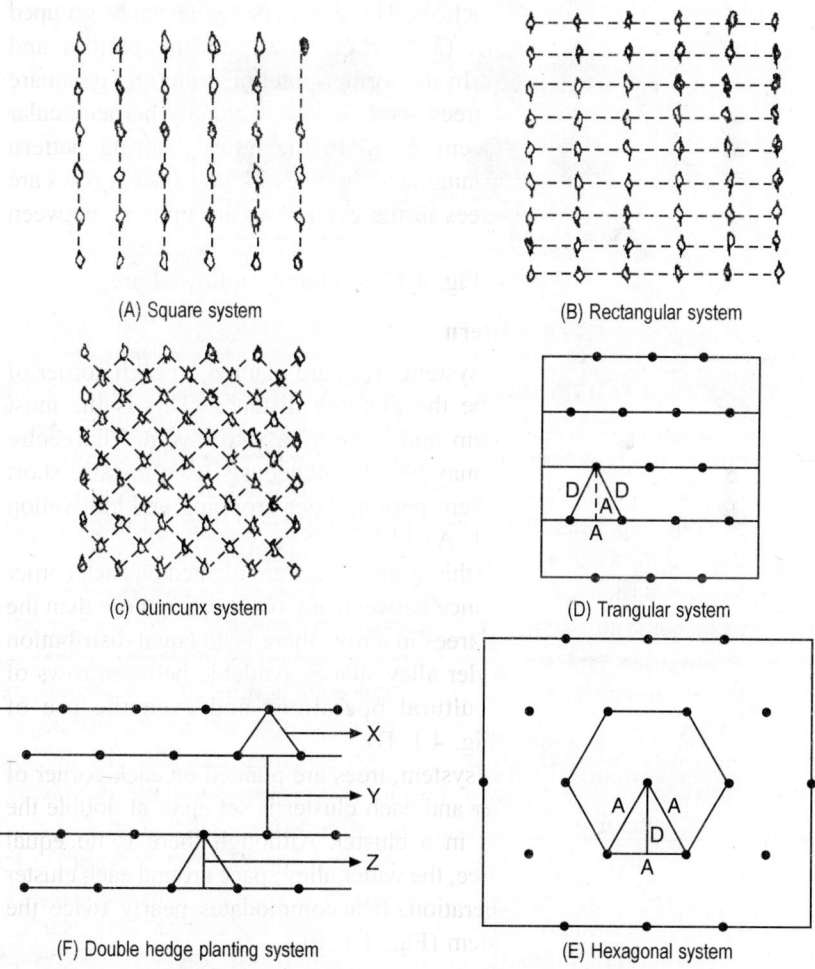

(A) Square system

(B) Rectangular system

(c) Quincunx system

(D) Trangular system

(F) Double hedge planting system

(E) Hexagonal system

Figure 4.1 System of planting in Horticultural Crops.

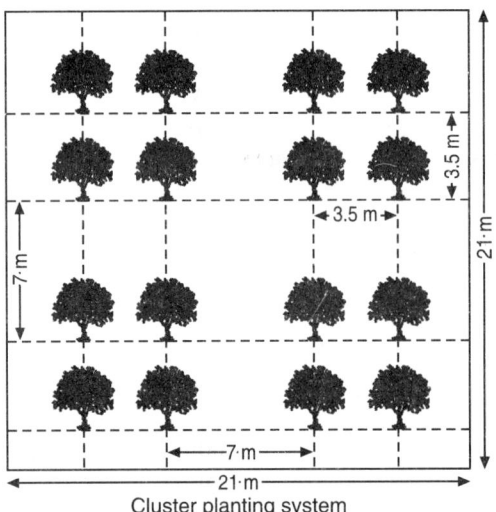

Cluster planting system

Figure 4.1F   Cluster system of planting.

5. **Diagonal or quincunx system:**   This is the square method but with one more plant in the centre of the square. This will accommodate double the number of plants, but does not provide equal spacing. The central (filler) tree chosen may be a short lived one. This system can be followed when the distance between the permanent trees is more than 10 m. As there will be competition between permanent and filler trees, the filler trees should be removed after a few years when main trees come to bearing (Fig. 4.1. C).

6. **Triangular system:**   The trees are planted as in square system but the difference being that those in the even numbered rows are midway between those in the odd rows instead of opposite to them. Triangular system is based on the principle of isolateral triangle. The distance between any two adjacent trees in a row is equal to the perpendicular distance between any two adjacent rows. However, the vertical distance between immediate two trees in the adjacent rows is equal to the product of 1.118 x distance between two trees in a row. When compared to square system, each tree occupies more area and hence it accommodates few trees per hectare than the square system (Fig. 4.1. D).

7. **Contour system:**   It is generally followed on the hills where the plants are planted along the contour across the slope (Fig. 4.2).

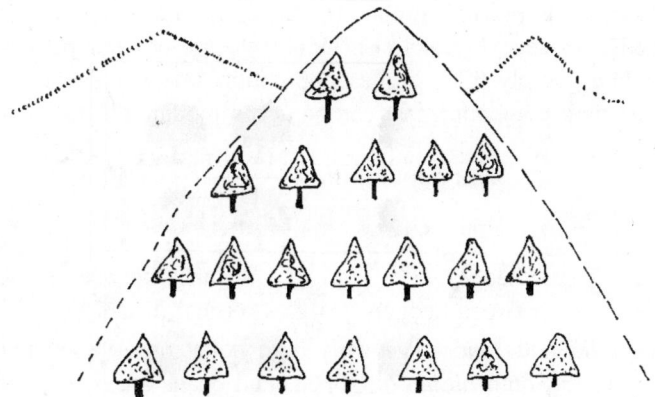

Figure 4.2 Contour system of planting.

It particularly suits to land with undulated topography, where there is greater danger of erosion and irrigation of the orchard is difficult. The main purpose of this system is to minimize land erosion and to conserve soil moisture so as to make the slope fit for growing fruits and plantation crops. The contour line is so designed and graded in such a way that the flow of water in the irrigation channel becomes slow and thus finds time to penetrate into the soil without causing erosion. Terrace system on the other hand refers to planting in flat strip of land formed across a sloping side of a hill, lying level along the contours. Terraced fields rise in steps one above the other and help to bring more area into productive use and also to prevent soil erosion (Fig. 4.3).

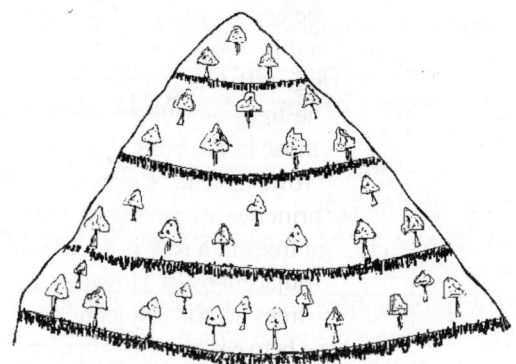

Figure 4.3 Terrace system of planting.

The width of the contour terrace varies according to the nature of the slope. If the slope becomes stiff, the width of terrace is narrower and *vice-versa*. The plant distance under the contour system may not be uniform. In

South India, tea is planted in contours either in single hedge system or in double hedge system (Fig. 4.1. F). Double hedge contour planting system accommodates nearly 22% higher population than single hedge system. Number of plant population that can be accommodated in this system is

$$\text{Plant population} = \frac{N \times \text{unit area}}{D(y + z)}$$

Where

N – number of hedges   D – distance between plants
Y – distance between hedges   Z – vertical distance between row

This system in tea helps to get early, high yield, conserve soil and suppress weed growth. In South Indian hills, peas and beans are sown under paired row system which is almost similar to double hedge planting system. The seeds are sown at 10 cm interval in each double rows of 30 cm apart with the distance of 1.5 m between each pair of rows.

Planting Distance

The minimum vertical distance between any two trees or plants is referred as the planting distance and this varies depending upon many factors. The principles in deciding the planting distances are the following:

1. Trees when fully grown, the fringes of trees should touch each other but the branches should not interlock.
2. Trees root will spread over a much larger area than top and there should be proper room for the roots to feed without competition.

Factors which decide the planting distance are the following:

1. Kind of fruit trees—mangoes are planted at a distance of 10 m × 10 m, guavas at a distance of 5 m × 5 m while papayas are planted at a distance of 2 m × 2 m.
2. Varietal factor—spreading nature of varieties requires wider spacing than upright growth types.
3. Rainfall—wider spacing should be given in low rainfall areas than the high rainfall areas for a kind of tree.
4. Soil type and soil fertility—closer spacing may be given in fertile soil.
5. Rootstocks—trees of the same variety grafted on different rootstocks will grow to different sizes and as such require different planting distances. e.g. Apple.
6. Pruning and training—trees trained on head system requires closer spacing than the other type of training system.
7. Irrigation system.

In general, if the spacing is too wide, it is obvious that the yield per unit area would be greatly reduced. Only in very exceptional cases would this be justifiable. Ordinarily, it is more profitable to plant the trees closer together and supply the needed water and nutrients. If the trees are too close together, the trees grow tall rendering pruning, spraying and harvesting difficult. There is root competition and inadequate nutrition and the trees as such give less yield and produce smaller fruits of poor colour. Cultivation also becomes difficult in the closely planted orchards. Hence, proper canopy management is vital through pruning and other techniques. Close planting results in a greater yield per unit area in the early life of the tree but less in the more important later years. Close planting is therefore a false economy.

The total number of trees per hectare for various important horticultural crops under (a) Square (b) Hexagonal and (c) Triangular system of planting are given below:

Crop	Planting distance (m)	No. of trees per hectare		
		Square system	Hexagonal system	Triangular system
Mango	10 × 10	100	115	89
Sapota	8 × 8	156	178	139
Clove	6 × 6	277	320	248
Acid lime	5 × 5	400	461	357
Coconut	7.5 × 7.5	177	205	159

It may be seen that hexagonal accommodates 15% more number of plants while triangular system accommodates 11% lesser number of plants than square system. The calculation of the number of trees per hectare when planted under square or *rectangular* system is very easy and is obtained by dividing the total area by the area occupied by each tree (a × a in square system or l × b in rectangular system). The theoretical and the actual number of possible trees which can be planted in an orchard depend upon the shape of the field. In practice, in large fields, the percentage difference between the theoretical number and the actual possible number will be less.

High Density Planting System

Planting of fruit trees rather at a closer spacing than the recommended one using certain special techniques with the sole objective of obtaining maximum productivity per unit area without sacrificing quality is often referred as 'High density planting' or HDP. This technique was first established in apple in Europe during sixties and now majority of the apple orchards in Europe, America, Australia and New Zealand are grown under this system. In this system, four planting densities are recognized for apples *viz.*, low HDP (< 250 trees/ha), moderate HDP (250–500 tree/ha), high HDP (500 to 1250 trees/ha) and ultra high HDP (>1250 trees/ha). Recently, super high density planting system has been also established in apple orchards with a plant population of 20,000 trees per ha. In some orchards, still closer planting of apple trees is followed (say 70,000 trees/ha) which is often referred as meadow orchards.

Advantages of HDP are

(i) Early cropping and higher yields for a long time; the average yield in apple is about 5.0 t/ha under normal system of planting and it is about 140.0 t/ha under HDP.
(ii) Reduced labour costs.
(iii) Improved fruit quality.

Characteristics of HDP are

(a) The trees of HDP should have maximum number of fruiting branches and minimum number of structural branches.
(b) The trees are generally trained with a central leader surrounded by nearly horizontal fruiting branches.
(c) These branches should be so arranged and pruned in such a way that each branch casts a minimum amount of shade on other branches.
(d) The height should be one and half times its diameter at the base. A key to successful HDP depends upon the control of tree size.

This is achieved by

(a) Use of size controlling root stocks. In apple, dwarfing root stocks and intermediate stocks like MM 106, MM 109, MM 111 are used to control the size of the plant. In pears, Quince A, Adam and Quince-C are commonly used as dwarfing root stocks. Vellaicolamban, a polyembryony cultivar of mango when used as rootstocks, it imparts compactness to the scion.
(b) Use of spur type scions—In temperate fruit crops like apple, the cultivars can be classified into a spur type or non-spur type. The spur types which have restricted annual growth are alone suitable for HDP.
(c) Training and pruning methods to induce dwarfness - under Indian conditions, apple trees trained under spindle bush, dwarf pyramid, cordon systems are found to contain the growth of the trees appreciably for HDP system.
(d) Mechanical device and use of chemicals to control size—Growth regulators such as daminozide, ethephon, chloremquat and paclobutrazol are extensively used to reduce shoot growth by 30 to

0%. This results in increased flowering in the subsequent years and may be useful in encouraging earlier commercial fruit production. Besides chemical manipulation, mechanical devices employing the use of spreaders and tying down the branches to make them grow from near horizontal to an angle of 45° from the main stem are also some of the standard practice to control tree size.

Planting system for HDP: The success of HDP depends upon the right choice of planting system. Generally, rectangular planting with single, double and three row plantings are followed. In single row planting the distance within the row is close, whereas the distance between the row is wide (4 × 2 m). In double row planting, a wider spacing is given after every two rows (4 + 2 × 2 m) whereas in three row planting, a wider spacing is given after every three rows (4 + 2 × 2 × 2 m). In meadow orchard system, a bed of 10 to 15 rows is closely planted (*nay* 30 × 45 cm) and separated by alleys of 2.5 m width between beds. This system is also called bed system.

Planting Season

The season of planting varies with different fruits and local conditions.

There are two seasons of planting in vogue in India.

- (i) Monsoon (June - August) and
- (ii) Spring (February - March)

Monsoon season is considered to be the best for planting evergreen fruit trees like Citrus, Mango, Sapota and Guava. If the trees are planted early in the rainy season they soon establish themselves and grow vigorously. Deciduous trees may be planted during the dormant period without shock. Care should be taken that planting is done before the growth starts, otherwise trees suffer severely and will be in poor condition to withstand the next hot weather.

Planting Methods

After locating the positions of the orchard trees, it is important that the trees are planted exactly where the stakes stood. It can be easily done with the help of a planting board. The planting board is usually of 1.5 m long, 10 cm wide and 2.5 cm thick with a central notch and one hole on either end, (Fig 4.4) the central notch and the two holes (one on either end) are in a straight line. The planting board is placed in such a way that the stake (tree marker) fits into the central notch. Two small stakes are inserted one in each end hole. The planting board along with the tree marker is then lifted straight up without disturbing the end stakes. A pit of about 1m cube or of the desired dimension the position of the tree marker is then dug.

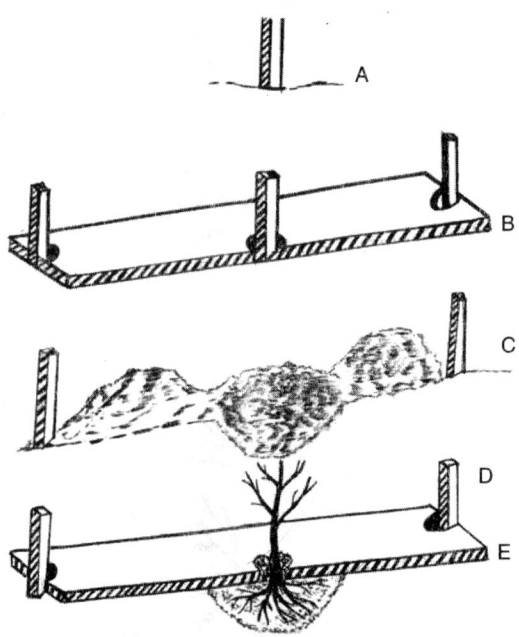

Figure 4.4 Use of planting board.

The pits are allowed to wither for few weeks before planting in some cases. The pits are then filled with top soil already mixed with red earth and well rotten farm yard manure. Irrigation is then applied to enable the contents of the pits to settle down properly. In the event of depressions taking place as a result of irrigation, more soil should be added to the pits so as to fill them to the level of land. The pits are then ready for planting.

Transplanting

The trees should be planted approximately where the original pegs were placed. This is achieved by replacing the planting board in position with the help of the guide pegs and the stem of the trees is brought to the central notch with the help of a hand hoe. One of the most common mistakes is that of planting the trees too deep. The plants should be set in such a way that the bud union remains slightly above the ground level. The trees in the field should be planted as deep as they stood in the nursery. The trees are irrigated soon after planting. This consolidates the soil and helps the roots to establish contact with it and to secure a supply of water quickly. A small basin may be made around the tree for this purpose. Planting if taken up during the rains, this basin should be demolished within a day or two so that water will not collect around the tree. This is more dangerous on heavy than light soils.

Spring winds cause damage to the growing plants by giving a constant shaking. To prevent this, plants should be staked when planted (Fig. 4.5). Some young plants are subjected to considerable injury, form sunburn particularly if they have been trained to single stem with no branches for 45 cm, or more from the ground. Such burn can be protected by wrapping them with paper or other material or by coating them with white wash. The latter is probably best, as most material wrapped around the trunk would be subject to termite attack.

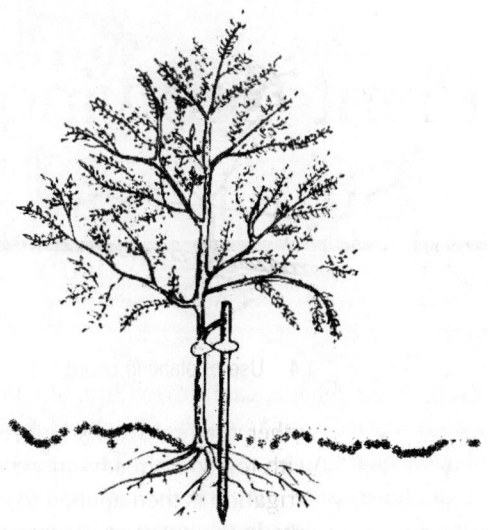

Figure 4.5 Correct method of tying tree sapling.

5

Plant Growing Structures

Some vegetable crops like cabbage, cauliflower, brinjal, chillies, tomato and few others are usually started in well prepared nursery beds and the young plants are transplanted later in the main field. This is the normal practice followed in most of the tropical countries where severe winter season is absent. But in countries having long cold winters and short growing seasons, these kinds of vegetables and some flower crops are grown in some plant growing structures. These plant growing structures are also used for starting solely for an early crop in certain areas where winters are not so severe. These plant growing structures include hot beds, cold frames, green houses or glass houses.

Advantages of starting plants in plant growing structures:

1. Increasing the length of the growing season and making it possible to grow long season crops in regions where summer is short.
2. Protecting the plants from unfavourable weather.
3. Obtaining higher yields of long season tender crops in places where summers are short.
4. Making it possible to produce an earlier crop by planting seed before it would be safe to plant in the open.

Hot Beds

The main use of hot beds is for starting plants to be grown in the main field and in certain cases to grow crops to maturity during non growing season. These hot beds are usually located near the farm buildings and near a good water supply. South and south-eastern exposures are preferable because beds will get more sunshine with these exposures than with others.

Generally the frames used in hot bed may be made of wood, cement, brick or stone, but the first two materials being the most common. Concrete frames are much more durable than wood and are cheaper in the long run. These frames may extend 30 to 45 cm above the surface of the ground on the backside and 15 to 30 cm in the front, thus giving a slope. For every one metre, a cross bar or a slide should be placed for the sash to rest upon.

Most durable wood should be used in making hot bed sash. The most commonly used size of the sash is 180 cm x 90 cm; as a large sash is too heavy to handle. The sash is usually painted before being glazed. These hot bed sashes need protection during cold weather and is provided by old matting, carpets or straw mats.

Hot beds have different heating systems; (i) Manure heated, (ii) Flue heated (iii) Hot water heated depending upon the sources through which heat is developed in the hot bed.

1. **Manure heated hot beds:**   Two parts of fresh horse manure is mixed with one part of straw which will give good result and preparation of this mixture is completed 10 to 14 days before the beds are actually needed. These mixtures are placed in a pit where we need actually the hot bed and optimum moisture condition is provided so as to allow the mixture to decompose and thereby it will begin to steam. A layer of 10 to 15 cm of good soil cover over the manure will serve as the hot beds to start the seeds.

2. **Flue heated hot beds:**   In this type of hot beds, the beds are frequently heated by hot air conducted through flues from a fire box located at one end of the beds. For a bed of these, pipes are placed below the bed at a depth of 40 to 45 cm. Two chimneys are constructed for each bed one in each corner of the fire box end. The height of the chimney varies from 1 to 1.5 m. The smoke and warm air from the fire box pass through the flues and circulate under the bed and finally escape through the chimneys.

3. **Hot water heated hot beds:**   It is the most satisfactory method as the temperature can be controlled more easily than any other method. But it is more expensive to install than the other methods.

4. **Electrically heated hot beds:** Heating of beds with electricity is more common in many regions where the vegetables are to be started. The electricity is passed through specially constructed cable, consisting of electrically insulated resistance enclosed in a flexible metal sheath. Though initial cost of an electric hot bed is high, the other advantages are that the beds are always ready for use and they are easily manipulated and readily removed.

Cold Frames

Cold frames are satisfactory for starting plants provided little protection is necessary. The cold frames are constructed in the same way as hot beds except that no pit is required and therefore, no heat is required in the cold frame except that provided by the sun. This is the main difference between cold frames and hot beds. Permanent cold frames are made of concrete and temporary ones are made of boards. They are covered with glass sash canvas or cloth. These cold frames are used (i) to start the plants in the spring, (ii) to harden the plants that have been started in the hot bed or green house, (iii) to grow certain vegetable crops like lettuce, celery, radish and beetroot to maturity.

Green Houses

A green house (they are also referred to as glass house in Europe) may be built either attached to the house or some other building or separate green house is advancement over hot beds or cold frames. The advantages of sowing in green houses over the rest of the plant growing structures are:

(i) Temperature can be controlled easily in green house than in other forcing structures.
(ii) Ventilation can be regulated better and with less danger of chilling the plants.

A small green house is useful to start plants during winter and early spring. This type of green house is very low in height (about one meter height) and is constructed by erecting frame structure, covered with hot bed sash.

The other type of green house which is used to grow the plants through-out their growth period is of very big size. The area of a green house may run from as small as ½ acre to 100 acres or even more. Inside the green house, raised benches or beds are provided to raise the plants. These green houses are constructed with a wood, a pipe or steel frame to which the wooden sash bars are fastened with the glass embedded in putty or some other glazing compound. The green house is heated by hot water or steam conducted through pipes which are distributed at predetermined distances. A thermostat should also be installed to control the heat. Now-a-days green houses are useful to force vegetables like tomato and cucumber in western countries and also ornamental plants like chrysanthemum, roses, carnations, African violet, hydrangea, poinsettia, etc.

Conservatory

The above mentioned plant growing structures *viz.*, hot beds, cold frames or green houses are exclusively useful only in temperate countries where winter is severe and the growing season is very much short. Therefore, the objective of green house in tropics (except in the hills) is not to protect the plants from cold but rather from the fierce sun's rays and strong winds, while also maintaining a fairly uniform and moist atmosphere. There are numerous kinds of ornamental plants with beautiful foliage or flowers or both, which cannot thrive in the open, exposed all the day time to sun and wind. The delicate ferns, graceful Anthurium, Alocasia, bright coloured Caladiums, wonderful orchids and several other plants require a reasonable amount of shade and protection from the sun and hot or cold breezes. Therefore, for successful culture of the above kinds of plants, a green house is essential in a tropic. This plant growing structure is also known as 'fernery' or 'conservatory'. A conservatory is usually built on a slightly elevated area. A short masonry wall, 75 to 90 cm high, can be constructed all round the fernery and enclosing it leaving gaps for entrance. The roofing is supported by stone pillars or iron or concrete posts and may consist of a strong frame work of iron girders and stout iron rods supporting a galvanized wire netting for the creepers to spread upon. Till the light creepers like *Antigonon leptopus* or *Ipomoea palmata* cover the roof, plaited coconut leaves may also be used to cover the top of the fernery. Humidity in the conservatory is increased by providing a small pool at the centre.

A glass house in the hill station of tropics provides ideal environmental conditions to grow tropical ornamental plants, orchids, cacti and succulents.

Plastic Green House / Poly Tunnels

A large amount of money is required for the construction of sophisticated green houses. As a consequence, it is beyond the reach of small or even medium growers. Since 1948, with the invention of plastic green houses, cultivation of plants under cover has become a reality to small growers. Depending upon the requirement the plants are grown under plastic green houses, poly-tunnels, row covers etc. More than 2 lakh hectares of land are now under plastic green houses and similar area under row covers in the world. For the construction of these simple structures wood, bamboo, steel pipes, mild steel rods, plastic pipes etc, are used as the frames and the roof is generally covered with UV stabilized polyethylene materials.

In certain polyhouses, exhaust fans or vents are fitted whereas in poly-tunnels one side of the plastic cover is lifted above the ground for ventilation and lowering the inside temperature during the day. In India, cultivation under plastic green houses and poly-tunnels has just been initiated. They do permit higher light intensity.

Advantages of Growing Under Cover are:

1. One can grow any plant in any place
2. Plants can be grown throughout the year
3. One can get blemish free products
4. Easy to protect from pests and diseases
5. Earliness
6. Reduced cost of cultivation and
7. Minimum water requirement.

Roses and gladioli when grown under polycover produce higher yield of quality flowers of export standards. Similarly, carnation growing for export purpose at Himachal Pradesh and Kodaikanal is possible only under poly green houses.

Net Houses

A net or shade house is a structure enclosed by agro nets or any other woven material to allow required sunlight, moisture and air to pass through the gaps. It creates an appropriate micro climate conducive to the plant growth. It is also referred as shade net house or net house.

Uses of Shade Net House

- Helps in cultivation of flower plants, foliage plants, medicinal plants, vegetables and spices.
- Used for fruit and vegetable nurseries as well as for raising of forest species etc.
- Helps in quality drying of various agro products.
- Used to protect against pest attack.
- Protects from natural weather disturbances such as wind, rain, hail and frost.
- Used in production of graft saplings and reducing its mortality during hot summer days.
- Used for hardening tissue culture plantlets.

A shade house structure comprises of two basic components i.e. frame and cladding material. The frame provides support for cladding material and designed to protect against wind, rain and crop load. Mild steel (ms) angle frame lasts upto 20 to 25 years, if anti rust treatment is done at regular interval, whereas bamboo structure can last upto 3 years. The agro shadenet as cladding material lasts for 3 to 5 years depending on the climatic condition. Shade nets are available in different colours with wide range of shade percentages *viz.*, 25%, 30%, 35%, 50%, 60%, 75% and 90%. The flower crops like anthurium, orchids can be grown under shade net houses.

Hydroponics

Intensive and profitable cropping in a green house condition is possible after the advent of a technique known as 'hydroponics'. The term 'hydroponics' was derived from two Greek words hydro (water) and ponos (labour), literally meaning 'water working'. It can be defined as the science of growing plants without the use of soil, but use of an inert medium such as gravel, sand, peat, vermiculite or saw dust, to which is added a nutrient solution containing all the essential micronutrients for normal growth and development. Since many hydroponics methods employ some type of medium. It is often termed as 'soilless culture'.

This technique was first developed on commercial scale by W.F. Greicke of the University of California as early as in 1930. He grew root crops such as beets, radishes, carrots, potatoes and cereal crops, fruits, ornamentals and flowers. With the development of plastics, hydroponics has become a reality for green house growers. In virtually all climate areas and in developed countries, hydroponic green houses are established with 10 to even 100 acres under a single stretch. In India, hydroponics did not reach until 1946. Research conducted at Kalimpong, Darjeeling district led to the development of a novel method of hydroponics, known as Hydroponics — the Bengal system which has the advantage of being inexpensive to install, simple to maintain and economical to operate.

Advantages of Hydroponics are

1. Much higher yields than those grown on soil.
2. It can be used in places where ordinary agriculture or gardening is impossible.
3. Quality of the produce is always superior.
4. Freedom from soil borne diseases and weed.
5. Less cost of management.

Essentially there are three methods by means of which plants may be grown without soil. They are water culture, drip culture and gravel culture.

Water Culture

It involves growing of plants with their roots in solutions, no inert medium being used for support. Nutrients are supplied in the solution in which the roots are immersed. The solution is kept in a bottle or a tank, the top is covered with a wire netting and above that is placed material such as peat moss or saw dust or rice husks to give support to the seedlings. The water level in the tank should be increased with increase in the age of plants since more air will be required. The great drawback in this system is the lack of suitable aeration to roots.

Drip Culture

This method is also known as sand culture. In this method, plants are kept in sand medium and are watered with a nutrient solution.

Gravel Culture

An improvement over the above two methods is gravel culture and is one of the most widely used hydroponic techniques. A gravel culture technique should have the following accessories.

(a) **Nutrient tank:** They must be water proof, acid resistant and of sufficient size to hold about 40% of the total volume of the benches that they fill. These tanks may be made of concrete, wood or metal.

(b) **Benches:** Gentle sloping benches made of concrete or wood may be preferred and it should have a 'V' bottom. The surfaces of the benches may be coated with an emulsified asphalt to prevent any possible leakage. PVC pipes of 75 mm are held in position with the holes or cuts facing downward.

(c) **Medium:** Crushed granite of irregular shape free of the fine particles less than 1/40 cm in diameter and coarse particles more than 1.5 cm in diameter is selected. The particles should not be of calcareous material in order to avoid pH shifts.

(d) **Solutions:** Many different formulae have been advocated but the following formula has proved satisfactory on many crops.

Chemicals	Per 1000 litres of water
Sodium nitrate	0.5 kg
Potassium chloride	0.5 kg
Ammonium sulphate	0.1 kg
Magnesium sulphate	4.5 kg
Mono calcium phosphate	2.5 kg
Calcium sulpahte	0.5 kg

The above chemicals should be weighed out individually and mixed in the tank. This solution is suitable for all crops for the first month and when the plants have become well established, usually 3 or 6 weeks, the concentration should be doubled, which means twice the amount of chemicals per 1000 litres of water. The pH of the solution should be checked twice a week and it should be maintained at 6.5 for most crops. To raise the pH a stock solution of 12.5 g of sodium or potassium hydroxide to one litre of water should be used. To lower the pH, a stock solution of 6.25 g of either concentrated sulphuric or phosphoric acid per litre of water should be used. The nutrient solution is pumped into the medium 3 to 4 times in a day. The plants are spaced in gravel as in soil and set in along with the ball of soil.

Recent improvement in soilless cultures are Nutrient Film Technique (NFT), Tube culture and Aeroponics. Nutrient Film Technique is a water culture technique in which the plants are grown with their root systems contained in a plastic film through which nutrient solution is continuously circulated. Tube culture is a modification of the NFT and water culture. The nutrient solution is pumped though PVC drain pipes of 10 cm thick covered with black polythene film through which the seedlings are set into the nutrient solution flowing along the bottom of the PVC pipes. Aeroponics on the other hand is growing of plants in an opaque trough or supporting container in which the roots are suspended and bathed in a nutrient bath mist rather than a nutrient solution.

6

Nutrition of
Horticultural Crops

Nutrient supply is one of the most important factors influencing growth and productivity of the horticultural crops. Sixteen nutrient elements are recognized as essential to all plants for their normal growth and development. These essential elements are carbon, hydrogen, oxygen which are derived from air, soil and water and nitrogen, phosphorus, potassium, calcium, magnesium, sulphur, iron, zinc, manganese, copper, boron, molybdenum and chlorine which are supplied by the soil in varying quantities either from the reserves in the soil or through application of manures and fertilizers. Nitrogen, phosphorus and potassium are used in large quantities by plants. These are, therefore, called 'major' or 'primary nutrients'. Calcium, magnesium and sulphur are required in relatively smaller but in appreciable quantities and are called 'secondary nutrients'. Iron, zinc, manganese, copper, boron, molybdenum and chlorine are required by plants in small quantities for their growth and development. Hence, these are referred to as 'micro nutrients'.

Each of the plant nutrients has a definite and specific function to perform in the growth and development of plants. A deficiency of any of these essential nutrients causes some abnormal conditions and upsets the growth of the plants. Similarly, application of nutrients to plants in excess of their requirements may be more harmful than what they are applied in insufficient quantities. This is particularly true of micro nutrients like iron, zinc, manganese, copper, boron and molybdenum which are required by plants in very small quantities. Excessive application or higher dose of micro nutrients than required produces toxicity which affects the normal growth and development of plants.

The nutrient requirement of the plant depends on the nutrient supplying power of the soil on which the plant is to be grown. When the soil does not furnish adequate quantities of the elements necessary for normal development of plants, it is essential that the required amount may be supplied through application of manures and fertilizers. This necessitates finding a method that will permit the determination of these deficient elements.

The mineral nutrition need of horticultural plants differ from those of annual crop species in a number of ways, many; of which are related to their perennial nature. In perennial species, there is a need to supply nutrients to current crop production as well as for the vegetative organs which persist for several years. Nutrient application in a year may have its effect on both the nutrient status and productivity in the following years and in subsequent years, plant responds to both direct and residual soil fertility.

To ensure high economic productivity and to sustain the available soil nutrient status at the desired level, correct dose of manures and fertilizers must be applied by use of reliable diagnostic tools to avoid nutrient shortages or excess. Several diagnostic methods are available to arrive at need based manuring.

The assessment of fertilizer requirement may be done by the following methods.

1. **Experiments under controlled conditions and field trials:** In this method, series of treatments containing various doses of fertilizers are selected and are assigned to the plots at random within a given area and are repeated a number of times to obtain more reliable results. Results of experiments are analyzed to find out the most economical fertilizer recommendations under a given set of conditions.

2. **Soil analysis:** Soil samples collected from the field are analyzed for the nutrient contents and their contents are rated as low, medium and high. Based on this rating, the required nutrients may be supplied through judicious application of manures and fertilizers.

3. **Nutrient uptake studies (Crop removal):** This gives a rough idea about which nutrient(s) are required more by the plants. For example, nutrient uptake studies conducted in many banana growing countries showed that a banana crop of 40–60 t per ha removes 250–300 kg N, 25–40 kg P_2O_5 and 800 to 1200 kg K_2O besides 150 kg Ca and 60 kg Mg. It indicates that the fertilizer schedule for banana should be such that it should supply more of potassium and less of phosphorus. Similarly, the nutrient uptake pattern by peas and beans showed that they remove lot of P than N or K.

4. **Leaf analysis:** Leaf is the centre of physiological activity in plants and leaf analysis to assess nutrient requirement of a given crop, is based on the assumption that within certain limits, there is a positive relation among doses of nutrient applied, leaf content of the element and the yield/quality of the economic produce. In a leaf, the nutritional concentration in zone of deficiency remains relatively uniform and the transition zone in zone of deficiency and adequacy serves as an ideal indicator of the nutritional status of the plant at any time during the growing season. The sampling of leaves should be done prior to the application of manures and fertilizers and the time and stage of leaf vary from crop to crop as indicated below:

Crop	Plant part	Age, stage and position and other
Mango	Leaf including petioles	4–7 months old leaves from middle of the shoot
Banana	Leaf lamina on both sides of midrib on centre of leaf petiole or midrib can be also taken	3rd fully opened leaf which is around 20 days old.
Citrus	Recently matured leaf	4 months old

Based on yield and fruit quality responses, the nutrient content so determined can be classified into five categories *viz.* deficient, low, optimum, high and excessive or toxic. Leaf analysis indicates the relative concentration of the nutrients in relation to its optimum. However, if it is intended to translate this into nutrient recommendations, i.e. how much manure and fertilizers are to be applied, adjustments are to be made in the existing schedule considering the soil fertility status.

Manures and Manuring

Manures are substances of organic or inorganic nature which are capable of supplying the various plant nutrients when applied to the soil. Manures are usually divided into two groups, *viz.,* organic and inorganic manures. Organic natural manures include excreta of animal matter such as blood, bones, flesh, wool, horn etc., and decomposed vegetation. They usually contain all the essential nutrients in various proportions. The inorganic manures usually supply one essential plant nutrient and hence are also called 'relative manures'. They supply nitrogen, phosphorus or potassium.

A. Important organic manures

1. Cattle manure or farm yard manure

The manures produced by horse, cattles or other animals are included in this category. It takes a long time to decompose, nearly a year before it becomes usable. It is more suited to light than heavy soils. They contain 0.6% of nitrogen, 0.35% phosphorus and 0.6% potash, however the percentage of these nutrients may vary depending upon the substances the animal feed, age of animals, conditions of animals and storage and handling including the kind of litter used. The manure is applied as a basal dressing by broadcast and immediately incorporated into the soil by ploughing.

2. Night soil

It is a powerful manure rich in nitrogen. The chief objection to its use is based on sanitary and sentimental grounds on account of its offensive odour when it is not sufficiently decomposed. On the dry basis, it contains 5.5% of N, 4.4% P and 2.0% K. As it is strong manure, regular watering should be done when it is used.

3. Guana

It is a well known manure rich in nitrogen (10.15%) and phosphorus (9.82%). It occurs as deposit in large quantities, principally off the islands off the coast of Peru and South America. It is the excrement of sea birds accumulating over several centuries. The pure form of Guana is practically exhausted now. Guana may be mixed with about six times its weight of soil.

4. Bones

They are rich in phosphoric acid and calcium. Steamed bone meal contains not less than 3.5% of nitrogen and 23% phosphoric acid. Bones are especially beneficial to soil deficient in lime.

5. Oil cakes

They are residues left after the oil is extracted from the seeds of Groundnut, Castor, Gingelly, Pongamia etc., and they contain 3–5% nitrogen and 1.5–2% of phosphorus. They are best applied to potted plants in the form of liquid manure.

6. Leaf mould

Withered and dry leaves and garden sweepings are thrown into a pit in a shady corner in the garden and covered with earth and watered copiously once or twice in summer to assist decomposition. Decomposition will be completed within a year. When well decomposed, leaf mould could be powdered and sifted through wire mesh. Leaf mould is rich in humus and is hence applied to both sandy and clay; soils. It is usually mixed with soil in the preparation of many pot mixtures.

7. Wood ash

It is rich in potash. Vegetables generally require liberal manuring with potash.

8. Coir compost

The coir pith, obtained as a waste material from coir industry, is composted with a fungus called *Pleurotus sojar caju*. This compost is having about 25:1 C/N ratio and hence is a richer source of organic manure and it improves the crop growth. They also contain good amount of macro (N 1.4%, P_2O_5 0.06%, K_2O 1.2%), secondary (Ca 0.98%) and micro nutrients such as Fe, Mn, Zn and Cu. Its unique property is to absorb and retain moisture about 5–6 times of its volume and hence it is a good material to apply for rainfed horticultural crops.

9. Vermi compost

Organic waste materials and animal dungs when fed with certain species of earthworm *viz., Eudrilus eugeniae, Eisenia fetida* and *Perionyx excavatus* the 'worm cast' or excrements of the worms form the needed organic fertilizer called 'vermicompost'.

It is rich in organic carbon content (47%) and humus substances which help in building soil structure and stimulating plant growth particularly that of roots. It can be applied to horticultural crops relatively in small quantity in the places of organic manures or organic cakes along with 50–75 per cent recommended quantities of NPK fertilizers.

The major roles of various organic manures are:
1. To serve as a source of major, secondary and minor elements
2. To build up soil organic matter and maintain fertility
3. To improve physicochemical and biological properties of the soil
4. To have residual effect
5. To reduce the incidence of pests and diseases
6. To improve the quality of the crop
7. To act as a chelating agent.

The soil organic matter can be increased by the addition of 'synthetic farm yard manure' which is popularly called 'compost'. It may be defined as the material resulting from the decomposition of plant residues under the action of bacteria and fungi. The final product is usually dark brown in colour and when correctly made, resembles ordinary farm yard manure both in its properties and appearance. By regulating the temperature and moisture and by introducing the required amount of nitrogen, phosphorus, potassium and calcium, the speed of decomposition and the nature of the product can be decided. Well prepared compost contains 0.75 to 1.00% N, 0.60 to 0.75% P_2O_5 and 1.00 to 1.50% K_2O.

Sometimes, green manure or green-leaf manures are ploughed into the soil for the purpose of incorporating organic matter, thus supplying humus as well as nutrients contained in them. The green manure crops are generally leguminous plants, raised for the sole purpose of serving as manure. About 45 to 60 days old crops can accumulate about 100 to 200 kg N per ha. The following are the commonly grown green manure legumes in India.

1. Sunhemp (*Crotolaria juncea*)
2. Daincha (*Sesbania aculeata*)
3. Pillipesara (*Sesbania speciosa*)

'Green leaf manuring' refers to the incorporation of the green leaves and other tender parts of the plants collected from the shrubs and trees grown outside the field and also collected from the waste lands and nearby forests into the soil. The popular plants are:

1. Gliricidia (*Gliricidia maculata*)
2. Sesbania (*Sesbania speciosa*)
3. Pungam (*Pungamia pinnata*)

The green manuring is generally recommended in orchard to build up the humus status of the soil, available plant nutrients and enhance the biological properties of the soil. Green leaf manures like *Gliricidia maculata* can be grown in the roadsides and bunds.

B. Important Inorganic Fertilizers

The following are the important chemical fertilizers with which the farmers may be familiar.

Nitrogenous Fertilizers

The following table illustrates the nitrogen content of different nitrogenous fertilizers.

S. No.	Name of the fertilizer	Nitrogen Content %	Form of nitrogen
1.	Sodium nitrate ($Na_2 NO_3$)	16.0	Nitrate
2.	Potassium nitrate (KNO_3)	12.5–13.5	Nitrate
3.	Ammonium sulphate ($(NH_4)_2 SO_4$)	20.6	Ammonical
4.	Ammophos – A	11.0	Ammonical
5.	Ammophos – B	16.0	Ammonical
6.	Ammonium nitrate	33.0	16.5% Ammonical, 16.5% nitrate
7.	Ammonium sulphate nitrate	25.6	19.0% Ammonical, 6.6% Nitrate
8.	Urea ($CO NH_2)_2$)	46.0	Amide
9.	Calcium cyanamide	20.6	Amide

Among the above, ammonium sulphate and urea are most popular among farmers in India. Ammonium sulphate is a fine crystalline salt containing no moisture and also it absorbs no moisture from the atmosphere. It is readily soluble in water and is retained in soil without any loss. Urea is cheaper per kg of nitrogen than other solid nitrogenous fertilizers available in India. Commercial urea is relatively pure white and crystalline. It is readily and completely soluble in water. This is suitable for all crops in all types of soil. The nitrogen present in it is converted to ammonical and nitrate forms which are suitable for absorption by crop plants.

Phosphatic Fertilizers

The commonly available phosphatic fertilizers are:

1. **Superphosphate:** This is the most extensively used one popularly referred as 'super'. It contains 16 to 18% of water soluble phosphate.
2. **Basic slag:** It is a by-product in the manufacture of steel, a black finely divided powder containing 8 to 18% of phosphoric acid which becomes available to plants slowly.
3. **Rock phosphate:** It contains 30 to 40% of P_2O_5, 3 to 4% fluorine and varying amounts of lime. It is best suited to acid soils when compared to neutral or alkaline soils. They are suitable source for reinforcing the compost or farm yard manure.

Potassic Fertilizers

The commonly used potassic fertilizers are muriate of potash (potassium chloride) and potassium sulphate. Potassium chloride contains 48 to 62% K_2O and 35 to 47% chlorine. Certain fruits and vegetales are sensitive to high chlorine content. In such cases, potassium sulphate may be preferred. It contains about 50% K_2O besides containing 18% of sulphur. Sulphate of potash (SOP) is having a low salt index and is virtually free of chloride and hence recommended for fruits and vegetables which are chloride sensitive.

Mixed Fertilizers

It is a mixture of more than one straight fertilizers, which can supply more than one plant nutrient elements.

The advantages of mixed fertilizers are:
1. Saving in time and labour in application.
2. Saving from cartage of too many straight fertilizers from too many places.

The disadvantages of mixed fertilizers over straight fertilizers are:
1. Specific needs of crops individual nutrient element cannot be satisfied.
2. Unit cost of the various nutrients contained in the mixed fertilizers will always be higher to the unit cost of nutrients contained in the straight fertilizers.

Mixed fertilizers are now available for important horticultural crops. The contents of these may vary depending upon the crops requirement.

Biofertilizers

Biofertilizers are carrier based preparations containing beneficial micro organisms in a viable state intended for seed or soil application and designed to improve soil fertility and help plant growth by increasing the number and biological activity of desired micro organisms in the root environment. They are supplement to chemical fertilizers and comparatively cheap. Biofertilizers are otherwise called as 'microbial inoculants'. The following microbial inoculants are being used to increase the growth and production of horticultural crops.

    (i) Inoculants of biological nitrogen fixing micro organisms
   (ii) Phosphobacterial inoculants and
  (iii) Mycorrhizal inoculants

Biological Nitrogen Fixation

Fixation of atmospheric nitrogen is effectively carried out by specific group of micro organisms either in free living condition (e.g. Azotobacter) or in symbiotic association with leguminous (e.g. Rhizobium) and non-leguminous (e.g. *Azospirillum*) plant systems. These micro organisms function at the prevailing atmospheric temperature and pressure.

The above microbial inoculants *viz.*, Azotobacter inoculant, *Azospirillum* inoculant and Rhizobium inoculants are available as carrier based (Peat soil based or lignite based) inoculants for different horticultural crops. The different methods of application of the above biofertilizers include (i) seed treatment or seed inoculation (ii) soil application or soil broadcasting and (iii) seedling dip or root bacterization in the case of transplanting vegetables to derive maximum benefits. By the above biofertilizer application, the fertilizer nitrogen can be reduced to the tune of 20 to 25%.

Phosphobacterial Inoculants

Some free living bacteria in soil belonging to the genera *Bacillus* and *Pseudomonas* and fungi belonging to the genera *Penicillium* and *Aspergillus* possess the ability to bring insoluble phosphates in soil into soluble forms by secreting organic acids. The phosphobacteria are also known to mineralize the organic phosphates through enzymatic actions. Pre-treatment of the vegetable and other horticultural seeds with phosphobacteria has been reported to help in reducing phosphate fertilizer requirements of the crops and increasing the yield.

Mycorrhizal Inoculants

Most of the horticultural plants have their roots colonized by soil inhabiting mycorrhizal fungi. These fungi known as Vascular Arbuscular mycorrhizal fungi (fungi VAM) live in plant roots and grow out into the soil and increase the uptake of soil nutrients of low diffusion, particularly phosphorus and also increase the uptake of water. Invariably VAM fungi stimulate plant growth through increase in P inflow into the host plant. Experiments have confirmed that the VA-mycorrhiza enhances phosphate uptake through physical extension of the root absorbing surface rather than by induced changes in root physiology or by chemical solubilization of insoluble soil phosphate.

However, the VAM inoculum is available as soil + root based inoculum since the culturing of the fungus is not possible under laboratory condition. Experimental evidences proved the role of VAM inoculation in bhendi, gourds, chillies, egg plant, sweet potato by increasing the growth and yield. VAM inoculation induces better growth and more stem thickness in acid lime facilitating early production of pre-immunized acid lime seedlings.

Thus, biofertilizers bring lot of benefits to the farming community by way of economizing fertilizers and thereby reducing the cost of cultivation besides improving nutrient status of the crop plants and improving soil fertility etc.

Time and Method of Application of Fertilizers and Manures

The crops are manured or fertilized in order to supply the nutrient elements which are not present in sufficient quantities in the soil. Their application results in higher yield only when they are applied at the proper time and at the proper place. The following are the broad principles that decide the time of application of fertilizers and manures after choosing the fertilizers to be used.

Nitrogen is required throughout the crop growth and all plants are found to absorb the nitrogen continuously from the soil throughout the entire growth phase. Their requirement may however be low at the initial and final stages. Besides, all the nitrogenous fertilizers are readily soluble in water and more loss is found to occur. Therefore, it is advisable to supply nitrogenous fertilizers in split doses instead of applying the entire quantity at one time. The fertilizers can be applied as basal dressing or top dressing whenever and wherever required.

Phosphorus is required in large amounts in the early stages of growth. Moreover all the phosphatic fertilizers should be applied before planting or sowing the crop. Sometimes the available phosphorus becomes unavailable due to fixation. Therefore, it is generally recommended that the entire quantity is applied as based for crop growth, but the release of this nutrient in the soil is very slow. Pottasium is required throughout the crop growth. Therefore, it is desirable to apply the entire quantity of potassic fertilizers before sowing or planting the crop.

Methods of Application

Solid manures and fertilizers are commonly applied to crop plants by the following methods:

1. **Broadcasting as a basal dressing:** The fertilizer is applied uniformly over the entire cultivated surface of the land. It may or may not be incorporated into the soil. The broadcasting can be done just before the last ploughing or planting or sowing depending upon the nature of materials and crops grown. Concentrated organic manures, bulky organic manures like cattle manure, various ammoniacal fertilizers and potassic fertilizers are applied as basal broadcast.

2. **Broadcasting as a top dressing:** This refers to the application of fertilizers by broadcasting when the crop is in the field. This is done to meet the immediate demands of the growing crops. Only nitrogenous fertilizers are usually used for top dressing.

Liquid Fertilizers May be Applied by the Following Methods

1. Starter solution

Solutions made up of water-soluble nitrogenous, phosphatic and potassic fertilizers in small quantities (concentration 0.05%) are used for the establishment of young plants. This solution is called starter solution. At the time of transplanting in the field, the roots of young plants are dipped in this solution. This results in quicker and more establishments of seedlings (e.g. tomato). One great advantage of it is that the nutrients reach all the plant roots.

2. Foliar application

Many nutrient elements can be absorbed through the leaves of the plants with very great rapidity and efficiently. Recently, specialty water soluble fertilizers are sprayed at various stages of fruit and other horticultural crops to increase the crop yield with quality produce. In all the cases, the concentration should never exceed 1%. Many of the nitrogenous fertilizers, some complete fertilizers and carriers of trace elements are recommended as foliar application. When compared to soil application, plants require less quantity of nutrients if supplied through foliar application. To be more effective, two or three spray applications at shorter intervals of time may be needed. Two or more trace elements may be combined and given as a single spray and sometimes this may be also combined with fungicidal or pesticide spray solutions. Foliar sprays employing urea to supply nitrogen have been successfully used for many crops (e.g. Brinjal, Bhendi etc.). One precaution to be observed with foliar spray is the use of correct concentrations of nutrients, otherwise higher concentration results in many problems to the crop plants.

Integrated Nutrient Management

Integrated plant nutrient supply (IPNS) or Integrated nutrient management (INM) system refers to the maintenance or adjustment of soil fertility and plant nutrient supply at an optimum level for sustaining the desired productivity through optimization of the benefits from all possible sources of plant nutrients in an integrated manner. Conceptually, the IPNS strives to achieve: (a) regulated nutrients supply for optimum crop growth and productivity, (b) maintenance or some times an improvement in soil fertility, and (c) minimum adverse impact on agro-ecosystem quality by means of striking a balance among various nutrient sources *viz.*, soil fertilizers, organic manures and bio-inoculants. This concept has been extensively followed in horticultural crops as it improves the soil physical, chemical and biological properties besides reducing the requirement of inorganic fertilizers.

7

Irrigation of Horticultural Crops

The water relations of the plant are of extreme importance both for vegetative growth and for fruit production. It is necessary for rapid growth, satisfactory crop and to maintain turgor in cells for maximum photosynthetic activity. In arid and semi arid zones, irrigation is a very important cultural practice. Even in humid areas where distribution of rainfall is not satisfactory, irrigation is essential during the drought period.

The need for irrigation and also the amount of water that should be supplied are influenced by the following factors:

1. **Annual precipitation:** If rainfall is high or low but irrigation facilities are available, intensive cropping can be followed. If irrigation facilities are not available and the rainfall is also poor, extensive cropping with drought tolerant crops can be followed.
2. **Period of moisture shortage:** In South India, the period from December–March is totally free of rainfall and during this period irrigation is a must even to perennial crops.
3. **Stage of the crop:** Irrigation requirements sometimes depend upon the stage of growth of the crop. For instance, fruit bearing mango trees are to be regularly irrigated at 10–15 days interval during the fruit development stage i.e. from fruit set to full development stage. But to obtain good flowering, irrigation must be stopped 2–3 months before expected flowering period.

4. **Type of crops and cropping:** Most horticultural crops have high moisture requirements. Some fruit trees have deeper root system and hence during the period of drought they suffer very little or not all if the subsoil moisture is at a high level. Moisture shortage during critical periods of cell enlargement will greatly affect in many succulent vegetable crops.

5. **Water use efficiency:** Water use efficiency (WUE) refers in general terms to amount of water used per unit of plant material produced i.e. in terms of dry matter produced or marketable crop. To maximize WUE, it is necessary to minimize the losses through run off, seepage, evaporation, and transpiration through weeds as well as to promote maximal growth by planting high yielding, drought tolerant varieties.

The frequency of irrigation is determined by the following factors:

1. **The nature of soil:** Fine texture soils hold moisture longer than soils of coarse texture. Deep soils hold larger quantities of water than shallow soil. Presence of organic matter content also increases the same. When the water holding capacity of soil is increased the interval between irrigation can be extended.

2. **Rate of absorption by plants:** Transpiration rate of the crop plants affects the rate of absorption of water and consequently influences the frequency of irrigation. Those plants with large leaf surface require more water than those with reduced leaf surface.

3. **The root system of the crop:** A shallow rooted crop requires more frequent watering than a deep rooted crop. There is no absolute method for determining the time to irrigate. Some growers can tell based on the external symptoms. The immediate symptoms of lack of water are wilting, drooping of leaves, curling of leaves, shrinkage of fruit etc. Since the feel test is difficult to describe and requires considerable skill, soil moisture meters like irritometers and Bouyoucos moisture meter are available which measures the moisture content of the soil.

SYSTEMS OF IRRIGATION

A. Surface Irrigation

Applying water to the soil without aerial application is known as surface irrigation. It depends on gravity for spread of water over the area. This system generally uses more quantity of water. Different systems of surface irrigation (Fig. 7.1) are:

1. **Flooding:** This is followed in wet lands mostly for banana. This is a wasteful method which will lead to stagnation of water and help weed growth.
2. **Check:** Check bunds of large areas enclosing a number of trees are provided with channels between two rows. This is more economical than flood system (Fig. 7.1(a)).

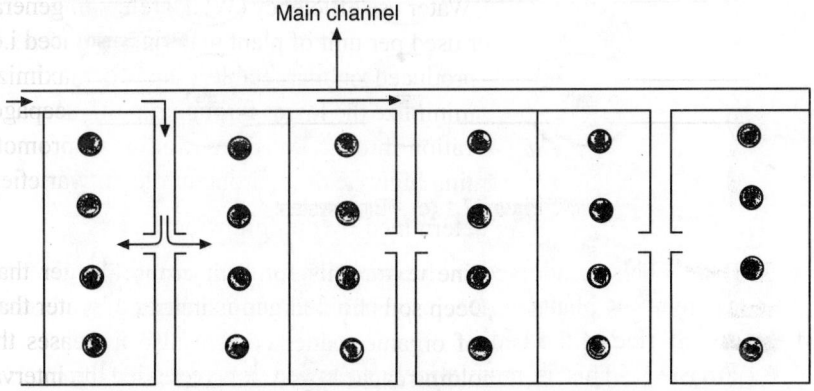

Figure 7.1 (a) Check method of irrigation.

3. **Basins:** This is widely practiced. The basins should be square or circular and should be sloping from the trunk to the periphery. This method is useful in young orchards, light sandy and alkaline soils. The size of the basin should be widened as the roots spread (Fig. 7.1(b)).

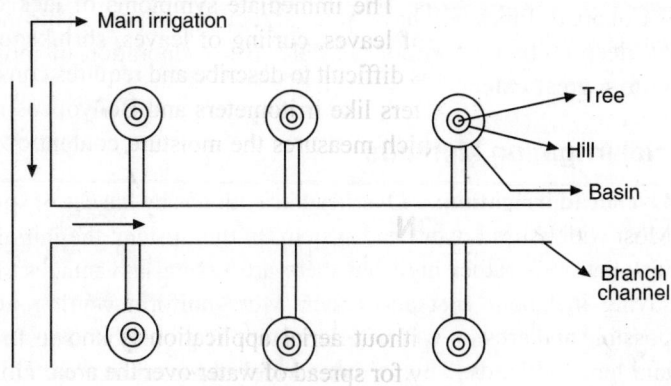

Figure 7.1 (b) Basin method of irrigation.

4. **Ring:** In this system, small ring bund will be provided around the trees or one single irrigation channel connecting all trees will be formed and around each tree the channel is widened to form basin (Fig. 7.1 (c)).

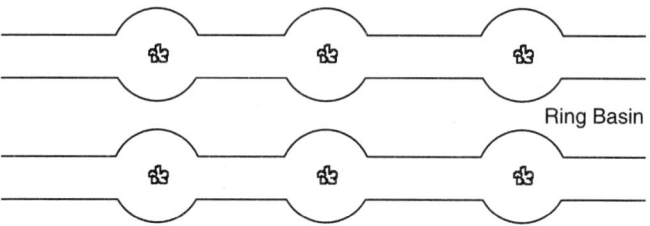

Figure 7.1 (c) Ring system.

5. **Bed:** This is adopted in heavy soils for fruit crops like banana, wherein 3–4 plants are enclosed in a bed and is irrigated by opening on one side of the bed.
6. **Furrow:** This is most widely followed for vegetable crops like tomato, onion, brinjal etc.

All the above different systems of surface irrigation do not ensure uniform distribution of water. It may be more near channels and less away from channels.

B. Sub Irrigation

This method supplies water from below soil through underground pipe or by ditches on one side. This is useful for green houses. Pipes are laid 45 – 60 cm deep and 6 m apart. Pipes will have holes at regular intervals. This method is costly and deep cultivation is not possible. But evaporation of moisture is prevented to a great extent.

C. Special Irrigation Methods

1. **Overhead irrigation:** Overhead irrigation is by the use of sprinklers. Most widely used over head system. In this system, the initial cost of installation is rather high but there are several advantages. There is saving in labour cost and water. More uniform wetting of soil is possible and erosion will be eliminated. This method is best for steep and terraced lands. This is more widely adopted in plantations (Fig. 7.1 (d)).

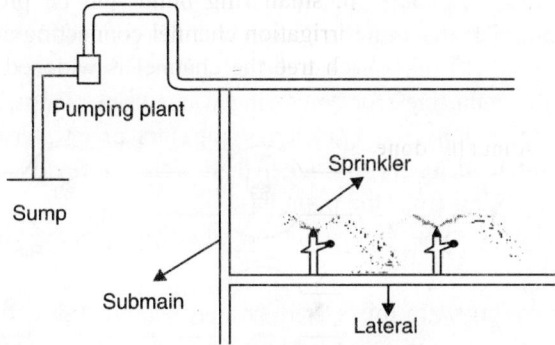

Figure 7.1 (d) Overhead irrigation.

2. There are also some disadvantages. Due to the influence of wind, there will be non-uniformity in coverage. In hot sun, droplets on leaves and fruits may cause sunburn. Certain diseases may spread easily.

3. **Drip irrigation:** Drip irrigation is known by various names like 'trickle irrigation' or 'high frequency irrigation' or 'daily flow irrigation'. This is a method of watering plants at a rate equivalent to its consumptive use so that plants would not experience any stress during the growing phase. In this the water is conveyed from a source under low pressure to the root zone of the crop only. The twin objectives of this method of irrigation are: (1) provision of optimum quantity of water to the crop for optimum production and (2) saving the valuable water from wastage thereby increasing the water use efficiency and the command area (Fig. 7.1 (e)).

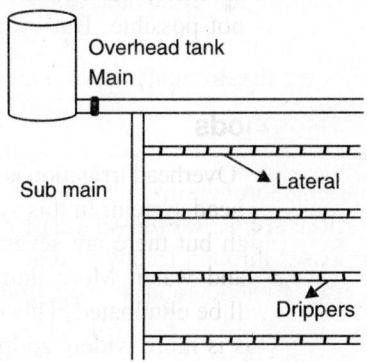

Figure 7.1 (e) Drip irrigation.

A typical drip irrigation system has the following components: (1) a water supply pump at the source of water, (2) filters, fertilizer mixing tank, (3) control system, (4) pressure regulators, (5) monitor valve/water meter, (6) head lines or main lines for conveying water from pumpset to the field where water is to be delivered, (7) laterals (pipe) to carry water to plant rows and (8) the emitters/drippers through which water is finally released at a distance of 5 to 25 cm from the plant base.

Drip irrigation system offers a number of advantages over other systems of irrigation like:

1. **Water saving:** In this system, water is applied directly to the root zone, eliminating wastage. The water saving may range from 30 to 70 per cent depending upon the area covered by the plants.

2. **Labour saving:** This system eliminates the need for constructing borders, bunds and labour intensive works associated with conventional irrigation techniques, thereby saving about 60 to 90 per cent labour.

3. **Use of lower quality water:** In this system, water is applied continuously and the root zone is kept wet constantly, therefore, salt concentration moves away from the root zone and thus it permits use of low quality water as compared to other systems of irrigation.

4. **Increase yield and plant vigour:** This system maintains soil moisture at optimum level eliminating water stress resulting in greater vigour, better establishment and high productivity.

5. **Reduced weed growth:** Since water is applied to the restricted area wide spread weed growth is inhibited due to restricted water supply.

6. **Saving of nutrients:** In this system, nutrients are directly applied to the root zone along with water, therefore, chances of leaching losses are minimized thus saving upto 30 to 60 per cent is expected.

However, the following are the disadvantages:

(a) Higher initial investment
(b) Clogging of drippers due to oxidants, bioxides and algae.

Fertigation

Application of plant nutrients through irrigation water is called 'Fertigation'. Advantages of fertigation are 1. saving in fertilizer application costs 2. reduction in nutrient losses through leaching, fixation, and denitrification 3. increase in nutrient use efficiency and 4. increased yield and quality of produce.

The success of fertigation depends upon (a) application of nutrients in the active feeder rootzone of plants in accordance to their relative requirement for different nutrients at different phenological growth stages (b) selecting the sources of fertilizers with reference to the soil physico-chemical characters, water quality, the need for combining the fertilizers and their compatibility (c) calculating the quantity of fertilizers per day for each growth stage and (d) allowing the dissolved fertilizers through drip system every alternate days or once in three days or once in 7 days depending upon the infiltration rate of the soil.

Common water soluble fertilizers like Potassium Nitrate, Mono Ammomnium Sulphate which have high water solubility can be used in fertigation. Common fertilizers like urea and muriate of potash can be also used, provided they need to be dissolved in the tanks one day prior to application. Mixing of ammonium sulphate with potassium chloride; calcium nitrate with any phosphates or sulphates; magnesium sulphate with mono or di-ammonium phosphates and phosphoric acid with iron, zinc, copper or manganese sulphate should be avoided to prevent the reduction in their solubility.

Fertigation studies conducted with fruit crops like banana revealed that fertigation with 75–100% of recommended dose of fertilizers at weekly intervals results in increased bunch weight, quality fruits etc. Based on this, fertigation technology is recommended to many horticultural crops.

8

Orchard Cultivation

Orchard cultivation refers to the careful management of the orchard soil in such a way that the soil is maintained in a good condition suitable to the needs of the tree with least expenses. This involves maintenance of the physical condition of the soil, its moisture and nutrient content. A good system of orchard cultivation should ensure:

1. Weed control and saving in moisture and nutrients
2. Very little disturbance to soil and preventing soil erosion and
3. Reduced cost of cultivation.

Methods of Soil Management Practices

1. Clean culture

This type of cultivation is extensively followed in India. This involves regular ploughing and removal of weeds. The clean culture has many disadvantages. They are:

(i) Humus will be completely depleted rapidly due to frequent cultivation.
(ii) Frequent intercultural operation causes injury to the feeding roots, the trees may be short lived or stunted in growth.
(iii) Clean cultivation aids in more aeration leading to the depletion of nitrogen.
(iv) Hard pan is created in the soil.
(v) Frequent cultivation causes more soil erosion.

The above mentioned defects in clean cultivation can be minimized by avoiding deep and frequent cultivation and also cultivation when the soil is too wet.

2. Clean culture with cover crops

This type of soil management involves raising of a cover crop or green manure after removing the weeds. If clean cultivation is attempted during the rains, considerable erosion is almost sure to occur. It is probably best to plant a green manure crop between the trees early in the rains and plough it into the soil towards the end of monsoon season. In India, green manure crops like sunnhemp, cowpea, daincha, lupins etc. are more commonly used. Legume cover cropping in grapes, mango, guava and other fruit crops is becoming a common practice in the management of orchards. Cowpea and French beans grow well under guava and sapota tree. In some places to prevent soil erosion, certain permanent cover crops like *Calapogonium muconoides, Centrosema pubescens* and *Pueraria phaseoloides* are raised in the alley spaces. They are leguminous crops, establish in a short period, dry up during summer to conserve moisture. With summer showers they come up again because of their profuse seeding habit and spread themselves as a vegetative mat by the time the heavy monsoon starts pouring in. Such permanent cover cropping is a common feature in rubber plantations of Kerala and Kanyakumari district.

3. Mulching

This is one of the important soil management practices adopted in certain countries. Crop residues like straw, cotton stalks, leaves, saw dust, pine needles, coir dust and other materials like polythene films or certain special kinds of paper are spread in the tree basins and in interspaces between trees. Main objective of mulching is to conserve soil moisture and to control the weed growth. The other advantages of mulching are:

1. Keeps soil cool in day; warm at night hours
2. Reduces surface run-off
3. Adds humus to the soil
4. Prevents soil erosion
5. Fruits are protected and kept clean since they fall of the mulches
6. It allows the absorption of more rain water and
7. It reduces irrigation frequency.

The following are some of the disadvantages:

1. Dry materials used as mulches encourage the risk of fire and consequent damage to trees.

2. Thick mulches may act as places for mice and rodents to live and multiply. They may cause damage to tree trunks and roots by eating the bark and burrowing to the land. The mulching materials should not be placed too close to the tree trunk and it should be spread in such a way that they give a good cover to the root system of the trees.

4. Sod

In this method, permanent cover of grass is raised in the orchard and no tillage is given. This type of orchard cultivation is followed in U.S.A and Europe. This may be useful in slopy lands for preventing soil erosion. But they compete for soil moisture and available nitrogen. The drawbacks of this system are the need for increased manuring and water application. They are harmful to shallow rooted trees. Hence, sod may be useful with deep rooted trees because soil moisture will be very low on the top layers.

5. Sod mulch

This is similar to sod with the only difference is that the vegetation is cut frequently and the cut material is allowed to remain on the ground. This is slightly better than the previous one, as the moisture loss is not so great as in sod. In both sod and sod mulch, more nitrogen should be applied to the fruit trees than usual application because the vegetation utilizes more soil nitrogen.

Intercropping

In young orchards, the question of how best one can use the soil between the trees arises. If the trees are properly spaced there is considerable area which will not be used by the permanent trees for several years. Similarly, in the case of other long duration horticultural crops like tapioca, turmeric, ginger and banana some area between adjacent plants will be remaining unocuupied by the main crop for few months. It naturally appeals to the grower to get some return from this vacant land especially when he is getting no return in the early periods. The practice of growing any economic crop in alley spaces of the fruit trees in the first few years or in the unoccupied spaces of the long duration crop in the early periods is referred as intercropping. They also act as a cover crop and the land benefits by the cultivation, irrigation, manuring given to the intercrops. The following important principles should be observed while growing intercrops:

1. Intercrops should not occupy the area where the roots of the fruit trees are concentrated.

2. Soil fertility should be maintained or improved when intercrops are grown.
3. Water requirements of the intercrops should not clash with those of the main fruit trees. The intercrop may require irrigation at a time when it would be detrimental to the trees.
4. Intercrops should be selected with reference to their effect on soil moisture. Grain crops remove excessive moisture to the detriment of fruit trees. The intercrops selected should not exhaust the soil water and nutrients and should not demand more water than is allowed for fruit trees.

Vegetables are the best intercrops when compared to millets. But whatever may be the intercrop grown, it should be kept well away from the main fruit trees and irrigated independently. The intercropping should be stopped when trees occupy the entire orchard space. Thereafter, green manuring or cover cropping should be only practiced.

Many growers prefer some quick growing fruit trees to grow as intercrops. A number of fruits crops are available for this purpose. In temperate regions, peaches are often grown between apple trees. Similarly, in properly spaced mango orchard, guava trees can be planted to bear in two or three years and will produce a number of crops before it is necessary to remove them. Such short-lived trees are known as **'fillers'**. Papaya, banana or phalsa may be well grown as fillers in orchards. The danger in using fillers is when they are allowed to remain in the orchard for too long periods. As normally root system makes a faster growth than the branches, the roots of the permanent trees come in contact with the roots of the fillers before there is any crowding above ground. Therefore, the fillers should be removed after a few years usually immediately after the main fruit trees have commenced bearing.

The recommended intercrops for some important horticultural crops are given below:

Crop	Age	Intercrop
Mango	Upto 7 years	Leguminous vegetables, Papaya (filler)
Grapes	Upto 8 months	Snake gourd or bitter ground in pandal
Apple, pears	Upto 5 years	Potato, cabbage
Banana	Upto 4 months	Sunnhemp, onion
Tapioca	Upto 3 months	Onion, beans, lab-lab, Black gram
Turmeric	Upto 3 months	Small onion, coriander
Arecanut	Upto 10 years	Pineapple
Coconut	Upto 3 years	Banana, tapioca, vegetables

Mixed Cropping

It refers to the practice of growing certain perennial crops in the alley spaces of the main perennial crops. The main advantage is the effective utilization of available area and increase in the net income of the farm per unit area. Extensive research conducted by CPCRI, Kasargod on mixed cropping in coconut and arecanut plantations showed that cocoa, pepper, cinnamon, clove and nutmeg can be grown as mixed crops in coconuts while nutmeg and clove as mixed crops in between four arecanut palms on alternate rows. In all the above cases, increase in yield (upto 10%) is obtained in the main crop due to synergistic effect of the crop combinations arising out of beneficial micro-organisms in the rhizosphere and the more availability of major nutrients in the active root zone of the crop mix as compared to the pure stand.

Multitier System of Cropping

Certain horticultural plants like coconut and areacanut are grown for about 50 years in a particular land. It takes nearly 4 to 7 years for the above trees to reach the bearing stage. Adequate alley spaces (nearly 75%) are available in between these trees and being the palm trees, their root system will not also spread beyond one metre in diameter. Hence, these vacant spaces can be profitably used for raising other crops, thereby increasing the employment opportunities and profit. This is the chief objective of the multitier system of cropping.

Intercropping and mixed cropping involve jointly multitier system of cropping and is defined as a compatible companion of crops having varying morphological frames and rooting habits, grown together in such a manner that their canopies intercept solar energy at varying heights and their roots forage the soil at different zones. The main principle here is that the land, water and sunlight should be effectively used. An ideal combination of crops for multitier cropping in coconut and arecanut plantations are given below:

Tier	Crop
First (Top)	Coconut or arecanut.
Second	Pepper trained over the trunk of coconut or Arecanut trees.
Third	Cocoa or cloves planted at the centre of four arecanut or coconut trees.
Fourth (ground)	Pineapple, ginger and dwarf coffee.

Organic Farming

Excessive use of chemical fertilizers and pesticides as a mean of intensive cultivation to boost up our food production has caused considerable damages to our soil health and the environment. This has been criticized recently by many environmentalists. This has focused the attention of several experts in ecologically sound viable and sustainable farming systems, known as organic farming.

It is production system which avoids or largely excludes the use of synthetically compounded inorganic chemicals. This system entirely relies on crop rotation, crop residues, animal manures, legumes, green manures, off-farm organic wastes, biofertilizers, mechanical cultivation etc. and aspects of biological pest control to maintain soil productivity and tilth to supply nutrients and to control insects, weeds and other pests. This system is often, referred as 'biological farming' 'regenerative farming' and 'sustainable farming' and 'eco friendly farming' etc.

Organic farming is essential because

1. Chemical fertilizers may have an adverse effect on soil life-they do not supply humus, have an adverse effect on physical, chemical and biological properties of soil
2. Chemical fertilizers are costly
3. Indiscriminate application of pesticides could lead to residues in horticultural crops which are consumed mostly in raw state by us
4. The continuous use of pesticides is ecologically unsustainable as pests acquire resistance.
5. There is a premium for the horticultural produces which are raised under organic farming.

Essential features of organic farming are

1. Use of organic manures like FYM, compost, vermicompost and coir compost etc. 2. Use of biofertilizers. 3. Use of green manures and grain legumes 4. Non-chemical weed management and 5. Use of botanicals and bio-control agents in the control of crop pests.

In India, organic farming in horticultural corps especially in vegetable and spices crops is slowly picking up, wherever premium prices are available for organically grown quality produces. In certain horticultural crops where productivity and total productions can not be compromised by following strict organic farming practices, a combination of inorganic and organic farming practices, known as 'integrated nutrient management (INM)' may be followed possibly avoiding known toxic inorganic chemicals.

9

Training and Pruning

Certain fruit plants if left to themselves grow wild and do not bear abundantly unless trained to a specific form. Not all species of fruit trees necessarily need pruning. Evergreen trees like mango, sapota do grow well naturally that they hardly require any pruning except the removal of dead woods. On the other hand most deciduous trees like apple, pear, peach and grapes have to be trained and pruned regularly. Fruit trees like citrus, pomegranate and guava need only intial pruning of a few branches to train to a desired form. Recently it has been found that evergreen trees like mango, sapota also need regular pruning to maximize its productivity especially under high density planting systems.

There is frequent failure to distinguish clearly between pruning and training. These two practices are often regarded as one and the same or atleast as inseparable. When a plant is tied, fastened, staked or supported over a trellis or pergola in a certain fashion or some of its parts are pruned with a view to giving the plants a frame work the operation is called 'training'. On the other hand, 'pruning' refers to the removal of plant's parts such as buds, developed shoots and roots to maintain a desirable form by controlling the direction and amount of growth. Therefore, these two operations can be distinguished on the following aspects:

S.No.	Training	Pruning
1.	It concerns form primarily	It affects functions only
2.	It determines plant's outline, its branching and frame work.	Assists more in what the tree does in respect to fruiting
3.	Generally concerned with first few years of the tree growth to determine its annual frame work.	Annually done

Objectives of Training

(i) To admit more light and air to the centre of the tree and to expose maximum leaf surface to the sun.

(ii) To direct the growth of the tree so that various cultural operations, such as spraying and harvesting can be done at the lowest cost.

(iii) To protect the tree from sun burn and damage.

(iv) To secure a balanced distribution of fruit bearing parts on the main limbs of the tree.

Details of Training

1. Height of head

The distance from the ground level at which the main or scaffold limbs branch from the trunk is known as the 'height of the head' (Fig. 9.1) and this has to be decided before training is done. Trees in which scaffold branches come out within 0.7 to 0.9 m is referred as *'low headed'* and those in which they come out from the trunk above 1.2 m or more is called *'high headed'*. High headed trees help in

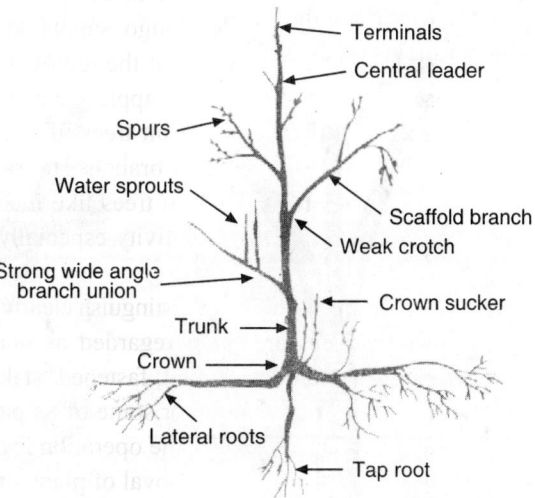

Figure 9.1 Major gross structure of plant.

easy orchard cultivation but in tropical climate, high headed trees are unsuitable as their exposed trunks are subjected to sunscald injuries. Low headed trees come into bearing comparatively much earlier and are able to resist stormy winds more effectively and permit easy cultural operations like pruning, spraying, thinning and picking.

2. Number of scaffold limbs

The number of main branches or scaffold limbs to be allowed while training varies from 2 to 15 or even more. Neither extreme is desirable. If there are only two or three main scaffold limbs, they are almost certain to form 'crotches', that they are likely to split and allow one or two branches to break down. These weak crotches may be avoided by training more and better spaced scaffold limbs. A fruit tree with 5 to 8 numbers of scaffold branches makes it mechanically strong and at the same time open enough to facilitate necessary orchard operations.

3. Distribution of scaffold limbs

The distribution is more important than the number. If the scaffold branches arise at closer interval *i.e.* 20 to 25 cm distances, they form bad crotches much sooner than when distributed at 45 to 60 cm distances of the trunk (Fig. 9.2). The crotch angle should be also considered in selecting scaffold branches. The cambium divides continuously forming new layers, therefore, branches which grow upright should be removed as the angle is 'y' shaped. The layers of cambium are not continuous and the branches may be easily broken. On the other hand, branches arising from the trunk at angles ranging from 45 to 90° are the strongest because the cambium is continuous without phloem inclusion (Fig. 9.3).

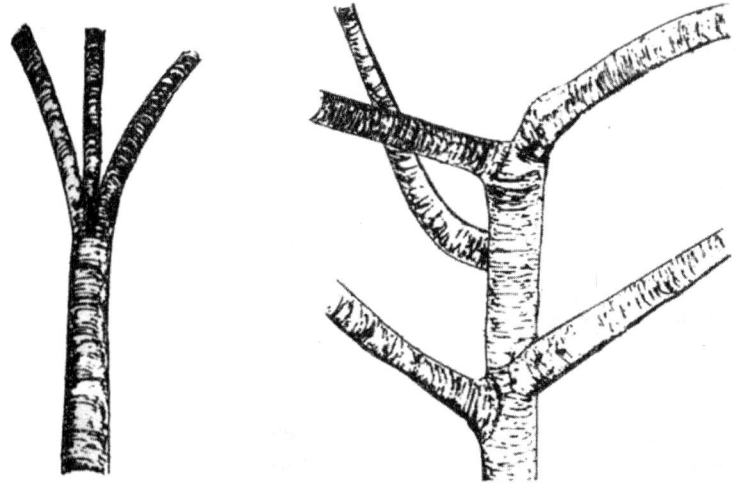

(a) Narrow-angled branches form a weak union and can split with the weight of the crops

(b) Good broad-angled strong branches

Figure 9.2   Branch angles

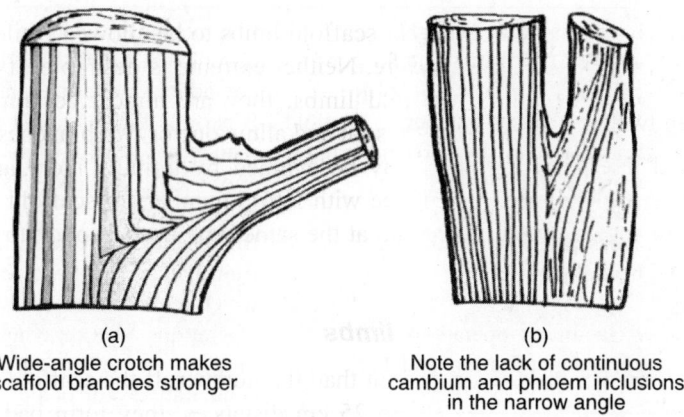

(a)	(b)
Wide-angle crotch makes scaffold branches stronger	Note the lack of continuous cambium and phloem inclusions in the narrow angle

Figure 9.3 Cross section of wide-angled crotch and narrow angled crotch.

Methods of Leader Training

The method of leader training should be suited to the normal growth habit of the fruit tree. The common systems of training followed are:

1. Open centre

In this system the main stem is allowed to grow only upto a certain height and the leader stem is pruned to encourage scaffold branches production. This system is also known as Vase-shaped system (Fig. 9.4).

2. Central leader

In this system a tree is trained to form a trunk which extends from the surface of the soil to the top of the tree. This system of training is also known as closed centered one (Fig. 9.4).

3. Modified leader

It is intermediate between the open centre and central leader. This is developed by first training the trees to the leader type by allowing the central axis to grow unhampered for the first four or five years. The central stem is then headed back and lateral branches are allowed to grow as in the open centre system (Fig. 9.4). It possesses practically all the advantages of the two other types of training system and few or none of their disadvantages.

The differences between the open centre and central leader system are given as.

S.No.	Open central system	Central leader system
1.	It admits more sunlight and enables the fruit to attain good colour	It is not possible to get a better coloured fruit as it does not permit more sunlight
2.	It forms lesser number of scaffold branches and they are more likely to split at the crotches	It permits the formation of number of scaffold branches and hence less likely to split at the crotches
3.	Trees are more fruitful in all branches	Branches at the top are more fruitful than the lower branches, which remain more or less shaded
4.	It facilitates operations like spraying, thinning and harvesting	Operations like spraying thinning & harvest are comparatively difficult

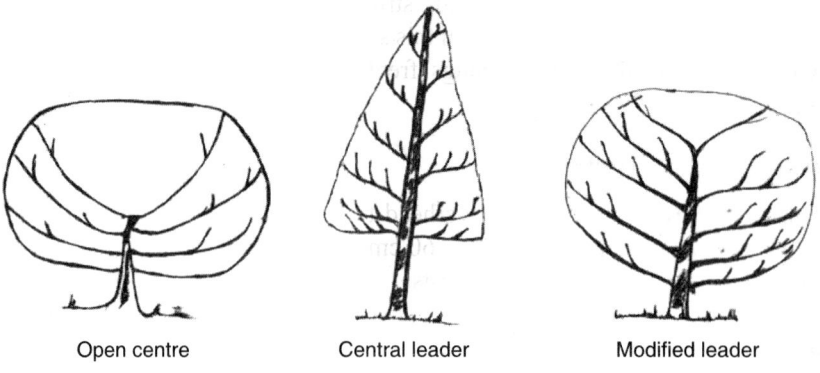

Open centre   Central leader   Modified leader

Figure 9.4   Common system of training.

Trees are trained to different forms with or without the support of certain structures. The following are some of the systems requiring the support of the structures:

1. Bower system

It is also called as 'Pandal' or 'Arbour' or 'Pergola' system. It is generally practiced in grapes and other cucurbitaceous vegetables like snake gourd, ribbed gourd, bitter gourd etc. In this system, the vines are spread over a criss cross network of wires, usually at 2.1 to 2.4 m above ground, supported by concrete or stone pillars or live support like *Commiphera* sp. The vine is allowed to grow single shoot till it reaches the wire net and is usually supported by bamboo sticks tied with jute thread. When the vine reaches the wires, its growing point is pinched off to facilitate the production of side shoots. The main advantage of this system is that it is capable of giving higher yield. But the main drawbacks are (i) most expensive system (ii) pruning, training and spraying operations are generally difficult.

2. Espalier system

Plants are trained to grow flat on trellis or on horizontal wires by training the branches perpendicularly to the main stem on both the sides and trained horizontally on to wires. Plants trained in this system are called 'espaliers'. An espalier with one shoot or two shoots growing in opposite or parallel directions are called a 'cordon' (Fig. 9.5).

3. Kniffin system

In this system, two trellis of wire are strung supported by vertical posts. The vines such as grape when trained in this system has four canes one along each wire and the bearing shoot hangs freely with no tying being necessary. (Fig. 9.6).

4. Telephone system

This system is also known as overhead trellis system. This system consists of 3 or 4 wires usually kept at 45 – 60 cm apart fixed to the cross-angle arms supported by vertical pillars or posts.

5. Tatura trellis

In this system, trees are trained to a multi-layered wire trellis. The trellis is V-shaped, supported by two long, stout poles embedded into the soil at angles of 60° from the horizontal. Five wires at 60 cm intervals are fastened to these poles. This system is being now followed for pome fruits, nut fruits and grapes. The trees are grown as double leader. Tree with each leader is inclined at an angle of 60° from the horizontal. The following are some of the training systems which do not require the support of any structure but will be trained to a particular shape.

(a) Head system

This system is usually followed in grapes. In this system, the vine is allowed to grow single stem with the help of stakes. After attaining a height of 1.2 m, it is cut back to produce side shoots. Only four laterals 75 cm above the ground in all directions are alone retained, the rest are thinned out. From these laterals, two secondaries are normally developed which bear one or two fruiting spurs. As it occupies lesser area per unit plant, it is suitable for high density planting (Fig. 9.7).

The main advantages are easy and simple to train, inexpensiveness to establish and inter cultivation is possible. The disadvantage is that yield obtained per plant is relatively low.

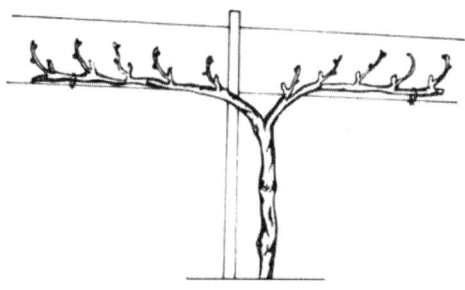

Figure 9.5 Cordon (espalier) system.

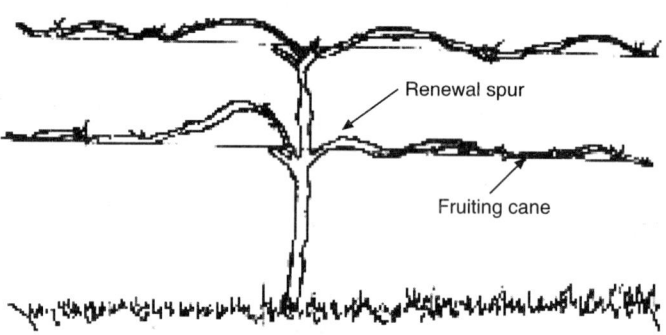

Renewal spur

Fruiting cane

Figure 9.6 Kniffin system.

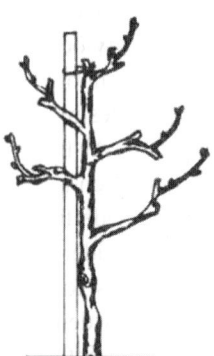

Figure 9.7 Head system of training.

(b) Palmette

This form of growing fruit trees consists of a tree with central leader having 3 to 4 tiers of two opposite branches trained at an angle of 30° from the vertical. The branches towards aisle or against the row are removed.

(c) Spindle bush

It consists of a central stem with a larger number of about 15–20 semi horizontal branches radiating from it. The branches are kept spindly and no secondary branches are encouraged on primary branches.

(d) Dwarf pyramid

It is also a central leader tree and fully grown tree consists of central erect stem with 20–30 branches radiating from it on all sides at an angle of 45°–90°. The branches are so trained that the lowest ones being the longest with the upper tiers being smaller, presenting 'pyramidal' shape.

(e) Head and spread systems

It is another variation form of modified leader system. In this system the primary branches are taken in different tiers, each tier being about 60 cm above the lower tier, instead of throughout the length of stems as in the conventional modified leader system.

The most of the above five systems are practiced in temperate fruit crops.

Tree training involves spreading and positioning of branches in various methods. In certain growing system, regular spreading is necessary for the first few years of tree growth and in some for the entire life of the tree. Branch spreading helps to obtain wide angled crotches and semi horizontal spreading branches which tend to bear early than upright branches.

Though branch spreading normally occurs in a tree as a consequence to pruning of the vertical shoots and the shoots growing towards centre and also by the heavy cropping in a branch, desired spreading effect is achieved by using 'spreaders' and 'branch tying'. A 'tree spreader' is a device which is fixed to a lateral shoot or branch in such a way to widen its angle and force it into a desired position. Materials made from 'thick wires', welding rods or from wood may be used as tree spreaders depending upon the length and strength required. Tying the branches to a desired shape is an excellent method of spreading the branches and widening their angles. One disadvantage with tying down branches is that it may interfere with the orchard cultivation.

Maintaining the Tree Form

After the trees have been trained to a particular form, precaution should be taken to maintain the desired frame of the tree. This is possible by avoiding the formation of weak crotches, water shoots and suckers. Weak crotches may be avoided by judicious training and having right number of scaffold branches distributed at the required distance or angles. 'Water shoots' or 'water sprouts' are extraordinarily vigorous vegetative shoots which grow from high points on the main branches in an upright direction at the expense of the parent branches from which they arise. They are erect much thicker than the normal branches and bear much larger coarse leaves. If such shoots are not removed, they soon close the centre of the trees and obstruct light. Citrus and pomegranate trees produce more often such shoots and these shoots should be removed as soon as they appear. On the other hand fruit trees like guava, apple and pomegranate often produce weak shoots in many numbers which arise from the adventitious buds on the roots or underground parts of stem of these trees. These shoots otherwise known as suckers should be promptly removed as soon as they appear.

Bearing Habit

Fruit trees may bear the fruits either terminally on a long or short growth, laterally on current or past season growth or adventitiously from any point on the trunk. The relative position of a fruit with reference to its potential bud giving rise to flower or inflorescence in the shoot is often known as bearing habit. Knowledge on the bearing habit is pre-requisite before resorting to pruning in any fruit crop. The position of flower or inflorescence on the shoot in relation to the growth of current season is characteristic of a species or variety. Position of fruit buds bears a relationship with the growth habit of a tree. Plants having terminal fruit buds do not possess enormous spreading habit and the trees are rather compact when compared to plants having lateral fruit bud bearing habits since they force the development of laterals below rather than beyond the flowers or flower clusters.

Different Kinds of Flower Bearing Shoots

Based on the position of fruit bud and the kind of flower bearing shoots, fruit trees can be classified into following groups: (Fig. 9.8)

Group 1: Fruit buds borne terminally and unfold to produce inflorescence without leaves e.g. Mango.

Group 2: Fruit buds borne terminally unfolding to produce leafy shoots that terminate in flower clusters, e.g., apples, pears.

Group 3: Fruit buds borne terminally unfolding to produce leafy shoots with flowers or flower clusters in the leaf axils, e.g. guava.

Group 4: Fruit buds borne laterally unfolding to produce flower parts only without any leaves, e.g. citrus, coconut, papaya, coffee.

Group 5: Fruit buds borne laterally unfolding to produce leafy shoots terminating in flower clusters, e.g. grapes.

Group 6: Fruit buds borne laterally unfolding to produce leafy shoots with flower clusters in the leaf axils, e.g. Fig, avocado.

Group 7: Fruit buds borne both terminally and laterally but unfolding to produce inflorescence terminally, e.g., walnut.

Group 8: Fruit buds always borne adventitiously in old trunk or shoots, e.g., jack, cocoa, Indian star gooseberry.

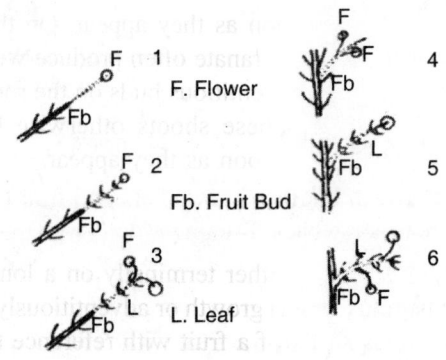

Figure 9.8 Bearing habits in horticultural crops.

Pruning

Commonly, trees are pruned annually in two ways. A few shoots or branches that are considered undesirable are removed entirely without leaving any stub. This operation is known as 'thinning out'. The other method which involves removal of terminal portion of the shoots, branches or limb, leaving its basal portion intact, is called 'heading back' (Fig. 9.9). Thinning out involving large limbs as in old and diseased trees is called 'bulk pruning'.

These operations are carried out to divert a part of the plant energy from one part to another. As trees grow older, they should receive relatively more of thinning out and less of heading back. Heading back tends to make trees more compact than thinning out. If a few of the several branches growing close together on the same parent limb are entirely removed or thinned out, the rest of the branches would grow more vigorously. Thinning out results

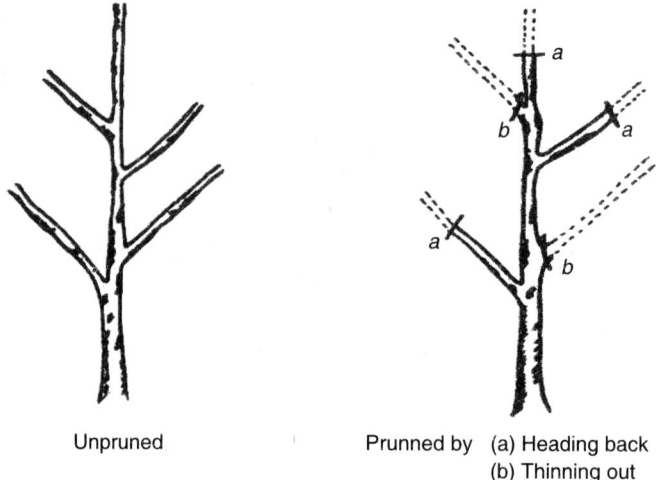

Unpruned   Prunned by   (a) Heading back
                                       (b) Thinning out

Figure 9.9   Methods of pruning.

in lesser new shoot growth but more new spurs and fruit bud formation than corresponding severe heading back. Pruning is done with the following specific objectives:

- To remove surplus branches,
- To open the trees so that the fruits will colour more satisfactorily,
- To train it to some desired form,
- To remove the dead and diseased limbs,
- To remove the water sprouts and
- To improve fruiting wood and to regulate production of floral or buds.

Season of Pruning

Little differences are likely to result from pruning at different times during the dormant seasons in deciduous fruit trees though in certain cases, earlier pruning causes earlier foliation in the spring. Late pruning during dormant periods is generally not advocated as it leads to more bleeding than earlier pruning. The exposed cut surface in certain cases may provide an excellent opportunity for infection by some pathogens. For this reason, winter pruning is usually preferred to spring pruning as bleeding will be excess in later period. Summer pruning may have a dwarfing effect or an invigorating influence. A light summer pruning may aid in colouration of fruit in certain species.

The amount of pruning or severity of pruning which is desirable for mature trees differs in different species. The minimum amount, which is common to all, is the removal of broken or diseased branches and those which cross against each other. Diseased branches should be completely removed from the base of the trunk. In other cases, annual pruning may be very light in the beginning but after some years it may become necessary to prune heavily. Otherwise, the trees may lack vegetative vigor and make very little growth. Under South Indian conditions, old non bearing mango trees are pruned to expose the centre portion to sunlight and also crowded terminal shoots thinned to one or two shoots during August-September. The pome fruits such as apple, plum, pears and peaches are pruned every year in December, January. Jasmines are pruned to 45 cm height from the ground level during the last week of November.

SPECIAL PRUNING TECHNIQUES

1. Root pruning

Root pruning has long been known as a recognized practice for the culture of dwarf fruit trees. A circular trench of 45 cm away from the stem is dug out annually and the roots are cut-off every year with a sharp knife. After pruning, the trench is filled with manures liberally. The tree is thus fed and watered artificially in a restricted area. As a result of this the circular mass of fibrous roots increases very slowly from year to year and the tree makes short but stocky and well ripened shoots. The diameter of the circular trenches may be extended slowly as years roll on. Each year 4 to 5 cm of the stumps of the previous year growth are pruned. This helps to increase the production of mass fibrous roots, dwarf the trees and bears abundantly. In Deccan Virdharba area of North India, root pruning is practiced to induce flowering of oranges in required season. However, this practice is not advocated every year to the fruit trees.

2. Ringing

It is one of the known practices to increase fruit bud formation in certain fruit crops (Fig. 9.10). The operation consists of removal of a complete ring of bark from a branch or the trunk. Ringing interrupts the downward passage of carbohydrates through the phloem and thus causes them to accumulate in the part of the tree above the ring. Scoring the bark and inverting a ring of bark also accomplish the same effect. Ringing is practiced on mango to force flowering in over vegetative trees which do not normally bear a satisfactory crop. This practice cannot be recommended for all fruit crops and it is found beneficial in promoting fruit set in certain vigorously growing grape varieties and they often result in large size fruits.

3. Notching

Notching is a partial ringing of a branch above a dormant lateral bud. The operation consists of removing a small narrow strip of bark just above and close to a dormant bud (Fig. 9.10). Notching above a bud increases the yield of fig trees in Pune. In apple, notching helps to produce a strong shoot. Notching a bud prevents the inhibitory influence of certain compounds on the bud, the supply of carbohydrates coming through the phloem from above and increase the supply of water and nitrogen from below through the xylem. The practice of notching below a bud (otherwise known as 'nicking') is effective in starting a floral shoot. It causes a greater concentration of carbohydrates in the bud and reduces the supply of water and nitrogen from below. Nicking helps to produce spurs from buds in apple.

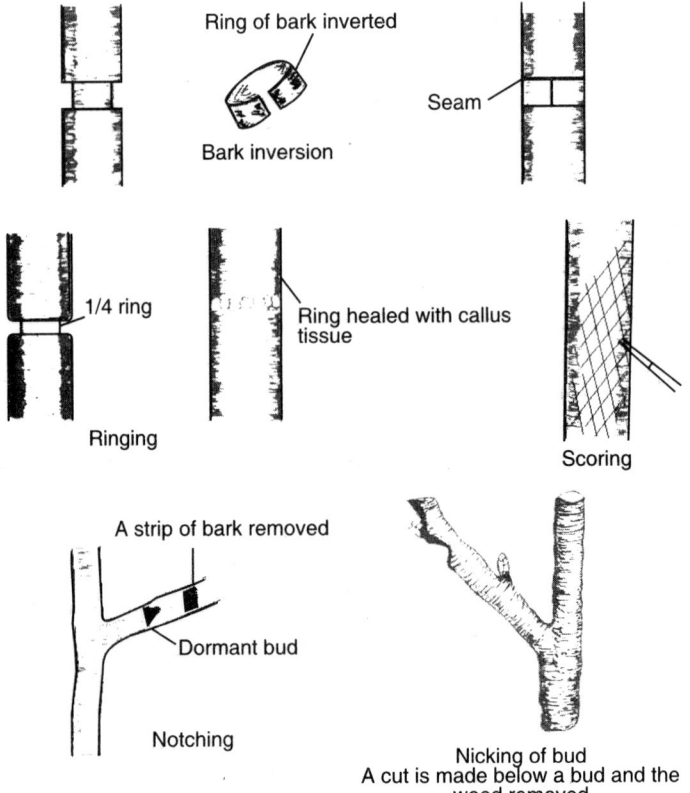

Figure 9.10 Special pruning technique.

4. Smudging

It refers to the practice of smoking the trees like mango, commonly employed in Phillippines to produce off-season crop. Smoke is produced by burning branches and fallen leaves on the ground and the smoke is allowed to pass through the centre of the crown of the trees. The tree is smoked heavily and continuously for a week. The smoking can be discontinued as soon as the terminal buds in the past season should begin to swell. Smudging of mango trees in India has not been found to induce early blossom.

5. Bending

Bending of branches is widely practiced in the Deccan for increasing fruit production in guava, especially in the erect growing-varieties. It's large branches in their erect upright position have a natural tendency to produce fruit bearing shoots near the terminal portion, while their lower portions remain more or less dormant. When such large branches are bent towards and tied and fastened with similarly bent branches of the neighbouring trees, the fruiting area is considerably increased due to the increased number of lateral branches produced on the lower parts of the main limbs. Bending practice is not recommended to those varieties of guava in which the branches naturally spread obliquely and droop giving rise to side branches.

6. Coppicing

This refers to the practice of complete removal of the trunk in trees like Eucalyptus and Cinchona leaving 30–35 cm stump alone. The coppiced stump starts producing many vigorous shoots in about 6 months time. Only 2–3 shoots are retained per stump and the rest ones are completely thinned out. These left out shoots attain coppicing stage in about 10 years depending upon the locations and other factors (Fig. 9.11).

7. Pollarding

This refers to the practice of removing the growing point in shade trees especially in silver oak in order to encourage side branches (Fig. 9.12).

8. Lopping

This refers to the practice of reducing the canopy cover in shade trees in order to permit more light.

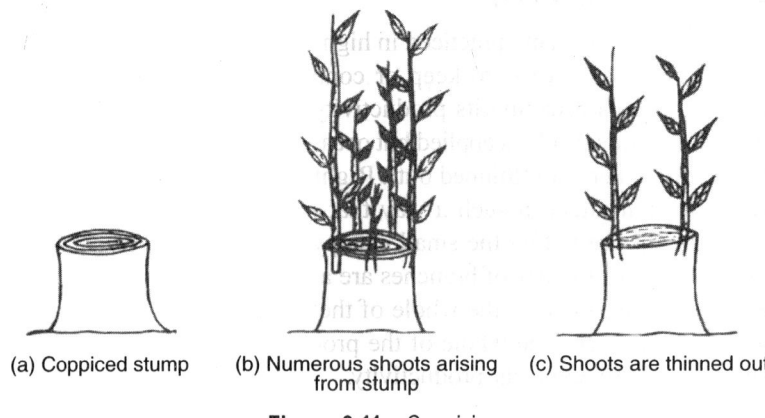

(a) Coppiced stump   (b) Numerous shoots arising   (c) Shoots are thinned out
from stump

Figure 9.11 Coppicing.

Figure 9.12 Pollarding.

9. Pinching

It usually refers to the removal of terminal growing point in herbaceous plant so as to induce more laterals or side shoots production. Pinching is normally practiced in flower crops like carnation, chrysanthemum to reduce the plant height and to promote axiliary branching.

10. Disbudding

The practice of removing unwanted flower buds in a cluster so as to encourage the remaining buds to develop into a large, showy, quality bloom is called disbudding. This practice is commonly done in cut flowers like carnation, chrysanthemum, dahlia, marigold and zinnia etc.

11. Containment pruning

This type of pruning is being practiced in high density plantation of temperate fruits. The main objective is to keep or contain the plant within the space allotted to it without affecting its productivity. In this technique of pruning, generally the 'heading back' is applied but occasionally extra vigorous branches originated at periphery are 'thinned out'. Right from the beginning, the plants are pruned to train them in such a way that most of the bearing woods ie. spurs and laterals are within the small canopy of the tree. When these trees start full bearing, 15 to 20% of branches are annually cut from the periphery. In this way, within 5-6 years the whole of the periphery is renewed. Thus in a period of 11-12 years, the whole of the productive area of the tree will be renewed without affecting its productivity.

12. Alternate differential pruning

It refers to light and severe pruning techniques followed to control the alternate bearing in certain fruit species. Since the physiology of plant is different in 'off' and 'on' years of production, special type of pruning techniques is required to induce regular bearing in these plants. During the 'off' year, the aim is to check the excessive flower bud formation and to encourage the production of many shoots. During the 'on' year, when there is excessive blossoming and formation of less number of fruit buds for the next year crop, the aim must be to reduce the number of fruit buds before the 'on year'.

Thinning

Fruiting is an exhaustive process to the tree especially if the crop is heavy. Therefore, the chief goal is to permit the tree to mature as large a crop as possible and yet conserve sufficient nutrients and carbohydrates for good shoot and spur growth, leaf development and flower bud formation for the next year crop. If the tree is permitted to mature an excessive crop, obviously, it becomes devitalized to the point where it not only produces an inferior produce but it becomes increasingly susceptible to diseases. In order to ensure proper balance between the amount of new shoot growth and flower bud formation, thinning is practiced in certain fruit crops. Thinning is the removal of a part of flower buds, flowers or fruits (before it matures on the tree) with the object of increasing the marketability of the remaining fruit and reducing the alternate bearing tendency.

The other objectives of fruit thinning are the following:

1. To increase the annual yield of marketable fruit,
2. To improve the fruit size,

3. To improve the colour of the fruit,
4. To improve the eating quality (T.S.S),
5. It reduces the limb breakage,
6. It promotes tree vigor and ensures more regular cropping,
7. It permits more thorough spraying and dusting of fruits during the late season application and
8. It ensures uniform ripening.

Time of thinning is also important. In certain fruit crops, thinning is done at blossom timing (blossom thinning) to prevent exhaustion. In other fruit trees, to prevent the overcrowding of the fruits, thinning is done at marble stage. In few other fruit crops, thinning should be done soon after the natural fruit drop of young fruits has started i.e. little before they have grown one-fourth of their normal size.

The amount of thinning of fruits depends upon crop set, response of the variety to thinning, nature of pruning given, age of the tree and the value of the increased size of fruit in the market.

Methods of Thinning

1. Hand thinning

The method is costly and advisable if diseases are present. Twenty per cent hand thinning of pea size berries is beneficial to certain varieties of grapes. Thinned inflorescence produces better quality fruits. Manual thinning is common in date palm in which removal of some branches or removal of some strands from each bunch or shortening the length of the strands is done.

2. Chemical thinning

Thinning of blossoms may be effectively and cheaply done by spraying certain chemical substances. Naphthalene acetic acid (NAA) is effective in thinning the crop when applied two or four weeks after petal fall stage in grapes. NAA at 100 ppm reduces the fruit setting from 67% to 50% in Anab-e-Shahi variety of grapes. In mandarin, NAA 600 ppm on marble sized stage is recommended to thin overbearing fruits so as to increase the size and quality of fruit.

10

Cropping

POLLINATION

Pollination refers to the transfer of pollen grains from anther to the stigmatic surface. It is of two types *viz.*, self-pollination and cross-pollination. If the transfer is from the stamen to the stigma of the same flower or to the stigma of another flower on the same plant or to the stigma of a flower on any plant of the same clonal varieties, then this type of pollination is known as 'self-pollination'. If it is effected without the aid of any outside agency, such as wind or insect, then this process is known as 'autogamy'. If the pollen is to be transferred to the flower of another individual or in the case of pomological varieties to the flower of another variety, this process is known as 'cross-pollination' Cross-pollination may be required for a number of reasons in horticultural crops.

1. Due to the dioecious nature (e.g. Papaya, Datepalm, Nutmeg) or monoecious nature (Cucurbits)
2. Due to the peculiar flower structure, (e.g. Brinjal, Delicious Apples, Vanilla).
3. Due to the dichogamy nature (Onion, Carrot, Sapota)
4. Due to the behaviour of bisexual flowers as functionally unisexual (e.g. Avocado and Allspice)
5. Due to self-incompatibility or self-sterility factors (e.g. Apple, buck wheat)

The pollen may be transferred from the anther to the stigma in a number of ways. In most plants, pollen is shed at or after anthesis and is transferred to the stigmas by insects or by mechanical means primarily the wind. In majority of the horticultural crops, pollination is effected by insects and pollination by wind is not a common one. Fruit crops like sapota, jack and aonla and nutcrops like walnut, chestnut and pecans are reported to be pollinated by winds. They normally produce large quantity of pollen grains which are light in weight so they can be carried to distance places. In these nutcrops, the stigma is feathery to facilitate wind pollination.

Honey bees, ants and many insects aid in cross-pollination. Their activity is greatly affected by weather conditions especially low temperature and rains. Honey bee keeping is reported to increase the fruitset and yield in many horticultural crops (e.g. apple, cardamom, coconut etc.). In apple orchards, each honey bee usually forages 2–3 trees and visit 50–100 flowers per trip, thus in a day, it visits about 5000 flowers. Hence, 10–12 colonies per hectare at a distance of 150 m are ideal for temperate fruit orchards. Recently, a chemical substance (proprietary product Name: Bee-Q) is used to attract the bees so as to increase the fruitset. It has been estimated that nearly double the number of the flowers are pollinated in cardamom if such chemical is used at the time of early and mid flowering phases.

Artificial Pollination

Self-incompatibility is common in apple and pear. This has been recently overcome with the 'recognition' or 'mentor pollen' technique. In this technique pollination is effected with a 2:1 mixture of Methanol killed or irradiated compatible pollen and self pollens which results in seed set. It is believed that the regulatory substance from the mentor pollen to the incompatible/incongruent pollen helps to overcome the crossing barriers. Another technique 'pioneer pollen' is also reported to increase the seed set in such fruit crops. In this method, pollination twice with compatible pollens at an interval of 1–2 days is done. It appears that the first applied pollens promote the activities of the pollens in the second application; hence the first applied pollen is called 'pioneer pollen'.

In emergent situations, when adverse weather conditions prevail or the orchard has inadequate pollinizers, hand pollination can be resorted to. This is much helpful in temperate fruit orchards and is being practiced in European countries and not yet in India. The followings are some of the methods by which artificial pollination is done:

- Artificially collecting the pollens, mixing it with the spores of *Lycopodium* (fern) and applying to the flowers with a soft brush.
- Placement of bouquets—in this method, branches of flowers of pollinizers are hung in the trees to be pollinated.

Fruitset

In an orchard, all the fruit trees do not bear equally or regularly, sometimes one fails to bear at the same time another tree of the kind under similar conditions produces a heavy crop. This problem may be due to failure to set the fruits, unfruitfulness and sterility. The following terminologies are useful in understanding the problem of unfruitfulness. Fruit setting refers to the initial setting of fruit at or just after the time of blossoming and to its remaining on the plant until maturity. A plant is said to be **fruitful** which not only blossoms and sets fruit but carries it through to maturity. Plants which are unable to do this are also as **'unfruitful'** or **'barren'**. **'Fertility'** on the other hand refers to the ability of the plant not only to set and mature fruits but to develop viable seeds. Inability of a plant to do this is known as 'infertility or 'sterility'. Fertile plants are necessarily being fruitful and all the plants need not be 'fertile'. **'Self fruitfulness'** indicates the ability of the plants to mature fruit without the aid of pollen from some other flower, plant and those plants are known as 'self fruitful plants'. 'Self fertility' refers to the ability of plant to mature viable seeds without the aid of pollen from some other flower. Such plants are known as 'self-fertile'.

The causes of fertility, unfruitfulness and failure to set the fruits be grouped into two main categories as those 'internal factors' and those 'external or environmental factors.

I. Unfruitfulness Associated with Internal Factors

These factors may be grouped into three as

- ✓ those due to evolutionary tendencies
- ✓ those due to genetic influences and
- ✓ those due to physiological factors.

A. Evolutionary Tendencies

Production of imperfect flowers as a cause of evolution may lead to unfruitfulness.

1. **Dioecious and monoecious nature:** Monoecious forms like melons are self fruitful but in dioecious forms like papaya pollinizers have to be provided. It is necessary to retain few of the staminate trees to ensure a good fruit set. Nutmeg which is a dioecious tree also requires pollinizers for good fruit set. The ornamental type of pomegranate flowers but fails to set fruit. This is because they are unisexual.

2. **Heterostyly:** In certain cases the structure and form of the flowers are peculiar in nature to prevent self pollination and make cross pollination more certain. If cross pollination does not take place, the plant is likely to remain unfruitful even though perfect sex organs have been developed. One such form is heterostyly, the style and stamens are at different height that self pollination is prevented. For example, in brinjal, four types of flowers are present depending upon the length of style *viz.*, long, medium, pseudo short and true short style types. The last two do not produce any fruits at all. In Delicious group of apples, upright position of the stamens combined with the spreading the petals permit the bees to collect the nectar without touching the pistil or anther.

3. **Dichogamy:** When the receptivity of the stigma and viability of pollen occur in different periods, it is known as dichogamy. This results in low production of fruits. If stamens ripen before the pistil, it is known as protandry (e.g. Annona) and if pistil ripens before the stamens, it is known as protogyny (e.g. Sapota). In Avocado, each flower opens twice and is closed in between. It behaves as a female flower during first opening time and during the second time, it behaves like male flower thus preventing self pollination.

4. **Abortive flowers or aborted pistils or ovules:** In Pecan nut the terminal portion of the pistillate clusters is almost defective and is shed at pollination time. The late flowers of strawberry are always abortive.

5. **Impotence of pollen grains:** Many varieties of grapes produce impotent or non-viable pollens though they appear to have perfect flower type.

B. Genetic Influences

These types of causes are more directly due to genetic factors and are inheritable.

1. **Sterility and unfruitfulness due to hybridity:** More widely if the parents for crossing are related, greater is the degree of self-sterility. A hybrid between pear and quince *viz.*, 'Purina' flowers and sets fruits freely but the fruits are seedless. Seedlessness in most of the commercial varieties of banana and pineapple are due to hybridity.

2. **Incompatibility:** It refers to the incompatibility between pollen and the ovule of the same plant or of the same variety. Pollen and ovule are fertile but they fail to effect conjugation. In apple, pear and plum incompatible varieties require another pollinizer for fruit setting.

C. Physiological Influences

1. Due to slow growth of the pollen tube. e.g. Cole crops.
2. Premature or delayed pollination. In tobacco flowers, if matured pollens are applied to the immature pistils, they germinate and enter the ovules but the flowers soon fall.
3. Nutritional conditions within the plant.

Nutritive conditions of plant just before or at or and just after the time of blossoming is important in determining the percentage of flowers carrying for setting and for maturity. Its effect may be on pollen viability or on fertility of the pistils. In apple, pollen from older trees are always less fertile than the pollens from young vigorous trees. In plums, exhaustion or weakening of the trees in one season by over bearing, drought and poverty of soil is associated with the production of many defective pistils in the following year.

II. Unfruitfulness Associated with External Factors

1. Nutrient supply

Sterility normally occurs in certain families due to over feeding. e.g. Graminae, Cruciferae and leguminaceae. It has been observed that Jonathan apple, which is 'self sterile', on rich soil is becoming 'self fertile' in poor soils.

2. Pollinizers

In temperate fruit crops like apple, adequate number of pollinizers are to be provided, otherwise it affects the fruitset.

3. Pruning and training

Pruning tends to produce more true hermaphrodite condition in variety 'Hope'. If pruning is not done the variety tends to remain sterile and produces aborted pistils.

4. Locality

Jonathan apple which is sterile in one location is reported to be 'self-fertile' in another location.

5. Season

'Ideal' is a hybrid grape cultivar which is 'self-impotent' early in the season but turns to be self-potent later on.

6. Temperature

High temperature at flowering will dry up the stigmatic fluid and prevent pollination. Tomato varieties do not produce any fruit when grown at high temperature i.e. when day temperature goes beyond 37°C.

7. Light

The development of stamens and petals in strawberry flowers takes place only when plants are exposed to long photoperiods.

8. Pests and diseases

Hopper in mangoes, mildews in mangoes and grapes etc. affect the development of fruit leading to poor fruitset.

9. Spraying the trees when they are in bloom

Spraying reduces the crop set during blossoming. Some of the fungicides have inhibiting effect on pollen grains. Bordeaux mixture at 200 to 10,000 ppm prevent the germination of pollen grains on the stigma.

Seedlessness in Horticultural Fruits

In the recent years, the crazy for seedless fruits is increasing among the consumers. The seedlessness results in certain fruits due to the phenomenon of 'parthenocarpy' which refers to the development of fruits without fertilization or even without the stimulus that comes from pollination. Parthenocarpic fruits are usually seedless and need not be always. If a parthenocarpic fruit develops even without the stimulus from the pollination, then this phenomenon, is referred as 'vegetative parthenocarpy' (autonomic) e.g. Banana and Japanese Persimmon. On the other hand, if a parthenocarpic fruit develops from the mere stimulus of the pollination (but fertilization won't take place), it is known as 'stimulative parthenocarpy'. The female flowers of triploid watermelon require the pollen grains of diploid varieties to develop into a seedless fruit. Diploid pollen grains give a stimulus for the ovary to develop into a fruit without seed. When Allahabad Round variety of guava was self pollinated it resulted in the development of parthenocarpic fruit due to the stimulation provided by pollen hormones. Black Corinth variety of grapes is seedless due to stimulative parthenocarpy. In grapes CV. Thompson Seedless pollination and fertilization take place but the embryo gets aborted subsequently resulting in seedlessness. This phenomenon of development of seedlessness is referred as 'steno-spermocarpy'. The seedless or parthenocarpic fruits are advantageous since there is a greater preference among the consumers for the seedless fruits owing to its quality and hence they fetch higher prices than the seeded fruits of the same kind (e.g. seedless grapes, guava or oranges). Besides, the problem of unfruitfulness may not arise if a fruit develops parthenocarpically and the grower is assured of his crop (e.g. banana). One drawback with the seedless fruits is that they are usually small in size (e.g. grapes) and irregular in shape (guava). Recently, in citrus particularly mandarin oranges, less seeded varieties have been developed through mutation breeding (gamma irradiation) where the isolated solid mutants produce sterile pollen grains which when pollinate the flowers result in lesser seeded fruits.

Induction of Seedlessness in Fruits

The seedlessness can be induced by the following methods:

1. Use of growth regulators

Application of GA at 8000 in lanolin paste on the cut end of the style of the emasculated flowers of guava resulted in the development of seedless fruits. Similarly, seedlessness in loquat was induced by spraying GA 100–200 ppm on the emasculated flowers.

2. By changing the ploidy level

It was first demonstrated in Japan that by developing a triploid watermelon (3n:33) by crossing tetraploid x diploid varieties, seedlessness could be achieved. The seedlessness in naturally available seedless guava varieties is due to autopolyploidy (triploid) and not due to parthenocarpic fruit development.

Parthenogenesis

In some plants, fruits develop parthenocarpically, still they produce viable seeds (e.g. Mangosteen and Strawberry). This phenomenon is referred as parthenogenesis. The seedlings of such fruits are genetically uniform. In certain cases, seeds develop parthenogenetically but they are non-viable. When female flowers of jack were pollinated with pollen grains of bread fruit, seeds formed in jack but they did not germinate as they were non-viable.

Fruit Drop

Fruit trees usually bear a large number of flowers and only a small percentage of which are enough to give a normal yield. For instance a single inflorescence of mango contains as many as 5000 flowers but an average of 5 fruits per inflorescence would provide a good to heavy crop, however, the actual percentage of fruitset will be much lesser. When the fruitset is much more than the trees can normally carry to maturity, there will be drop of fruits at various stages of development as an adjustment of tree to its resources. Such a drop is natural and beneficial to the trees and it will prevent exhaustion of the resources and breaking of branches by over bearing.

This drop occurs at three stages as indicated below:

First drop

It occurs shortly after flower opening. Usually flowers with aborted pistils drop off at this stage. Lack of pollination, low stigmatic receptivity, defective flowers, poor pollen transference, and occurrence of incompatibility are some of the causes attributed for this drop.

Second drop

It occurs a fortnight later than the first drop. This drop includes unfertilized flowers and some fertilized flowers. Fertilized flowers also drop off at this stage as a result of adjustment in the trees between nutritional factors and fruitset.

Third drop

This drop occurs when the fruits are 'marble size' due to the formation of abscission layers in the young fruit stalks. This drop is generally common in most deciduous fruits and as it coincides with the month June, this drop is also known as 'June drop'. This natural thinning of fruit helps the trees to produce remaining fruits with good size. The above drops which occur in three stages are beneficial to the fruit tree.

Pre-harvest Drop

Another kind of fruit drop which is a loss to the grower is a 'pre-harvest drop'. In this case, dropping or shedding of fruits takes place before harvest. At this stage, half and three-fourth developed fruits are shed due to many causes. This is a loss to the growers and is a serious problem confronted by them especially in apples, mango and citrus fruits.

In all the above cases *viz*., dropping of flowers, flower clusters or fruits are preceded by the formation of a layer or several layers of cells at the base of petiole or pedicel or peduncle and at the right angle to this axis. These cells are more or less rectangular in shape and are not cemented together tenaciously and not crossed by fibers or vascular tissues. Consequently breakage or abscission takes place at that point. The formation of this layer of cells is a natural phenomenon taking place as natural maturity is reached. It may be hastened by certain environmental conditions.

Causes of Fruit Drop

1. Mechanical

Wind and hailstorms cause fruit drop.

2. Climatic factors

Climatic factors such as high temperature, low humidity and very low temperature hasten the formation of abscission layers, consequently the fruits drop. It has been observed in South India that shedding of fruits in mango will be more if the temperature is high and humidity is low.

3. Physiological factors

Abnormal fluctuations of soil moisture favour heavy fruit drop.

4. Nutritional

Lack of available nitrogen and similar other nutritional factors may cause fruit drop. The shed is more in weak shoots than in the strong ones and also more in young trees than in medium or old trees.

5. Cultural practices

Deep digging or deep ploughing during the fruit development phase will injure the roots and cause the fruits to be shed. Drought or lack of irrigation especially in mango during the third drop stage increases the dropping per cent.

6. Pathological causes

Incidence of pests and diseases will cause more shedding of fruits. For instance, high incidence of diseases like powdery mildew and anthracnose and pests like hopper and mealy bugs in mango favour more fruit drops.

7. Varietal factor

Within a kind of fruit, the varieties differ among themselves in the extent of fruit drop. In one study, it has been found that under similar conditions the extent of shed varied from 0.9% in Willard variety to 32.5% in Jehangir variety of mango.

Prevention

The pre-harvest drop may be reduced by controlling the causes to a certain extent. Proper and timely culture such as irrigation and manuring, plant protection, provision of pollinizers and wind breaks will help to prevent or reduce the amount of fruit drop.

A definite relationship between the auxin content and the abscission of fruits during various stages of development has been established in apple fruits. In the final stages of the fruit growth, a rapid decline in auxin content is correlated to degeneration of endosperm causing preharvest fruit drop. This led to the thought that high concentration of auxins supplied exogenously may inhibit fruit drop.

The possibility of reducing the preharvest drop by means of plant regulator sprays has been well established in many fruit crops. Naphthalene Acetic Acid (NAA) and its related compounds (10 to 20 ppm) are very effective in reducing the drop of fruits in pome fruits such as apples and pears. 2.4-Dichloro phenoxy acetic acid at very low concentration i.e. < 20 ppm checks preharvest drop in citrus fruits. 2.4-D and 2.4.5-T at 5 to 10 ppm when sprayed on Kodai orange decreased fruit drop and increased fruit retention. NAA at 10 ppm was effective in reducing the fruit drop in litchi and GA at 60 ppm reduces the fruit drop in jamun. Great care must be exercised in the use of hormones as it is possible to cause much damage to the foliage and new growth.

Alternate Bearing

Alternate bearing or biennial bearing is an age old horticultural problem and is also designated as irregular bearing. Essentially, these terms imply that the tree carries optimum load of crop in one year but it fails to flower and produces a satisfactory crop in the following year. This phenomenon is observed in most of the perennial tree crops in varying degrees. Any perennial plant, which carries a heavy crop load in the current year is likely to bear less in the following year. However, this phenomenon is distinctly expressed in horticultural crops like mango, olive, coffee, tamarind, apple, and certain cultivars of plum, pears and coconuts (Cv. Malaysian Semi Tall). In this type of bearing, the heavy crop year or season is referred as 'on year'. If an 'on year' is regularly followed by an 'off year', then it is called as 'biennial bearing' and or 'alternate bearing'. Irregular bearing on the other hand refers to rhythm of a heavy crop year followed by more than one year 'off crops'. The disadvantages of biennial bearing are:

- When compared to a regular bearer, the average yield per tree in an alternate or irregular bearer is always less.
- The quality of fruits may be inferior during 'on year' cropping as compared to the regular bearer.

- More breakage of heavy loaded branches is likely to occur in an 'on year' and the trees become more sensitive to frost damage than a regular bearer.

Causes of Alternate Bearing

1. **Genetic causes:** Certain trees which belong to the particular species or genus or family express this tendency always. For instance, apple (*Malus domestica*), pear (*Pyrus communis*), plum (*Prunus domestica*) and apricot (*P. armeniaca*) belong to the family Rosaceae have a more tendency to express this phenomenon. Within a species also, distinct cultivar differences exist for varying tendencies towards alternate bearing. In Mango, cultivars like Bangalora, Neelum, Rumani, Mallika and Amrapali are the regular bearers while cultivars like Mulgoa, Dashehari and Pairi are biennial bearers in habit.

2. **Bearing habit:** Terminal flower bud formation is often regarded as all explanation for alternate bearing habit in apple. The formation of spurs (which is a terminal bearer) is strongly stimulated in the 'off year'; hence in the following year (on year) the crop bears heavily.

3. **Age of the tree or the age and size of shoots:** In alternate bearing cultivars of apple, this tendency is found to be more in young trees than in older trees. In mango, it has been found that the bearing shoots which attain about 8–10 months maturity alone tend to produce the flower buds in those varieties which tend to express this phenomenon.

4. **Carbon/Nitrogen ratio:** In mango, it has been found that higher starch reserve, total carbohydrate and Carbon/Nitrogen ratio favour flower bud formation. Starch reserves and carbohydrates are extensively used as a 'sink' during heavy cropping year and hence bud formation is drastically affected leading to an 'off year' immediately to follow an 'on year'.

5. **Endogenous hormonal factors:** The problem of biennial bearing in mango appears to be closely associated with the fruit development process and inhibitory influence of the developing fruits on vegetative growth. The developing seeds are found to exert inhibitive effects on flower bud formation in mango and apple. In citrus, particularly in sweet oranges, alternate bearing is caused by the fruits, inhibiting flower formation on the wood on which they are borne and that is brought out by a flower inhibiting substance diffusing from the fruits.

11

Harvesting, Handling and Storage of Fruits

One of the most important periods in the life of fruits is from the time it is picked until it is disposed to the consumer. This period is commonly referred as 'post harvest life'. Fruits and vegetables are living organisms and highly perishable commodities. These are affected by a number of factors leading to the post harvest spoilage. The total losses in fruits are estimated to be 20–30% amounting to nearly Rs. 8000 crores annually depending upon the fruit varieties and post harvest handling systems. This may vary from fruit to fruit; 14% in apple, 20–80% in banana, 40–100% in papaya and 35–100% in potato and 60–85% in onion. The magnitude of these losses in developing countries like India is comparatively higher due to a variety of reasons; mainly lack of infrastructure needed for post harvest management of the perishables and poor utilization of the fruits and vegetables by processing industry.

The various causes for losses are

1. Loss of moisture, causing wilting/ shrinkage
2. Loss of stored energy
3. Loss of food constituents
4. Physical loss through pest and disease attack
5. Loss due to physiological disorder
6. Fibre development

7. Greening
8. Root/shoot growth
9. Seed germination
10. Other causes–Micro organism, Insects, Rodents and action of enzymes.

POST HARVEST TECHNOLOGY

The post harvest technology may be defined as those techniques (methods) which reduce the post harvest losses and help economic utilization of crops (including waste) to the maximum as fresh produce or to make nutritious, safe and stable products. To reduce the losses, we must understand the biological and environmental factors involved in deterioration of the perishables and secondly to use the post harvest technological procedures to achieve the same. Basically, enzymes, micro organisms, pests and the environmental factors are responsible for spoilage of these crops. From this point of view, any method which prevents enzymatic activity of plant or microbe and their activities would increase the post harvest life. Thus, post harvest technology involves multi disciplinary approach with horticulture, plant nutrition, pathology, microbiology, biochemistry and physiology, food science and technology and engineering.

PRE-HARVEST FACTORS

Many pre-harvest factors decide the post harvest life of horticultural produce.

1. Maturity and maturity indices

The post harvest quality and storage life of fruits are controlled by maturity. If fruits are harvested at a proper stage of maturity, their quality is excellent. Vegetables are harvested as and when they attain maximum size and yet are tender. Over maturity in root crops causes sponginess and pithiness. Their harvesting should not be delayed. Delay in harvesting of onion and garlic reduces their storage quality.

Maturation is the stage of development leading to the attainment of physiological maturity (when a plant or plant parts will continue ontogeny even if detached). On the other hand, horticultural maturity refers to the stage at which a plant or plant parts possesses the pre-requisites for utilization by the consumer for a particular purpose.

Methods to Determine the Proper Time to Pick the Fruits

1. Maturity tests

Following are the rough but ready maturity tests of fruits employed to pick the well matured fruits:

(a) Colour change is one of the criteria to judge the maturity of fruits. The change of peel colour from green to yellow is the main criterion to test maturity in mangoes. Similarly in papaya change of colour at apical end of the fruit indicates the full maturity stage. In the case of pineapples nearly 25% of the fruit surface should have turned to yellow colour.

(b) Increase in size.

(c) Softening of the tissue of the fruits. e.g. Figs and grapes.

(d) Ease of detachment from the stalk. e.g. Sapota and annona.

(e) Shriveling of fruit stalk. e.g. Watermelon.

(f) Sound by tapping-jack and watermelon when ripe produce hollow and dull sound on tapping but produces metallic sound if unripe.

(g) Days from full bloom i.e. **DFFB** is based on the apparent effect of temperature on flowering and fruit development which suggests the use of degree days or heat summation for the more accurate prediction of harvest data. Accumulated heat units for a fruit crop's optimum maturity is computed by calculating the time in relation to temperature above a certain minimum base temperature. For example, in apple the base temperature is fixed as 50°F and a day with an average temperature of 68°F (20°C) would provide 18 degree days F (10° days C) of heat units. A day with an average temperature of 41°F (5°C) would provide zero degree days of heat units. Based on this, heat requirement of certain fruit crops is worked out as below:

Fruit	Cultivar	Cut out or base temperature	Degree days
Apple	Red Delicious	50°F	1659–1705
Grape	Thompson Seedless	50°F	1600–2000
Mango	Banganapally	18°F	1426±5
Banana	Cavendish	9.8°C	1930

2. Accurate tests

(a) Colour charts

Charts are prepared for indicating colour at different stages of maturity. By referring to this ready chart, one can easily judge the correct stage of maturity.

(b) Penetrometer

It is an instrument which indicates or measures the softening of tissues as an index of maturity. It chiefly helps in determining when fruits are too soft and ripe to storage rather than when picking should begin. Firmness of the flesh can be assessed by removing a thin slice of the skin and flesh with a knife and using a special hand operated tester which records the kilogram of pressure (kg/sq. cm) for the plunger to penetrate the flesh.

(c) Sugar/acid or Brix/acid ratio

This is based on the principle that acid content reduces and sugar increases on ripening. The fruit growers should bestow more attention and considerable care during the picking season to reduce to a minimum level of careless handling of fruits by pickers.

1. Picking must be commenced from the lower branches of a tree advancing towards the top in order to reduce dropping of fruits to the minimum.
2. As far as possible, dropping of the fruits from the tree should be avoided to avoid any possible physical damage.
3. During picking, care must be taken to avoid any possible damages to the branches especially to the spurs as the subsequent cropping depends upon them.
4. Picking early in the morning is always best. Picked fruits should be kept in shade and excluded from sun. After picking, the fruits must be kept in the coolest place available which is well ventilated to arrest respiration and break down as much as possible.
5. There should not be any bruises in the fruits while picking as it will lower the marketable quality.
6. If picking is done in mid-day or hot weather, fruits should be kept in a shed overnight to cool.
7. The time of picking depends upon the following factors:

   (i) **Variety:** Certain varieties will ripen only in trees and certain others will ripen even when they are harvested 3/4th maturity stage.

   (ii) **Purpose:** For which it is meant, for example, tomatoes meant for table purposes can be harvested at red ripe stage whereas for distant market they may be harvested during breaker's (colour changing) stage.

   (iii) **Distance to market:** The closer is the market; ripe fruits may be picked and distant is the market, mature green stage may be harvested so that when the fruits reach the point of consumer, it will be ripe. In order to maintain good quality, storage capacity and commercial value, it is highly important that they should be picked at the proper stage of maturity. Prematurely picked fruits are generally small, poorly coloured, sour, tough and inferior in quality. Therefore, it is important that fruits should be allowed to remain in the tree as long as it is possible as it continues to increase in size when it remains upon the tree.

Handling

Handling includes all processes from picking to delivery or disposal at the consumer point. This includes the treatments given for getting the fruits ready for the market *viz.*, packaging and wrapping, ripening and storage. One of the important treatments is the dipping the fruits in antiseptic solutions like 1–2% caustic soda to remove the dust and infestation of scale insects and washing with 1–1.5% of Hydrochloric acid to remove any spray residue and to improve the appearance.

Pre-cooling

It refers to the rapid removal of the field heat from the freshly harvested fruits and vegetables in order to slow down ripening and reduce deterioration prior to storage and shipments. Different methods are adopted to precool the fruits, the important ones are (i) air cooling in which the fruits are kept in a cold room, (ii) hydro cooling-dipping of the fruits in cold water or by spraying cold water on the fruits and (iii) vacuum cooling—a costlier technique in which the atmospheric pressure is reduced so as to reduce the pressure of water vapour in chamber which results in evaporation of water from fruits which bring down the temperature. Vacuum cooling causes about 1% weight loss in the produce.

Grading

Grades or grading refers to the assortment of the fruits into different groups based on certain characters. This includes colour, firmness, soundness, free from blemishes and size of the fruit. Grading is a good market practice which improves the mutual confidence of salesman and consumer.

In India, grading is mostly done on the basis of size. But in the developed countries, grading is a rule and for example, the U.S. Department of Agriculture prescribes the following grades to apples in the U.S.A.

(i) U.S. Extra fancy
(ii) U.S. Fancy
(iii) U.S. No.1
(iv) U.S. Commercial

In India, grading of fruits like apple, plum, pear and mango varieties like Alphonso, Rumani, Bangalora and Sathugudi is done by Agmark mainly based on size only. For vegetables like tomato and brinjal grading is being followed to some extent for specialized city markets. The Indian standards Institute has specified four grades *viz.*, Super A, Super, Fancy and Commercial for tomato and three grades *viz.*, Super, Fancy and Commercial for brinjal.

Wrapping

Covering the fruits after harvest with any material in order to improve its post harvest life is known as wrapping. The materials commonly employed as wrappers are tissue papers, waxed paper, pliofilm, cellophane paper, aluminium foils and alkathene paper etc. Wrapping has the following advantages:

(i) It minimizes the loss of moisture in shriveling,
(ii) It protects against the spread of diseases from one to the other,
(iii) It reduces bruises,
(iv) It reduces damage during transport or in storage and
(v) It makes the fruit more attractive.

Care must be taken to see that wrap is not too impervious to the passage of oxygen and carbon-di-oxide. Pre-packing of banana fruits is done in 100 gauge polythene bags under room temperature and cold storage.

Waxing

Another treatment given to the fruits during handling is waxing. Waxing of fruits helps in reducing the moisture loss, improving the appearance of fruits and reduces the incidence of storage diseases. Wax emulsion is prepared by melting microcrystalline paraffin or cranaube wax along with emulsifiers. Boiling water free from hardness is slowly added to the molten ingredients and thoroughly stirred in order to make a stable emulsion. The harvested fruits are dipped in dilute wax emulsion for a minute and then these fruits are completely dried for 10–15 minutes. Care must be taken that excessive quantities of wax are not applied to the fruit. If the wax is not sufficiently diluted or if it is so thick, CO_2 and O_2 cannot readily pass through to and from the fruit. Mango fruits treated with wax emulsion containing 8 to 12% solids have one or two weeks longer storage life than the untreated ones.

Packaging and Packing

The term packaging encompasses both the direct or primary packaging around the product and the secondary and tertiary packaging, the over packaging such as over warts, cartons and crates etc. Proper packaging is essential otherwise the spoilage of fruits and vegetables are more in our country.

A packing material should be sturdy and it should protect the fruits in transport, more specifically it must be economical. The materials that are generally used in India for fabrication of package of fruits and vegetables are bamboo, wood, gunny bags, plastic films, fibre and plastic corrugated boards etc. Bamboo baskets and wooden crates of different shapes and sizes are used for a number of perishable commodities. Mupots, gunny bags and palmyrah mats are also used for a variety of purposes. Bamboo baskets are though relatively cheaper, they have many disadvantages like (i) the low dimensional stability and inability to withstand stacking load (ii) they are not strong enough to withstand rough handling. Packaging of grapes in mud pots is quite common in South India. It is often observed that during transport, the mud pots break and the contents get damaged. Though the mud pot has its own advantages as a container for grapes and such other fruits, it has to be handled very carefully thus affecting the speed of handling. In some cases like mango, pineapple, banana etc., a straight load is practiced in certain regions. For example, banana in bunches are loaded without any packaging into the railway wagons or trucks and transported from Maharashtra to Delhi. Similarly, mangoes are transported from South to North and pineapples are shipped from North East India and Kerala to different regions. In these cases, it has been observed that the losses due to spoilage are considerable.

Prepackaging

Prepackaging is generally defined as packaging the produce in consumer size units either at producing centre before transport or at terminal markets. Packaging of fresh produce in consumer unit packs protects the produce against the damage and excess moisture loss. The packaging material used should have the following properties:

(a) Sufficient permeability to oxygen, carbon dioxide and water vapour
(b) Desired protective physical properties
(c) Transparency.

The permeability requirement depends upon rate of respiration of the produce, the package bulk density and temperature of the storage. The prepacking of fruits and vegetables has multiple advantages:

1. It reduces the transportation cost by eliminating unwanted and inedible portion of fruits and vegetables.
2. The space required for shipping and storage is less.
3. It has a better eye appeal as the produce is prepacked in attractive film and the quality of the produce can be seen from outside without opening the pack.
4. Prepackaging reduces the shopping time of the consumer as the produce is graded before being prepacked.
5. By increasing the shelf-life both at room temperature and refrigerated conditions, prepackaging helps the grocer to market his produce over a longer period and thereby, avoid losses due to spoilage.
6. Prepackaging has a very quick turnover because of the recent development of automatic machine.
7. It saves labour costs and also makes the produce easy to handle.

Among the different types of packaging films, polyethylene film finds the maximum use. The shelf-life of asparagus is extended to 5–6 days or upto 2 weeks under ideal conditions when wrapped in special plastic film. Polyethylene bags of 100 gauge thickness even without ventilation could be used as prepackaging of brinjal, okra. Perforated polyethylene bags are also found to be suitable as prepacking material for packaging of Alphonso mangoes.

Cushioning Materials

The cushioning materials used for packaging fruits and vegetables are dry grasses, paddy straw, leaves, saw dust, paper shavings etc. The properties of good cushion materials are

1. It should have a resilient property.
2. It should have the ability to dissipate the heat of respiration of produce.
3. It should not carry any infective pathogens or it should not injure the soft fruit in any way.
4. It should be physiologically inactive.

Recently pulp tray, comb partition, cell pack etc. are used as the cushioning materials.

The expansion of horticultural industry has created a crisis with respect to the availability of wooden packaging cases. With the existing scare resources of wood, the ever increasing demand for wood cases cannot be met. Hence, considerable work has been done by different agencies on the introduction of alternative packaging. This included wire bound wooden cases, hard board, ply board, fibre board, pine needle case, corrugated fibre boards (CFB) and plastic board etc. The CFB boxes can be fabricated from draft paper made from bamboo, long grasses and many other types of agricultural residues like paddy and wheat straw, cotton and jute stick and recycled paper and cardboard. On account of severe shortage of fuel wood, timber boxes to a large extent find their way for use as firewood while all the CFB cartons are recycled as pulp or paper. CFB is also the suitable package material for cut flowers since they have isothermic properties. The CFB cartons offer added advantages like (1) Minimal bruising damage-research conducted showed that apples in CFB cartons had only 5% loss as against 30% in conventional packing (2) easy handling and stacking (3) more economical transport (4) can be turned quickly into highly precise and accurate size (5) can be appropriately punched, ventilated, printed at low cost (6) made pilfer-proof and reveal tampering at a glance (7) offer the most acceptable packaging in the international markets (8) collapsible and occupy less volume for storage of empties (9) can be used under cold storage conditions after giving water-proof treatment (10) can be made by reinforcing with hessian or nylon fibre.

Another important alternative to wood for packaging is plastics. Various plastic materials which can be used to package fruits and vegetables are

(a) Polypropylene boxes
(b) Molded expanded polystyrene boxes
(c) Stretch film
(d) Film wrap and
(e) Polyethylene net

Among them, polypropylene corrugated box has very good properties and can be used for horticultural produce, especially for apple. One main added advantage of this material is that this can be reused quite a few times.

Pelletization

Loading and unloading are done manually in India. Due to low unit load, there is a tendency to throw, drop or mishandle the package, damaging the commodity. This loss can be considerably reduced by using pellet system. However, this requires the standardization of box dimensions. For each commodity it should be worked out. Once this is accomplished, mechanical loading and unloading become very easy with the fork-lift system.

Storage of Fruits

Production of horticultural produce is limited by certain climatic conditions and through proper storage of fruits; the availability of the produce is extended. In this way, the reasonable prices are ensured to the fruit growers and the dumping of fruits (glut) in the market is avoided. Consumers on the other hand are able to get fruits at reasonable price.

Factors influencing storage: In general horticultural produces contain large amount of water as compared to cereals. Fruits and vegetables contain about 85% and 95% moisture content respectively. All the fruits and vegetables are living tissues and the act of harvesting upsets the balance of living processes existing during the growth. The physiological processes like photosynthesis and water uptake stop in the harvested produce but the transpiration (loss of water) continues. The development of certain diseases can progress unhindered and the chemical changes progress rapidly (starch to sugars). Any storage procedure should aim at slowing down, regulating or preventing these changes taking place.

Following factors are associated with storage of fruits.

1. Temperature

Fruits are alive and they continuously carry on respiration even after harvest. The faster a fruit respires faster it ripens. Therefore, the basic objective in the storage should be to keep this destructive process minimum. This rate of respiration can be controlled by temperature. Low temperature can also slow down the respiration considerably and prolong the storage life of fruits. Each fruit has a specific critical temperature which has to be determined for successful storage.

2. Humidity

As fruits and vegetables relatively contain a large percentage of moisture, maintaining a high humidity during storage will minimize transpiration. Otherwise the produce will shrivel and lose quality. A relative humidity ranging from 85-95% is essential.

3. Storage atmosphere

In addition to oxygen and carbon-di-oxide, other gases may affect the storage of fruits or damage the fruits. Ethylene gas which is released during ripening of fruits will hasten ripening of other fruits also. CO_2 is continuously produced by the respiration of the produce kept in the storage and may accumulate to a dangerous extent. This storage atmosphere can be improved by ventilation which means introduction of outdoor air into the storage room. Ideal storage temperature and relative humidity for some of the important crops are given in Table 11.1.

Table 11.1 Storage requirement for important fruits and vegetables

Name of the fruit/vegetable	Storage temperature °C	Relative humidity %	Approximate storage life
Apple	0.2.0	85–90	4–8 months
Grapes	0.2.0	80–85	2–4 weeks
Peaches	0.2.0	80–85	2–4 weeks
Plums	0.2.0	85–90	4–8 weeks
Bananas	12.5–14.5	85–95	3 weeks
Mangoes	9.0–10.0	85–95	4 weeks
Oranges	6.0–6.5	85–90	18 weeks
Papayas	4.5–6.0	80–85	5 weeks
Guavas	9.0–10.0	85–90	3 weeks
Tomato-ripe	0–4.5	85–90	10 days
Tomato-mature green	10.0–15.0	85–90	30 days
Peas	0–0.5	90–95	2 weeks

Methods of Storage

1. **Air cooled storage:** Fruits are placed in racks in the insulation building. By keeping the ventilators open at night to admit the cool night air and shut down during the day to keep off the warm dry air, the storage rooms can be kept cool. Fruits should not be heaped but spread at racks. Affected fruits should be removed.

2. **Air cooled storage-refrigerated with ice:** A few storages are filled with ice and salt and equipped with fans for quick cooling of the fruit during warmer periods.

3. **Treating with wax emulsions:** In the absence of cold storage facilities, certain fruits and vegetables can be stored by this method. Fruits are dipped for a minute in a wax emulsion and dried completely for 10 to 15 minutes.

4. **Refrigerated storage:** It has many advantages. (1) Prompt cooling of products. (2) Maintenance of optimum relative humidity (3) an even holding temperature with 1.0 to 2.0°C fluctuations. (4) No need to depend upon the external conditions of air, their movement etc. and (5) Most efficient and scientific way of storing. Refrigerated storage involves the use of **refrigerant**. It is a liquid that evaporates or boils at temperatures and relatively at low pressure. When the liquid refrigerant boils and changes to gas in the storage room coils, it absorbs heat from the stored product. The gas with the heat is then piped outside storage room where it is compressed and passed through a condenser which absorbs and dissipates the heat and then changes

the gas back liquid for repeated use. Refrigerants commonly used are Freon, ammonia and methyl chloride. Freon is most popular, odourless, and not toxic but leaks are difficult to locate. Ammonia gas is toxic if it escapes in large quantities. The commonly used temperature and relative humidity to store fruits and vegetables are furnished in Table 11.1.

5. **Controlled atmospheric storage (CAS storage):** By reducing the O_2 supply available to the fruit and by increasing the amount of CO_2 around the fruit or by lowering the temperature, the respiration rate can be slowed down in fruits and vegetables. A combination of all the above principles is involved in the controlled atmospheric storages. Sometimes, oxygen is replaced by nitrogen in the storage room. Apple can be stored in CA storage with 10% of CO_2 and 11% O_2 with a temperature of 4°C. A modification in the controlled atmospheric storage is the use of sub atmospheric pressure to store the horticultural produce. This method is known as 'hypobaric storage' (HBS). In this method, all partial pressures of the various gases including water vapour are reduced. This atmosphere reduces the respiration rate, ethylene synthesis (a gas which hastens ripening) and the rate of other metabolic processes. Therefore, shelf life of many fruits and vegetables including ornamental flowers get extended.

Recently in modified atmospheric storage method maintaining the relative humidity at 90 to 95% is recommended for the storage of green vegetables and other root and tuber vegetables to prolong the storage life. This method reduces the loss due to decay and moisture evaporation.

6. **Preservation by irradiation:** Potato, onion, garlic and other root crops lose their commercial value by sprouting or rooting during storage. These processes can be delayed by irradiating the produce with gamma rays. Potatoes treated with 10 KR of gamma rays prolong the storage life upto one year with least loss in weight. In small onion, gamma ray treatment keeps the bulb fresh upto 150 days in storage. This irradiation treatment also checks the infectivity of storage pathogens. Irradiation at low dose levels can delay ripening and over-ripening of tropical fruits like bananas, mangoes and papayas, when treated mature but unripe. Various studies conducted at BARC have revealed that at ambient temperature, shelf-life of irradiated mangoes (0.25 kGy) is enhanced by 5–7 days and there is an increase of about ten days in shelf-life of bananas when irradiated at 0.25 to 0.35 kGy.

7. **By use of chemicals:** There are certain chemicals other than fungicides which increase the shelf-life of fruits by delaying the ripening

and senescence. Potassium permanganate, an ethylene absorbent proves effective in the complete absorption of ethylene from banana held in sealed polyethylene bags. Mangoes dipped in 1000 to 2000 ppm of Maleic hydrazide delays ripening. Storage life of citrus fruits is prolonged by the application of 2,4-D and 2,4,5-T. The storage potential of fruit is largely dependent on the level of calcium. Any other nutrients which disturb the calcium content also adversely affect the shelf life of the fruits. The higher levels of N, P and Mg and low levels of K and B leads to the calcium deficiency in fruits and reduce its storage life. Calcium treatment delays ripening and senescence and improves the quality of fruits and vegetables. Pre-harvest spray of Calcium chloride (0.6%) and Calcium nitrate (1%) helps in the enhancement of shelf-life of mango and guava. Exogenous calcium application is known to get incorporated into protopect in molecules in the middle membrane and retard hydrolysis during post harvest ripening, inhibit fruit softening and extend storability of fruits.

8. **Evaporative cooling (cool chambers):**   Evaporation of water produces a considerable cooling effect and this evaporative cooling occurs when air that is not already saturated with water vapour, is blown across any wet surface. Cooling chambers work on the principles of evaporative cooling. This structure essentially contains a single layer of bricks as floor and a double layer of bricks as side wall with an interspace of 7.5 cm wide filled with river sand. The top of the storage space is covered with plaited vetiver roots or gunny cloth in a bamboo structure (Fig. 11.1). Once the cool chamber is saturated with water, sprinkling of water once in the morning and once in the evening is enough to maintain the temperature and humidity. These storage chambers maintain a very high humidity of about 95% throughout the year and it can reduce the temperature during summer months and increase the temperature during peak winter months. These chambers are ideal for storage of fresh fruits and vegetables for a short period. Firm ripe mangoes can be stored upto 10 days as against 4 days under room temperature. Similarly, citrus fruits like grapefruit and orange can be stored upto 60–90 days as compared to 8–15 days in ambient room temperature.

Ripening of Fruits

Ripening transforms a physically mature but inedible plant organ into a visually attractive taste and smell sensation. It marks the completion of development and commencement of senescence with life of a fruit and is normally an irreversible event. Ripening can be achieved by the application of ethylene.

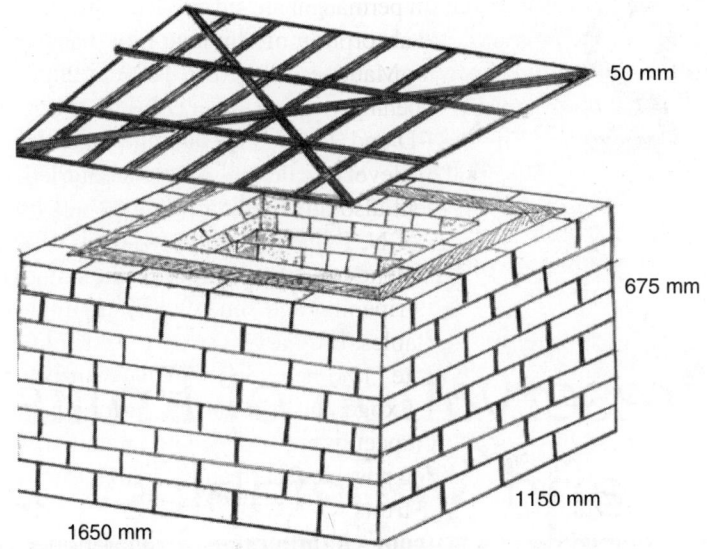

50 mm

675 mm

1150 mm

1650 mm

Figure 11.1   Cool chamber.

Accurate quantity of ethylene should be used in the ripening room at regular intervals. A concentration of CO_2 above 1% delays ripening. Hence, thorough ventilation is essential. By use of Ethephon commercially known as ethrel or CEPA (7 fluid ounces release 1 cft of ethylene), making it alkaline using caustic soda (3 g of caustic soda for 20 ml of Ethephon). Calcium carbide can also be used for ripening (100 g for 100 kg fruits).

DEGREENING

Degreening is the process of decomposing green pigments in fruits usually by applying ethylene or other similar metabolic inducers to give a fruit its characteristic colour as preferred by consumers. It is applicable to banana, mango, citrus and tomato. The time required to degreen a fruit depends upon the degree of natural colour break and maturity. The higher the green colour and more mature a fruit is, the less time is required to reduce the chlorophyll to a desired level.

Degreening is carried out in special treating rooms with controlled temperature and humidity in which low concentration of ethylene (20 ppm) is applied. The ethylene should be supplied from a gas cylinder. These rooms are thoroughly ventilated to keep the CO_2 level below 1% which does not allow higher colouring.

12

Preservation of Fruits and Vegetables

Most fruits and some kinds of vegetables are available during specific seasons at a cheaper rate and during the other seasons they may not be available or available in a limited quantity and hence they may be costly. Their availability can be extended by cold storage but it is highly expensive. Therefore, it is essential to preserve the fruits and vegetables by some means so that they can be made available in a form throughout the year. The spoilage of fruits and vegetables are caused by microorganisms like moulds, yeasts and bacteria. So, any preservation technique aims at preventing the above type of spoilages. Several types of preservations are used and are discussed briefly here.

1. Drying

Sun-drying or mechanical drying of fruits and vegetables involves complete removal of the moisture from them. They are then treated with sulfur fumes to maintain colour and also to avoid spoilage by microorganisms. Dates, grapes, jack, peaches, beans, bitter gourd, garlic, onions, potatoes etc., are preserved by this method.

2. Freezing

When fruits and vegetables are brought suddenly to a very low temperature at which all the chemical reactions stop. But it requires rather expensive equipment and shipping and storing under refrigeration.

3. Preservation by heat

It consists basically of the application of heat in varying degree to the food in closed containers for a sufficient long time to sterilize contents before they are hermetically sealed. This method of preservation by heat is also known as processing. Canned fruits and vegetable come under this category.

4. Preservation by sugar

No organisms can grow in a substance which contains sugar concentrations above 65% as microorganisms get desicated at higher concentrations by osmosis. Jams, jellies, marmalades and crystallized fruits are preserved by virtue of the above said principle.

5. Preservation by salt

A product which contains about 15% salt makes it most unfavourable for the growth and multiplication of microorganisms. Salt acts both by osmosis and as a poison on microorganisms. Pickles are preserved by this method.

6. Preservation by chemicals

Chemicals like benzoic acid and sulphur-di-oxide at specific concentration help to preserve the fruit juices, squashes or cordials. Benzoic acid in the form of 0.06 to 0.10% sodium benzoate is sufficient and is effective against yeast and its action is effective in the presence of CO_2. Potassium metabisulphite (KMS) when added liberates sulphur-di-oxide which reacts with water to form sulphurous acid. The concentration of SO_2 is recommended around 350 ppm. It is more active against moulds, spores and bacteria.

7. Preservation by fermentation

Fruit juices containing sugar are fermented to produce alcohol. These fermented substances keep off the spoilage organisms due to the presence of alcohols. e.g., wine making. Methods and techniques involved in the preparation of selected products like jam, jelly, squashes, syrup and pickles are discussed below.

Jam

Jam is more or less a concentrated fruit pulp possessing a fairly heavy body form. It is rich in flavour. Pectin in the fruit gives it a good set. It contains not less than 68% soluble solids and thus high concentration of sugar facilitate preservation. A jam manufacturer can choose fresh fruit, frozen, chilled or cold stored fruit, fruits or fruit pulp preserved by heat, sulphited fruits or fruit pulp and or dried fruits. Fresh fruits make good jams, the difficulty being regular supply. Good fruits free of any damage should be alone selected and for better jellying effect, green fruits should be mixed with the ripe fruits. Over ripe fruits should not be used.

Preparation of the fruits for jam making:

1. The fruit must be washed thoroughly to remove any adhering dust or dirt.
2. Leaves, stalks and other undesirable portions of the fruits must be removed.
3. Fruit is then subjected to preliminary treatment which varies from fruit to fruit. For instance, apples are peeled, cut into pieces and then pulped. Strawberries are crushed between rollers but in stone fruits like peaches, they are lye peeled (i.e. dipping the fruits in boiling water few minutes to facilitate the removal of the skin) and stones are then removed and the pulp is cut into pieces.

Addition of sugar

Cane sugar is generally used. Preparation of sugar to be added depends upon the degree of ripeness and acidity of the fruit. To get a minimum of 68.5% of sugar in the jam, generally 55.0 g of sugar is added to every 450 g of fruit. The finished jam should contain 30 to 50% invert sugar or glucose to avoid crystallization of cane sugar during storage. If the percentage of invert sugar is less than 30, cane sugar may crystallize out and if it is higher than 50%, the jam becomes gummy and sticky. Corn syrup or commercial glucose may be used to avoid crystallization.

Addition of acid, colour, favour

Generally citric, tartaric or maleic acids are used to supplement the acidity of fruits for jam making, if the fruits are deficient in acids, because appropriate combination of pectin, sugar and acid is essential to give a set to the jam. A good result is obtained if the pH of mixture is about 3.0. Only permitted edible colours should be used, if necessary and these should be added towards the end of the boiling process. Ordinarily, jams do not require addition of flavours if desired, may be added when jam boiling is nearing completion. Fruit and sugar mixture are boiled to concentrate the soluble solids to about 68.5% and also to bring about necessary degree of inversion of sugar. Determination of soluble solids is easily carried out by using a hand refractometer, while boiling is in progress. A sample of the boiling mixture is taken and is immediately cooled. A portion of the cooled material is squeezed through a muslin cloth directly on to the prism of the refractometer and the percentage of total soluble solids is read directly on the scale.

Packing

Soon after the end point is reached, the jam should be cooled in cooling pan to about 93°C and filled into jars at this temperature either mechanically or by hand. The surface of the jam in the jars should be covered with a thin disc of waxed tissue paper and allowed to cool. The jars should be stored in a dry place. Jams can be prepared from fruits like apple, tomato, papaya, strawberry etc.

Jelly

In jelly making, pectin is the most essential constituent. This pectin is present in the cell walls of fruit. While boiling, precipitation of pectin causes jelly formation. Precipitation takes place only when pectin, acid, sugar and water are in definite equilibrium range.

Fruits for jelly

Many fruits are rich in pectin as well as acid and are thus well suited for jelly making. A fruits rich in pectin but deficient in acid may be combined with a fruit deficient in pectin but rich in acid or *vice versa* to make a good jelly, the only drawback being that the flavour of the jelly is affected. Apple, grape, guava, lemon, orange, plum are examples of fruits rich in pectin and acid.

Selection of fruits

The fruits should be sufficiently ripe (but not over ripe) and should have good flavour. Fruits are washed thoroughly with water to remove any adhering dirt. Such selected fruits need not be peeled as in the case of guava and apple while fruits like lemon and oranges require peeling before they are used for extraction of pectin.

Extraction of pectin

Only a minimum quantity of water should be used to the fruit for a simple extraction of pectin. If necessary a second or even a third extraction may also be taken and these extracts may be mixed with the first one. The amount of water to be added would depend on the kind of fruit and usually it is added at the rate of 1.5 kg to a kg of apple and 2.5 kg to a kg of oranges or guavas. In the case of grapes, no water is added, the fruit being boiled in its own juices. Fruit juices when extracted in cold do not contain the required amount of pectin and hence do not form a jelly. To get the desired pectin requirements, the fruit should be cooked. The cooking time also varies from fruit to fruit as given below:

Name of fruits	Time in minutes
Apples	20–25
Grapes	5–10
Plums	15–20
Oranges	45–60
Jaman	20–25
Guavas	30–35

The pectin extract obtained can be clarified by allowing to settle overnight and the supernatant pectin liquor separated or it may be filtered through a muslin cloth. For boiling, copper and iron kettles should not be used for obvious reasons *viz.,* reaction of these metals with the acids and salts of the fruit. Aluminum vessels are satisfactory for this.

Pectin requirement

Usually about 0.5 to 1.0% pectin of suitable quality in the extract is sufficient to produce a good jelly. If the pectin is in excess of this, a firm and tough jelly is formed and if it is less the jelly may fail to set. Pectin, sugar, acid and water must be present approximately in the following proportions:

Pectin	1.0%
Sugar	60–65%
Fruit acid	1.0%
Water	33-38%

To the volume of the pectin extract, equal quantity of sugar is added. The sugar should be sprinkled on the fruit extract while it is boiling and should be thoroughly mixed by stirring to ensure complete dilution. During boiling, the scum which rises to the top is removed.

Cooking of Jelly

The mixture is boiled for about 20 minutes and the end point is that the final brix reaches 65° brix. The end point can be also determined easily by flake test. In this case some portion of a jelly is taken in a large spoon and cooled slightly. It is then allowed to drop. If it drops like syrup, it requires further concentration or if it falls in the form of flakes or a sheet the end point has been reached. Jelly is then cooled slightly and poured into hot and dry containers.

Marmalades

They are similar to jelly in all respects, the only difference being they contain the peels of the fruits. The shredded peels previously boiled in water to remove the bitter principles are used to boil along with pectin-sugar mixture. Marmalade preparations are common for like oranges.

Squash

Preparation of grape squash or pineapple squash is given below:

The stalks are removed first and the fruits are washed in cold water. In the case of pineapple, the skin is removed and cut into pieces and they are washed. The juice is extracted in a pulper. Required quantity of sugar is dissolved in water. Add citric acid and they are heated to boil. This sugar syrup is strained with a muslin cloth and cooled. Juice is added to the syrup and mixed well. Sodium benzoate (for grapes) or potassium meta bisulphite (for pineapple) is dissolved in a little quantity of the juice separately and mixed well with the squash. Essence is then added and stirred well. This squash is then filled into already sterilized bottles leaving about 2 to 2.5 cm head space and the bottles are then sealed with a cap. These bottles are to be labeled and stored. This can be diluted with water in the ratio of 1:2½ and served. Normally squash keeps well for 1 to 1½ years without much change in colour and taste.

Squashes can be prepared from oranges, grapefruit, lemon, lime, mango, jamun, passion and peaches. Fruit juice is often clarified by adding gelatin and tannin in proper proportions. Such clarified fruit juice is mixed with sugar, water, colour (if necessary), preservatives and the mixture is then filtered using a filter paper. Such clear solution obtained is known as **'cordial'** and lime juice cordial is very often prepared. Sugar is added in syrup form. The quantity of preservative added to the original juice is taken into account while adding the preservative to the cordial. Colour should be mixed in very small quantities.

Syrups

Syrups of strawberry, pineapple, oranges, mulberry etc., are very popular as summer drinks in different parts of India. They are generally prepared from extracts of natural material or by using artificial flavours and colours. These are added to heavy sugar syrups of 70°–75° brix. The syrup is prepared by heating sugar in water to which a little acid is added to invert the sugar.

Pickles

The preservation of food in common salt or vinegar is called pickling. Spices and oil may also be added. Pickles are good appetizers and add to the palatability of a meal. They aid digestion by stimulating the flow of gastric juices. Salt, vinegar and lactic acid are the three important ingredients used in pickling. These substances when used in sufficient quantities act as preservatives either singly or collectively.

Salt

Vegetables do not ferment when covered with a large amount of salt. Spoilage is prevented by adding sufficient amount of common salt bringing its final percentage in the material to about 15 to 20. At this high concentration mould and lactic acid forming bacteria do not grow. This method of preservation is applicable only to vegetables which contain very little sugar.

Vinegar

It also acts as a preservative in pickles. The final percentages of acid should not be below 2. To avoid dilution of vinegar by water from their tissues, the vegetables are generally put in strong vinegar of about 10% concentration.

Lactic acid

Lactic acid forming bacteria can grow even in the presence of 8–10% of common salt and when this lactic acid is formed in sufficient quantity, the lactic acid bacteria ceases to function and any further change in the composition of the material is prevented provided it is air-tight.

Pickling process

Pickling is done in two stages *viz.*,

1. **Dry salting:** The vegetable is prepared, washed thoroughly in running cold water, drained and weighed. For every 100 kg of the prepared vegetables 3 kg of salt is used. The vegetable is placed about a few cm and a small quantity of salt and another layer of vegetables thick higher is placed. This process of keeping the vegetables and sprinkling the salt is continued till three-quarters of the barrel full. The top of the vegetable is sealed with a cheese cloth and placed over a clean stone to give some weight to the vegetables placed in. The container is kept in a warm dry place and fermentation is allowed to proceed. In a short time, juice from the vegetable forms the brine converting the whole

mass. Fermentation then starts and is usually complete, in about 8–10 days. This may be confirmed by tapping the containers gently.

2. **Fermentation in brine:** Soaking of the vegetables in salt solution of predetermined concentration for a certain length of time is called brining. This treatment is given only to vegetables like cucumber which do not contain sufficient juice to form brine with dry salt. Brined vegetables will keep in vinegar for a long time.

In India, several kinds of pickles (mango, lime, chilli, cauliflower, beetroot, onion etc.,) are sold in the market. Oil pickles, which contain some edible oil, are also highly popular.

13

Plant Growth Regulators

Plant growth regulators or plant regulators are the organic chemical compounds which modify or regulate physiological processes in an appreciable measure in the plants when used in small concentrations. They are readily absorbed and they move rapidly through the tissues when applied to different parts of the plant. 'Plant hormones' or 'phyto hormones' are also regulators but are produced by the plants in very low concentration and these hormones move from the site of production to the site of action. Therefore, the difference between the plant regulator and plant hormone is that the former one is synthetic while latter one is natural from the plant source. Many of these chemical compounds have been manufactured and tested and it is seen that the behaviour and response of each crop is highly variable according to the composition and concentration of the compounds. In other words, these compounds are specific in their action. These synthetic plant regulators are put into several uses in horticulture which are discussed below:

1. **Propagation of plants:** The most common use of plant regulators in horticulture is to induce rooting of cutting in many horticultural plants. Certain kinds of plants may not successfully root under normal condition and with the aid of plant regulators; they can be easily made to induce rooting. The most commonly employed compound is Indole butyric acid (IBA) which is the most effective one and the other compounds are Naphthalene acetic acid (NAA), Indole acetic acid (IAA) and 2,4-Dichloro phenoxy acetic acid (2,4-D).

Certain of these chemicals can be used in a powder form, mixed with talc. The cuttings may be moistened with water at their lower ends and then dipped in the powder and planted afterwards, lower end going beneath the soil. Some patented products like Seradix A and B can be used in the form of powder. Some chemicals can be used as solutions after dissolving them in alcohol or water as the case may be. The lower end of the cuttings may be soaked in such solution of low concentrations say 10, 25, 50 parts per million (ppm) for 12–24 hours (soaking method) or at higher concentration from 500 to 2000 ppm for a minute or less (quick dip method). The concentrations differ according to the type of cuttings *viz.*, herbaceous, semi hard-wood and hardwood cuttings. Another usage of plant regulators in propagation of plant is in aiding rooting of air layering. When the ring of bark is removed from the stem, the growth regulator like IBA or IAA in powder or in lanolin paste is applied at the distal end of the bark removed portion.

Another use of plant regulators in the field of propagation is the stimulation of growth in the nursery plants so that it is possible to obtain graftable rigid rootstocks in mango and citrus within short time, say three months, otherwise, normally these rootstocks take one year to reach graftable size. Gibberellic acid is one such plant regulator which is used to induce vigorus growth of nursery plants. When this chemical is applied to the growing tips, this chemical will induce rapid cell division and cell elongation resulting in very rapid growth.

2. **Control of flowering:** The plant regulators are used for the regulation of flowering in certain crops. In pineapple, flowering is irregular and harvesting becomes a problem and hence to regulate the flower production, plant regulators are used. The treatment generally consists of pouring a required quantity 50 ml of the solution containing 0.25 to 0.50 mg of the alpha naphthalene acetic acid (NAA) in the central core of the plants. In *Jasminum grandiflorum* the flowering period is extended by the application of cycocel at 500 ppm.

3. **Fruit setting:** Synthetic growth regulators induce greater fruit set in various edible varieties of plants. Naphthoxy acetic acid, 2,4-Dicholoro phenoxy acetic acid and para chloro phenoxy acetic acid are the growth regulators which are successful in increasing fruit set. These chemicals are mainly used as spray solutions on the trees at required concentrations. The effective range of concentrations is between 25 and 50 ppm. It has been found that in chillies spraying of NAA at 10 ppm at 60th and 90th day after planting is beneficial for good fruit setting.

4. **Induction of parthenocarpy:** The other important effect of synthetic growth regulators is the production of seedless fruits. The chemicals are usually applied in the form of lanolin paste or spray solutions. Chemicals used for this type of action are 2, 4-Dichloro phenoxy acetic acid, 2, 4, 5-Trichloro phenoxy acetic acid, alpha naphthalene acetic acid and gibberellic acid. Application of gibberellic acid (GA) at 100 ppm induces complete seedlessness in the grape varieties *viz.*, Anab-e-Shahi, Pachadraksha etc. The Poovan cultivar of banana develops seeds in their fruits in Trichy tracts of Tamil Nadu and this malady is known as **'Kotta vazhai'**. The fruits are made seedless by spraying 2, 4-D at 25 ppm concentration in the bunches when the last hand has opened.

5. **Control of pre harvest fruit drop:** Losses resulting from pre harvest drop of fruits have long been serious problem. In apples and pears, pre harvest fruit drops can be checked by the application of 2,4-D and 2,4,5-Trichloro phenoxy acetic acid. Pre harvest fruit drop in citrus is controlled with 2,4-Dichlorophenoxy acetic acid at a concentration of 8 ppm. Naphthalene acetic acid and its sodium salt is the active ingredient in most of the commercial hormone preparations for preventing fruit drop. Timing of spray is important, in that the effectiveness of the spray should last until the expected drop occurs.

6. **Blossom thinning:** Blossom thinning is practiced to reduce the heavy setting of fruits in one particular season. The advantage is mainly to improve the size and grade of fruit in that season and also to prevent the exhaustion so that it can bear an equally good crop in the next season. In apples, plums and peaches heavy flowering in one season leads to exhaustion, next year it may produce no crop or poor crop. Therefore, when the trees are in full bloom, the growth regulator like 3-chloro-isopropyl-N-phenyl carbonate at 250 to 300 ppm is applied to obtain required amount of blossom thinning. NAA is an effective chemical in thinning the crop in grapes when applied two to four weeks after petal fall stage. Thinning effect will be good in varieties like Khandari, Pachadraksha, Seedless and Anab-e-Shahi when applied at 50 to 100 ppm.

7. **Fruit ripening:** The plant growth regulators can be employed to hasten or delay fruit ripening. Plant regulators like 2,4,5-Trichloro phenoxy acetic acid (2,4,5-T) at concentrations of 25–100 ppm has been found to hasten the ripening in some varieties of plums and peaches. In banana, ethrel treatment at 2500 ppm induces ripening in 24 hours. Application of 2,4-Dichlorophenoxy acetic acid at 16 ppm concentration delays ripening in Washington Novel oranges. These

chemicals are sprayed in water solutions on the fruit before harvest or after harvest or the fruits can be dipped in emulsions or in solutions.

8. **Weed control:** Chemical weed killers are now extensively used as hand weeding is very laborious and expensive. Derivatives of 2, 4-Dichloro phenoxy acetic acid are commonly used for this purpose. Sodium salts of 2, 4-D is used as a spray solution at 0.1% to kill the broad leaved plants only.

9. **Modification of sex expression:** Plant regulators can be employed to modify the sex expression in certain crops. In cucurbitaceous vegetables the production of male flowers will be always more in number than the female flowers and this sex ratio can be narrowed down by the application of ethrel at 100 to 250 ppm if sprayed four times at weekly intervals commencing from 10-15 days after sowing. This growth regulant not only increases the number of female flowers to male flowers, but also produces female flowers at earlier nodes. Certain plant regulators are employed to induce male sterility in crop plants so that such male sterile plants can be used as female plants in the hybridization programme. This process dispenses the expensive work *viz.*, emasculation. Complete male sterility in bhendi can be obtained by spraying with 0.4% of Maleic hydrazide (MH) and 0.4 to 0.5% of F.W 450 (Mendek). A single spray one week prior to floral bud initiation offers male sterility for 10 days and a subsequent spray at floral initiation extends the effect to 22 days.

10. **Control of dormancy:** Plant regulators can be employed to break dormancy or prolong the dormancy. Sprouting of potato tubers and onion bulbs is a common phenomenon in storage. Pre harvest spray of Maleic hydrazide (MH) at 2000 ppm given just before 15 days of actual date of harvest prolongs dormancy in the above storage organ by inhibiting the sprouting. In fruit trees of apple, plums and figs, early flowering is induced by spraying Dinitro orthocresol at 0.1% in oil emulsion. Seed treatment of tomato with GA at 1.0 ppm breaks the dormancy and increases the percentage of germination.

11. **Production of latex flow in rubber plants:** Application of NAA, 2,4-D, 2,4,5-T and ethrel at the tapping panel increases the flow of latex in rubber. These stimulants are useful especially in older trees before they are felled to completely exhaust the flow of the latex.

12. **Arresting the plant growth:** Periodical pruning of hedge plants and centering in tea (removal of the terminal shoots to encourage shoots) are laborious processes, as well as costlier operations. Growth regulators like Maleic hydrazide (MH), CCC (2-chloro ethyl trimethyl

ammonium chloride), and ethrel at appropriate concentration normally retards the growth rate in plants thus arresting the unwanted plant growth.

13. **Increasing the fruit size and quality:** Dipping the inflorescence in a beaker containing 50 ppm of gibberellic acid especially during calyptra falling stage increases the fruit size in grape and the quality of the treated fruits are always superior than the normal fruits.

14

Plant Protection

Innumerable pests and diseases attack plants and cause considerable losses. In this chapter, a general account of important causative organisms and their control measures are discussed.

INSECT PESTS

The term 'pest' is used very broadly to insects, other invertebrates like nematodes, mites, snails, slugs etc. and vertebrates like rats, birds, jackal etc., that cause damage to crops, stored produce and animals. Among them insect pests are more important and they inflict injury to plants either by directly or indirectly in their attempts to secure food and almost all portions *viz.*, the root, stem bark, shoots, leaves , buds, flowers and fruits. They cause injury to the plants by anyone of the following methods:

1. **Injury by chewing:** Insects which cause this type of damage chew off external parts, grind them up and swallow them. Larvae of different kinds of insects like moths, butterflies, beetles, weevils etc., cause the above type of damage.
2. **Injury by piercing and sucking:** The insects which cause this type of injury remain outside and with the mouth parts pierce through the epidermis and suck the sap. Aphids, psyllids, thrips, leaf hoppers, mealy bugs and scale insects cause injury by sucking and hence they are generally known as 'sucking pests'. As a result of sucking, a

general chlorosis of leaves and withering and drying of the affected portions are observed in the case of damage by aphids.

3. **Injury by sucking:** Thrips cause silvering or whitening of leaf surface due to the removal of cell contents below the epidermis. Sometimes, premature shedding of developing fruits is caused by these sucking pests.

4. **Injury by feeding internally:** The internal feeders cause damage by remaining within the plant tissues during a part or all of their destructive stages. This is accomplished by the adults thrusting their eggs into the tissue by their ovipositor or by the larvae eating their way in, after they hatch from the eggs. The internal feeders include borers, grubs or weevil, leaf miners and gall insects. When the larvae feed on the wood or pith of the plant which may be generally large enough to contain the body of the pest, they are referred to as borers. The larvae may bore into the terminal shoots and cause death of the shoots. Grubs of weevils are also borers in flower buds and fruits including nuts and seeds. The larvae of those bore into flower buds and cause shedding are usually called bud worms. Leaf miners are also larvae but being very small in size, they live in between the two epidermal layers of the leaves and feed on the food materials inside. Certain insects in their immature or adult stage cause the formation of special plant deformities known as galls and these galls provide shelter and food to the insects. Due to the formation of galls, the growth of the plants may be impaired and also the setting of fruits and seeds.

5. **Injury by subterranean insects:** Insects which are found in the soil live by feeding on the roots of plants by chewing or boring or sucking the sap or forming galls. Apart from the above type of damages, they may make harvests more difficult or reduce the quality of the produce or may be responsible for spreading many plant diseases caused by bacteria, fungi, MLO and viruses.

Methods Adopted for Controlling Pests

The control of insect pests falls under following heads:

(a) **Legislative:** By which the Government prevents the import of produces infested with insects, which if introduced into this country, would become local pests (e.g. potato tubers with nematodes).

(b) **Biological method:** The successful control of a pest species by means of another living organism that is encouraged and disseminated by man is called so. It is inexpensive and as long-term control, causes no pollution and poses no risk to human health. Biological agents are available in nature abundantly. Several pathogens including viruses such as nuclear polyhedrosis virus (NPV) and granulosis virus (GV), bacteria like *Bacillus thuringiensis*, *fungi* like *Metarhizium*, protozoa like *Schizogregarine* cause diseases in insects to destroy them. This method has been successfully used to control many important pests in a number of economic crops.

The other biological method of control involves the use of parasitoids and predators. A parasitoid is an organism which completes its life on a single host and ultimately kills it. A predator, on the other hand is a free living-organism and kills the host (prey) immediately and requires more than one prey individuals to complete its life. If the parasitoid attacks the egg stage of the host, it is then called egg parasitoid, (e.g.) *Trichogramma chilonis* on bhendi borer. When they attack at the larval stage of the host, it is then called larval parasitoid, (e.g.) *Apanteles plutella* on diamond back moth caterpillars in cruciferous vegetables. The predatory group of insects capture and consume another insects as their food (e.g.) green lace-wing, *Chrysoperla cornea* whose grubs and the maggots of Syrphid flies dramatically exert control over several aphids in many crops. The adults and grubs of ladybird beetles such as *Coccinella septempunctata*, *Menochilus sexmaculatus*, *Brumoicles suturalis* and *Scymnus nubilus* play important role in the population regulation of several sucking pests and defoliating insects.

(c) **Cultural methods:** The control of insects through adoption of ordinary farm practices at appropriate time in such a way that the insects are either eliminated or reduced in population is called the cultural method of control. Proper crop rotation or tillage operations may help to keep down the insect population.

Some early crops are sown in narrow strips around a major crop to serve as a trap for the pests that might be common to both. For instance, sowing of mustard in every twenty fifth row of cabbage crop will help preventing higher incidence of diamond back moth in cabbage and cauliflower and the preferred mustard plants can be cut and destroyed when the pest appears. This practice is called trap cropping.

(d) **Mechanical method:** Mechanical control is one by which the insect population is directly hit by mechanical devices or manual operations. Mechanical devices include using fly and maggot traps, setting light and bonfires to attract adult moths and beetles. Manual methods involve hand picking of egg masses, larvae and killing them. Mechanical exclusion consists of the use of devices by which insects are physically prevented from reaching the produce (e.g.) wrapping of individual pomegranate fruits with butter paper to save it from the attack of Anar butterfly, *Virachola isocrates*.

(e) **By use of insecticide:** Insecticide is a substance or mixture of substances used for killing, repelling or otherwise preventing insects. The insecticide is referred as a 'repellent' if it prevents the pest species in attacking its host, an 'attractant' if the pest species is attracted to source, trapped and an 'antifeedant' if it inhibits feeding on the host. The insecticides are available in any one of the following formulations:

(i) **Dusts:** The toxicant is diluted by mixing with or by impregnating a suitable finely divided carrier. The carrier may be organic flour clay. The toxicant in a dust formulation ranges from 0.5 to 25% (e.g.) endosulfan 4 D, malathion 5 D.

(ii) **Granular or pelleted insecticides:** In a granulation the particle is composed of a base such as an inert material or vegetable carrier impregnated or used with the toxicant which is released from the formulation in its intact form or as it disintegrates giving controlled release particles in the formulation generally possess a size range of 0.25 mm to 2.38 mm diameter. The formulations contain 2 to 10% concentrations of the toxicant (e.g.) carbofuran 3 G, Phorate 10 G.

(iii) **Wettable powders:** It is a powdered formulation which yields a rather stable suspension when diluted with water. The active ingredients in such a formulation ranges from 15 to 95% (e.g.) BHC 50 WP, sulfur 25 WP.

(iv) **Emulsifiable concentrate:** The formulation contains the toxicant, solvent for the toxicant and an emulsifying agent (e.g.) endosulfan 35 EC, dimethoate 30 EC, fenvalerate 20 EC.

(v) **Concentrated insecticide liquid:** The toxicant at highly concentrated level is dissolved in non-volatile solvent. An emulsifying, agent is not added here (e.g.) monocrotophos 36 WSC, phosphamidon 85 WSC.

(vi) **Fumigants:** A chemical compound which is volatile at ordinary temperatures and sufficiently toxic is known as a fumigant (e.g.) ethylene di bromide, methyl bromide, aluminium phosphide etc.

Plant Protection Appliances

The important methods of applying pesticides are dusting and spraying. The dusting operation allows the dust particles when falling free either slowly to settle down due to gravity or drift for long distance due to wind. The appliances that are used for applying dust formulations of pesticides are called dusters. They are either manual or power operated.

The spray fluid may be a solution, an emulsion, or a suspension toxicant. To achieve an effective control of pest, the toxicant is well distributed and to meet this requirement the spray fluid is blown down to fine droplets. The spraying machines may be either hand operated or power operated ones. Commonly employed manual operated sprayers are knapsack sprayer (hydraulic or pneumatic) and pneumatic hand sprayer. Rocker sprayers are useful for tall trees and pneumatic hand sprayers are helpful to spray in gardens. The power operated mist blowers are useful in field to cover more area in a limited time. A spray volume of 150 to 200 l water is necessary to cover one hectare of land with power operated high volume sprayers but high volume hand sprayers like knapsack sprayers require about 450-500 l water to cover one hectare. Insecticides should be applied in the morning or evening hours when the weather is calm or else they will fall on unwanted areas and also may not hit the target. Before applying insecticides, it must be ensured that there are no pollinators (like-bees) in the area; for the same reason insecticides should not be applied during blossoms when bees are likely to be at work. After application of insecticides a time lag (7–10 days in case of organophosphorus compounds and 20–30 days in case of organochlorine compounds) should be given before consuming the produce. During this period, the insecticides will get degraded and become non-toxic.

Non-insect Pests

Besides the different kinds of insects which damage crops, mites, rats, birds and nematodes cause damage to crop plants.

(a) **Mites:** Mites possess four pairs of legs as against insects which do have only three pairs of legs. In recent years the mites have become major pests. They cause damage by way of sucking the cellular materials by forming severe deformities. The chemicals which are used to control the mites are known as 'acaricides'. Sulphur, ethion, dicofol, phosalone and propargite are commonly used as acaricides at the rate of 15–20 ml per 10 litres of water.

(b) **Plant nematodes:** Plant nematodes are small organisms which live in soil around the roots of plants. They are about 0.1 to 1.0 mm in length. They are confined to the top 20 to 25 cm of soil, sometimes even to a depth of 3 to 4 m. They spread from one field to another through percolating water and agronomic practices like ploughing and weeding which involve transport of soil. Most of the symptoms of damage by plant parasitic nematodes are non-specific and often likely to be confused with those caused by other pathogens or soil factors like poor drainage, lack of soil nutrition etc. Some of the commonly observed symptoms are:

1. Stunting and wilting
2. Leaf curl
3. Browning or bronzing of leaves
4. Distortion of leaves, stems
5. Brown lesions in roots
6. Knot-like galling of roots

Control of plant parasitic nematodes is difficult, but nevertheless, necessary for obtaining profitable yields. Crop rotation with a non-host crop or application of large quantities of green leaves or grasses as mulches or summer fallowing and use of resistant varieties will reduce the incidence to some extent. Commonly used nematicides are DD mixture, dibromoethane, dibromo chloropropane, thionazin and aldicarb. In Tamil Nadu, nematode infection is a devastating problem in banana, citrus, potato and in vegetable crops like tomato, chillies and brinjal.

Integrated Pest Management (IPM)

IPM is a new system approach which has been necessitated primarily out of the growing concern about the undesirable side effects of large scale use of inorganic insecticides and often failure of the same to provide for suppression of pests at economic level. Attempts to totally suppress the pests by insecticides may lead to the following problems:

- development of resistance to chemicals in pest population
- outbreak of secondary pests
- resurgence of treated populations
- unacceptable residues on food and forage products and associated legal complications
- destruction of beneficial insect predators, parasitoids and pollinators, hazards to personnel involved in insecticide application, domestic animals and wild life; and
- expense of pesticides, involving the cost of materials, labour and maintenance of equipments.

Thus in any IPM programme, the ecological factors are exploited, the control methods are so designed that they are compatible with natural mortality factors in order to optimize control.

II. Plant Diseases

Fungi, bacteria, viruses and flowering plants cause infectious diseases. The different group of plant diseases caused by the above causative agents are discussed below along with their control measures.

Damping-off: The pathogen infects the collar region of the plants which results in collapse and death of young seedlings. Generally, fungi like *Pythium, Phytophthora, Fusarium* and *Rhizoctonia* are responsible for this. Damping-off at pre- emergence stage kills the seedlings before their emergence over the soil surface. If it manifests at post-emergence stage, the infestation is usually initiated at ground and soft, water soaked areas appear on the young stems. With advancement of the disease, the stem gets constricted and finally collapses. High moisture and thick sowing of the seeds favour the incidence of this disease. This is very common in vegetable crops like tomato, brinjal, chillies and cucurbits and ornamental plants in nurseries.

The control measures comprise of the use of seed protectants (like Thiram or Captan 2 g/kg of seed) to avoid pre-emergence infection. Drenching the soil with any copper fungicide at the rate of 2.5 g per litre of water at 15 days interval will control post-emergence incidence of this disease.

Rots: Rotting of plant parts is caused by a large number of fungal and bacterial pathogens. Fruits, bulbs, tubers and roots are the main targets of attack. The rots may be distinguished as dry rots and soft rots based on the degree of tissue-disintegration, lesser in the case of former one and more in latter one. Rhizome rot of ginger, root rot of turmeric, bud rot of arecanut and coconut palms and stem rot of papaya are some of the examples of rot diseases of economic importance. Most of the fruits suffer with soft rot diseases which are more common during post harvest period. In all these cases, small water soaked areas appear on the host surface which enlarge rapidly and the affected tissues turn soft and pulpy. Rot affecting the stems and roots can be minimized by maintaining proper drainage, Drenching the soil with 0.25% copper fungicides will also check infection. Rot affecting the fruits in storage in the godowns and warehouses may be checked by maintaining proper sanitary conditions and spraying with antibiotics like streptomycin (0.1%).

Downy mildews: In the lower surfaces of the infected leaves a white or grey downy growth is normally seen. This disease is more common in grapes. This is caused by the fungus *Plasmophora viticola*. This can be controlled by spraying 1% Bordeaux mixture or any other copper fungicides.

Powdery mildews: Powdery mildews are easily recognized by the white powdery coating on the plant. Powdery mildew infected plants show stunting and distortion of leaves, surface necrosis of infected tissues, decline in growth and defoliation. This disease infects crops like grapes, apples, cucurbits, roses and few ornamental annuals. This disease is caused by fungi belonging to the genera *Erysiphe, Uncinula, Podosphaera, Sphaerotheca, Phyllactinia and Leveillula.* Powdery mildews are controlled effectively by sulphur dust and certain other fungicides. Wettable sulphur (2 g/lit), Karathane are also recommended to control this disease. For cucurbits, sulphur dust has to be avoided, instead, wettable sulphur alone to be used. Tridemorph has also been found to be effective against powdery mildew in many crops.

Sooty mould: It occurs as a black thin crust covering the leaves, young stems and blackening the fruits too in particular kinds of trees and plants, such as the citrus, mango, sapota, guava which are attacked by sucking pests like green bugs. Though the fungi causing sooty mould are only saprophytic, growing upon the excreta cast by the insects feeding upon leaves, they are harmful since they reduce the surface area available for photosynthesis and thereby weaken the trees. This disease is caused by the fungus *Capnodium sp.* This mould can be prevented by spraying starchy solutions which will clear the mould from the affected parts as it dries off.

Rust: This disease can be identified by rusty yellow or brown or dark blotches on the epidermis of the stem and leaves. Rusts are harmful and they are difficult to eradicate once they infect the plants. Attention to plant sanitation, good cultivation and preventive spray are the useful protection measures. Among the vegetables, it is common in beans and peas. Dusting of finely ground sulphur powder will effectively control this disease, if it is dusted even before the appearance of the disease. Spraying mancozeb (2 g/lit.) is also effective.

Part 2

PLANT PROPAGATION

15

Plant Propagation

Plant propagation refers to the multiplication or perpetuation of individual or group of plants which have specific value to human kind. The methods of propagation in various plant species can be broadly grouped under two categories *viz.*, sexual propagation and asexual or vegetative propagation.

SEXUAL PROPAGATION

Sexual propagation of plants is by means of seeds. The seeds are the fertilized ovule containing embryos resulting from the union of male and female gametes during fertilization. The embryo in the seed gives rise to the new plant during germination. The development of a male and female gametes and how fertilization takes place are described below.

MICROSPOROGENESIS

Pollen is produced in the anther. In the early stage of development, the anther consists of a small group of meristematic cells. As it matures, four groups of microspore mother cells develop with it. Each microscope mother cell divides meiotically to produce four microspores, each of which contains the haploid (n) number of chromosomes. Each microspore develops into a pollen grain after undergoing a mitotic division producing two nuclei. The two nuclei of the pollen grain are the 'tube nucleus' and the 'generative nucleus'. The generative nucleus subsequently divides to produce two male gametes or sperm cells before the pollen is shed (Fig. 15.1).

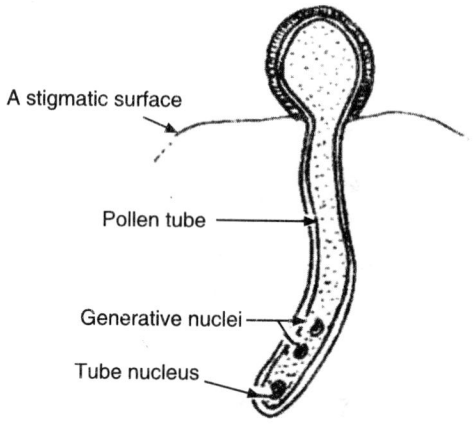

A stigmatic surface

Pollen tube

Generative nuclei

Tube nucleus

Fig. 15.1 A germinating pollen grain.

MEGASPOROGENESIS

The formation of a seed begins with the development of an ovule. In the young ovule, all the cells that compose the nucleus are identical. However, one of the nuclear cells, usually just below the epidermis near the top, differentiates from the surrounding cells, eventually forming the embryo sac (megaspore) mother cells. Generally this is 2n or diploid tissue. The nucleus of this mother cell undergoes two successive meiotic divisions, forming a row of four cells called 'megaspores'. They are genetically 'n' or haploid cells. Generally, three of the four cells nearer the micropylar end disintegrate and the remaining cell develops into the mature embryo sac. The nucleus of the remaining megaspore divides meiotically thrice to form the eight nucleated mature embryo sac. Thus, each mature embryo sac contains eight nuclei (each n) consisting of one egg cell, two synergid cells, three antipodal cells and one primary endosperm cells with two nuclei (Fig. 15.2).

During pollination, pollen is transferred from the anther to the stigma where it germinates and grows down the style until it reaches the embryo sac. The pollen tube penetrates the tissue of the stigma, grows down the style and enters the ovary usually through the micropyle. The tip of the pollen tube then ruptures and the sperm nuclei move towards and fuse with the egg effecting fertilization and form the zygote (2n). The second sperm nucleus unites with the two polar nuclei in a double fertilization forming the primary endosperm (3n). After fertilization of the ovule, the embryo and endosperm continue to grow and differentiate, ultimately forming the seed.

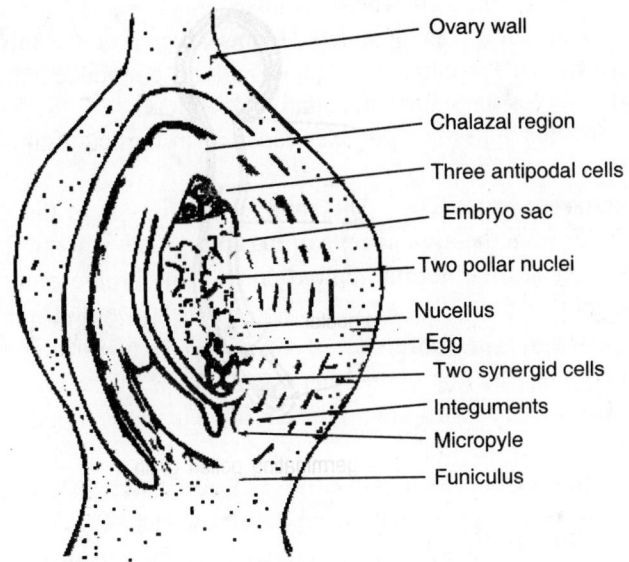

Fig. 15.2 An ovary showing an ovule with a mature embryo sac and its eight nuclei.

Usually, in the development of fruit and seed, the following relationship can be established:

Ovary	Fruits
Ovule	Seeds
Integuments	Testa (seed coat)
Nucellus	Perisperm
Two polar nuclei + sperm nucleus	Endosperm 3n
Egg nucleus + sperm nucleus	Zygote–embryo (diploid 2n)

The developed seed has three parts (a) embryo, (b) food storage tissues which may be the endosperm, the perisperm or cotyledon and (c) seed coverings. The seeds in which the endosperm is large and contain most of the stored food are called as 'albuminuous seeds' while the seeds in which the endosperm is lacking or reduced to a thin layer surrounding the embryo are referred to as 'exalbuminous seeds'.

APOMIXIS

In some species, embryos are produced not as a result of meiosis and fertilization but by certain asexual processes. The occurrence of asexual reproductive process in the place of the normal seed reproductive process of reduction division and fertilization is known as 'apomixis'. Such seedling plants produced in this manner are known as 'apomicts'. Plants which produce only apomictic embryos are known as 'obligate apomicts' and those produce both apomictic and sexual embryos are called 'facultative apomicts'.

There are four types of apomixes reported in horticultural crops. They are

1. **Recurrent apomixis:**   The embryo develops from the diploid egg cells or from the diploid cells of the embryo sac without fertilization. The egg has the normal diploid number of chromosomes as that of the mother plant. E.g. *Parthenium, Rubus, Malus, Allium.*

2. **Non-Recurrent apomixis:**   The embryo develops directly from the haploid egg cell or some other haploid cells of the embryo sac and hence haploid plants are produced. It occurs in *Solanum nigrum* and *Lilium sp.*

3. **Nucellar Embryony or Adventitious Embryony:**   The embryo arises from a cell or a group of cells either in the nucleus or in the integuments and hence they are diploid in nature having the genetic constitution of its mother plant. It differs from recurrent apomixes in that such embryos develop outside the embryo sac besides a normal embryo developing from the embryo sac. e.g. Citrus and certain varieties of mango.

4. **Vegetative apomixis:**   In some species, the flowers in an inflorescence are replaced by vegetative buds called bulbils which sprout and produce new plants while still on another plants. e.g. *Allium sativum, Agave, Furcreae* and *Dioscorea bulbifera.*

Significance of Apomixis

1. Apomictic seedlings are identical with its mother plant, similar to those propagated by vegetative means.
2. Such seedlings, when used as rootstocks, provide uniformity to the scions when grafted.
3. Apomictic seedlings are free from virus diseases.

POLYEMBRYONY

The phenomenon in which more embryos are present within a single seed is called polyembryony. It may result due to (a) nucellar embryony e.g., Citrus (b) development of more than one nucleus within the embryo sac (in addition to the egg embryo during the early stages of development) leading to multiple embryos (e.g. conifers).

Occurrence of polyembryony is widespread in all citrus species but the number of embryos per seed varies from species to species. In rough lemon, it varies from 3 to 5; in mango certain cultivars are reported to be polyembryonic with the number of embryos ranging from 2 to 10, the seedlings 1 to 7 and the germination per cent form 40 to 87. Polyembryonic seedlings can be identified from its true seedlings by their uniformity and vigorous in growth. The greater vigour is probably due to the elimination of viruses.

Advantages of Sexual Propagation

1. The plants raised by seeds are long–lived.
2. They are hardy with deep root system.
3. The possibilities are there to obtain chance seedling, the performance of which are better than their parents, e.g., mango variety (i) Chinna Suvarnarekha (ii) Mundappa.
4. The polyembryony produces true to type nucellar embryonic seedlings which could be used as rootstocks for uniform performance, e.g., mango-Vellaicolamban, Olour and Bappakai; citrus – all *Citrus spp.* except *C. grandis.*
5. Seed propagation is necessary when vegetative propagation is unsuccessful or difficult or expensive (e.g. Papaya).
6. Exploitation of hybrid vigour is possible only when the hybrids are multiplied in the first instance through sexual propagation although subsequent fixing of heterosis is effected through vegetative propagation, e.g., Co-1 sapota (Cricket Ball × Oval), PKM-1 mango (Chinna warnarekha × Neelum) and PKM-2 mango (Neelum × Mulgoa).

Disadvantages

1. The progenies are not true to type and so they become inferior, because in the commercial orchards it is necessary to have uniform growth and yielding capacities besides the quality attributes.
2. Choice tree or any hybrid trees cannot be perpetuated true to type by seed.
3. Seedlings have a long juvenile period. Oranges when raised by seedlings take 7 to 10 years for bearing while a budded plant comes to bearing within 3–4 years.
4. Seeds lose viability in a short period. In crops like citrus, cocoa and rubber the seeds must be sown afresh, *i.e.*, immediately after extraction.

Importance of Sexual Propagation

1. Seed propagation is necessary when the vegetative propagation is unsuccessful or difficult or expensive e.g. papaya.
2. When seedlings are required in large number, seed propagation is the only mean, e.g., forest trees.
3. The rootstocks used for grafting mango and sapota and for budding peaches and plums are raised only by seed.
4. To take advantage of the polyembryony seedlings.

SEED FORMATION AND MATURITY

Seed develops along with the fruit and reaches full size and maturity when the fruit ripens. The development of seeds in the fruit can be characterized by two distinct stages: (1) ripening stage-associated with rapid increase in dry weight as well as fresh weight resulting in an increase in seed moisture content and (2) maturation stage-during which the seed looses water but continues to increase in dry weight. Most of the seeds accumulate food materials during both the stages. The embryo acquires a strong resistance to desiccation during maturation stage so that it can tolerate desiccation efficiently with potentiality to resume growth when hydrated during germination. Hence, seed should be extracted only from ripe fruits. Seeds gathered from immature fruits may not germinate under favourable conditions and may lose viability more quickly than fully matured seeds.

SEED STORAGE

Seed should be normally stored in relatively dry conditions at low temperature. Some seeds such as citrus, rubber, cocoa should be sown immediately after extraction as the seeds are short-lived. Seeds of most of the plants retain their viability longer when stored at a relatively low than at a higher temperature. The storage conditions that maintain seed viability are those which slow respiration and other metabolic processes without injuring the embryo. The most important conditions for achieving this are reduced moisture content of the seed, reduced storage temperature and modification.

SEED GERMINATION

A seed consists of an embryo and its stored food supply (endosperm) surrounded by protective seed coverings. During seed germination cell metabolism increases, the embryo resumes active growth, seed covering ruptures and the seedling emerges. The water that is imbibed during germination softens the seed coverings and causes hydration of protoplasm. Enzymes become active now and begin to digest reserve substances like fats, proteins and carbohydrates in the storage tissues such as cotyledons, endosperm to simpler substances. These compounds are then translocated to the growing points of the embryonic axis to be used for growth and the production of new plant parts.

As germination proceeds, the structure of the seedlings soon becomes evident. The embryo consists of an axis bearing one or more seed leaves or cotyledons. The growing point of the shoot, the plumule, is at the upper end of the embryo axis above the cotyledons. The seedling stem is divided into the section below the cotyledons as the hypocotyl and above the cotyledons as the epicotyl. The initial growth of the seedlings follows one of the two patterns. In the first type *viz., epigeous germination,* the hypocotyl elongates and raises the cotyledons above the ground. In the second type, the *hypogeous germination,* the lengthening of the hypocotyl does not raise the cotyledons above the ground and only the epicotyl emerges (Fig 15.3).

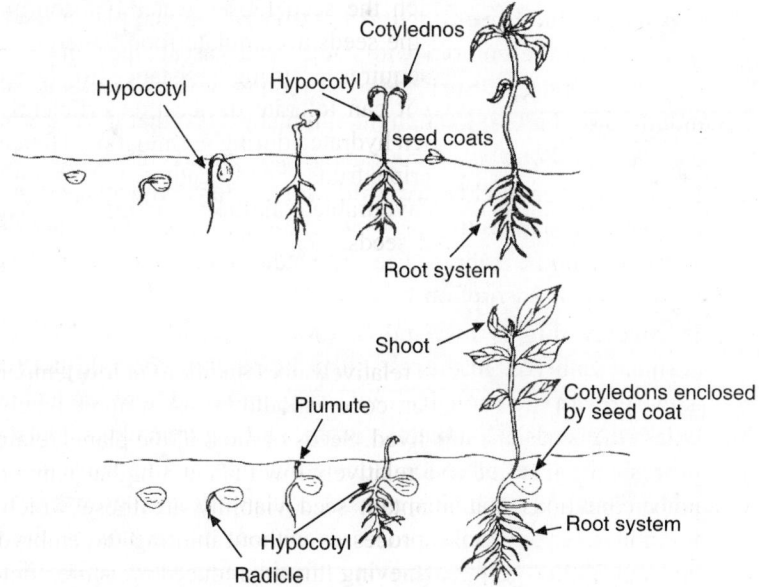

Fig. 15.3 Epigeal germination and Hypogeal germination in higher plants.

DORMANCY IN SEEDS

If the seed can germinate immediately upon the absorption of water without a barrier to germination, the embryo is said to be *'quiescent'* or *'non-dormant'*. But in certain cases the seeds do not germinate readily even when they are provided with all conditions required for germination. Such seeds are known as *dormant seeds*. The dormant seeds do not germinate because of any condition associated either with the seeds itself or with existing environmental factors such as temperature and moisture. The following are some types of dormancy associated with horticultural crops.

Types of Dormancy

1. **Seed coat dormancy:** In certain species of plants belonging to the families like Leguminoceae, Malvaceae, Cannaceae, Convolvulaceae, the seed coats or other tissues covering the embryo are hard and are impermeable to soil and oxygen, thus preventing the germination. In certain species, the seed coats (e.g. pits of stone fruits or shells of walnut or other nuts) are apparently permeable to water and gases but they are so hard to resist the embryo expansion. Hence, germination does not occur in such case.

2. **Dormancy due to rudimentary embryos:** Some plants shed their fruits before the embryo within the seed has attained the maturity stage to germinate. Such embryos require several weeks to several months after harvest to attain its full maturity so that it can germinate e.g. *Ilex sp, Pinus sp, Viburnum sp*, palms, orchids.

3. **Dormancy due to chemical inhibitors:** In certain species, specific chemical substances that prevent germination occur in the seed coats, endosperm or the embryo. These are reduced or eliminated by leaching with water or adsorption by soil.

4. **Dormancy due to internal factors:** It is due to physiologically dormant embryos. In this case, the dormant embryos do not resume active growth eventhough all environmental conditions are favourable, unless the seeds are subject to moist, chilling treatments. During this process, the level of endogenous growth promoting substances (e.g. gibberellins and cytokinins) increase while the level of growth inhibiting hormones (e.g. Abscissic acid) decreases, thus removing the block and permitting germination. e.g. freshly harvested seeds of apple, pear, peaches, apricot, rose and grapes do not germinate due to the above factor. They require 'after ripening' during which the physiological changes occur in the dormant seeds, permitting the germination to take place.

5. **Double dormancy:** Seeds of some species (e.g. *Cercis occidentals*) exhibit seed coat dormancy and embryo dormancy.

Seed Viability and Longevity

Seed viability means the presence of life in the embryo, while longevity refers to the length of time up to which the seeds will retain their viability. Seeds have been broadly classified into two major groups *viz.,* orthodox and recalcitrant, based on longevity *vis à vis* seed moisture content and response to drying. The orthodox seeds can be safely dried to low moisture content and the storability of such seeds improves with the lowering of seed moisture. Seeds of most field crops, seasonal vegetables and flowers are orthodox in nature on the other hand, many horticultural crops produce seeds which lose viability; when dried to moisture contents below a critical level as irreversible ultra structural damages are caused to the seed. Such seeds are known as recalcitrant seeds. Because of this factor, the longevity of horticultural seeds is relatively shorter, ranging from few days to few months only as indicated below:

Species	Longevity
Mango	80 days
Jackfruit	1 month
Mangosteen	8 weeks
Cinnamon	One week
Cocoa	4 months

Viability can be tested by germination tests, excised embryo test and tetrazolium test. Among them, the tetrazolium test is more reliable and easy to do. One per cent aqueous solution of 2,3,5-triphenyl tetrazolium chloride (pH 6-7) is taken in a petridish and water soaked seeds are placed in it and kept in dark, warm place. A viable seed takes red coloured stain while a non-viable seed remains colourless.

Techniques of Seed Propagation

Testing the purity: It is done to find out whether seed is true to name and is adulterated with weed seeds and other inert matters.

Pretreatment of seeds: Special treatments are necessary for seeds of many plants to germinate even when conditions appear to be favourable.

Mechanical scarification: It includes breaking or scratching the seed coats mechanically to modify the hard or impervious seed coats. This can be done easily by revolving the seeds in a drum lined with sand paper.

Soaking in water: Generally seeds will be soaked in hot water for a few seconds and then soaked for 24 to 48 hours in cold water which make the seed coat to get soften and wash off the inhibitors. e.g. wattle seeds. In some case, the seeds are soaked in running cold water for a period of 8–12 hours which help in removing the inhibitors. e.g. beet root.

Acid treatment: Soaking the seeds for a few minutes (15–60 seconds) in concentrated hydrochloric acid or sulphuric acid modifies the hard or impermeable seed covering. At the end of treatment period, the seeds are washed to remove the remnant acid. e.g. *Mucana bracteata.*

Cold stratification: During stratification, seeds are exposed to abundant moisture, ample oxygen and a relatively cool temperature. It consists of placing the seeds in a moist medium of sand, peat or vermiculite and holding at a temperature slightly above freezing. The time varies between 1 to 4 months depending upon the type of seeds. This permits the physiological changes within the embryo to occur. e.g. Peaches.

Dry storage: It promotes the after ripening in certain seeds which are dormant when freshly harvested. Freshly harvested seeds of many annuals and herbaceous plants fail to germinate until after a period of dry storage. Such post-harvest dormancy may last from few days to several months.

Treatment with chemicals: Soaking in potassium nitrate (0.2%), gibberellic acid (200 to 500 ppm) or thiourea (0.2%) solution prior to sowing has been found to stimulate germination of different kinds of seeds. For instance, soaking of seeds in gibberellic acid stimulates the germination of many citrus species *viz.,* Trifoliate orange, Rangpur lime, Sweet orange, Sour orange etc. Cardamom seeds when presoaked for 10 minutes in 25% acetic acid, 25% nitric acid or 50% hydrochloric acid show improvement in the germination from 18 to more than 90% in all the cases.

Seed Sowing

The seeds may be sown in raised beds in open or sown in containers like seed pan or seed boxes. The soil in the bed must be porous and light; may be made by incorporating one part of loamy soil, one part of sand and two parts of finely sifted cattle manure.

The seed pans may be of 10 cm in height and 30 to 45 cm in diameter and the seed boxes have a 45 cm length and 10 cm depth. These may be filled with a good soil mixture. Proper spacing and maintaining optimum moisture level in the soil media are essential to produce healthy seedlings.

ASEXUAL PROPAGATION

Asexual propagation or vegetative propagation or clonal propagation refers to the multiplication or perpetuation of any plant from any vegetative parts of plants other than the seed. Vegetative propagation is possible as the vegetative organs of many plants have the capacity for regeneration. For instance, stem cuttings have the ability to form adventitious roots; root cuttings can generate new shoot system while leaves can regenerate new roots and shoots.

Vegetative propagation is also known as clonal propagation since this kind of propagation is the rule to maintain a clone. A '*clone*' may be defined as genetically uniform materials derived from single individual and propagated exclusively by vegetative means such as cuttings, divisions or grafts. Bartlett pear, a clone originated as a seedling in England during 1770 has been propagated ever since by vegetative means. The leading cultivars of tea in South India namely Sundaram and Pandian have been exclusively propagated by cuttings ever since they were identified as superior seedlings in a plantation.

Advantages of Vegetative Propagation

1. The progenies are true to type of the cultivars.
2. Vegetative propagation is the only rule where no seed is formed or germination of seed is very slow or no viable seed is formed (e.g. banana, pineapple and roses).
3. Certain rootstocks have the capacity of resisting or tolerating the adverse environmental factors such as frost and adverse soil factors like salinity and alkalinity e.g. Frost resistance–*Poncirus trifoliate* (Trifoliate orange). Almond and Myrobolan plum roots tolerate excess boron.
4. The ability of certain rootstocks to resist pests and diseases can be advantageously exploited. Apples when grafted to rootstocks like Merton 778, 793 and 779 are resistant to wooly aphids.
5. Vegetatively propagated plants are generally dwarf in stature than the seedlings. Dwarf trees facilitate pruning, spraying and harvesting besides more number of plants can be accommodated in a unit area.
6. To replant an undesirable existing tree; either with reference to its quality or susceptibility to pests and diseases or pollination requirement, is expensive and time consuming. The defect can be overcome easily by vegetative propagation through grafting or budding of desirable scion to the existing tree by top-working technique.
7. Many plants are propagated by vegetative means because of the speed and ease of multiplication. For instance in date palm, which can be grown to maturity from large off-shoots, the size of which would require several years if to be produced from seeds.

Disadvantages of Vegetative Propagation

1. Vegetative propagation is sometimes more expensive than seed propagation.
2. Vegetative propagated plants are comparatively short-lived.
3. The chances of carrying viruses from one generation to subsequent generation are more.

Genetic Variation in Asexually Propagated Plants

Genetic variation in clonally propagated plants occurs due to gene and chromosome (mutational) changes which can occur within somatic cells and if followed by mitotic division, may lead to a permanent change in the clone. Some of these changes are of horticultural curiosities or economically important. When mutations that occur within a plant often affect only a segment of the meristem and as a result, give rise to sectors or layers of the mutated tissues. Thus, a plant propagated from such meristem is composed of two or more genetically distinct tissues growing adjacent to each other, and such plants are called 'chimeras'. Plants with variegated leaves as found in Coleus, Crotons, Euonymous, Bougainvilleas are example of chimeras.

Depending upon the extent of mutated cells lying adjacent to non-mutated cells in a meristem, three kinds of chimeras are recognized.

1. **Sectorial chimera:** In this type, the growing point of the shoot is composed of two genetically different tissues situated side by side occupying distinct sectors of the stem. Leaves and lateral buds arising from such a shoot may be composed of two tissues combined in various ways, depending upon their location.
2. **Periclinal chimera:** In this type, tissue of one genetic composition occurs as a relatively thin skin, usually with one or several cell layers in thickness over a genetically different core. This is most common and relatively stable chimera type. Such chimeras will revert back if they are propagated by seed or root cuttings arising from the adventitious roots which originally arise from non-mutated layer.
3. **Mericlinal chimera:** This is similar to periclinal chimera except that the outer layer of a different tissue does not extend completely around the shoot occupying a segment of the circumference.
4. **Budsport:** If a branch which shows changes from the rest of the plant in one or more inheritable characters and that can be perpetuated by vegetative means is called a budsport. Many of these budsports are only chimeras and are detected when a lateral shoot develops from a mutated area. Many horticultural cultivars originated due to budsport only. Deep red coloured 'Starking' and 'Richa Red' apple cultivars are sports of the original Delicious apple cultivar only.

Methods of Vegetative Propagation

The various propagation methods include cuttage, layering, graftage and division or separation.

Cuttage

Cuttage may be described as a method of propagation of plants by the detached vegetative plant parts which when placed under conditions favourable for rejuvenation will develop into a complete plant similar in all characteristics to the parent from which it was taken. A cutting may be defined as any vegetative plant part, when detached from the parent, is capable of regenerating the missing organ or organs. According to the plant part from which a cutting is prepared, they can be classified as root cuttings, stem cuttings, leaf cuttings and leaf bud cuttings.

Root Cuttings

Root cuttings may be made from the true roots of any plant species. Their use is limited to plants with roots capable of producing shoots or shoot primordia. Root cuttings of 10 to 25 cm long are planted horizontally in soil or moist sand and watered regularly. The adventitious buds will sprout to produce shoots (e.g. seedless breadfruit).

Stem Cuttings

According to the nature of the wood used in making the cuttings, stem cuttings are of four classes *viz.*, hard wood, semi hard wood, soft wood and herbaceous cuttings:

(a) **Hard wood cuttings:** Most of the fruit plants are propagated by hard wood cuttings. Cuttings are prepared during dormant season from the wood of the previous season growth. In certain fruit crops like Fig and Olive, hard wood cuttings are prepared from two years or still older woods. These cuttings have ample supply of stored food to nourish developing shoots and roots.

Hard wood cuttings should be about 15 to 25 cm long, with at least 2 to 3 nodes. Basal end of the cuttings should be slanting and just below the basal node while the top cut end should be straight and 2 to 3 cm above the last node. These cuttings may be planted straight or slantingly in rooting media with or without leaves keeping about 3/4th of length buried in them. Grapes, rootstocks of roses, pear are propagated by hard wood cuttings.

(b) **Semi-hard wood cuttings:** This type of cuttings is generally from evergreen species during summer from new shoots just flush of growth has taken place and the wood is partially matured. Leaves are retained on this type of cuttings usually on the top side (e.g. Duranta, Hibiscus, Crotons etc.).

(c) **Soft wood cuttings:** This type of cuttings is also known as green wood cuttings. These cuttings are taken from woody plants prior to lignification when the tissues are still relatively soft. The best material of this kind has some degree of flexibility but it is enough to break when bent sharply. Soft wood cuttings root easier and quicker than the other types, but require more attention (e.g. Jasmine, Hibiscus etc.).

(d) **Herbaceous cuttings:** This type of cutting is made from succulent herbaceous plants such as geranium, chrysanthemum, coleus or carnations. This differs from soft wood cutting in that these plants will not develop woody tissues. These will root relatively in a shorter period under proper conditions.

(e) **Leaf and lead-bud cuttings:** Leaf and closely related parts are used to propagate many common green house plants. The red begonia may be propagated by severing some of the radial veins or vascular bundles of the leaf and placing it in top of a moist propagation medium. The severed tissues are capable of regenerating both roots and shoots, with the subsequent production of new plants.

A whole leaf of the common Bryophyllum produces a plant from each of its lobes (Fig. 15.4). In other plants like Sansivieria, long tapering leaf is cut into sections of 5 to 8 cm long. These leaf pieces when inserted 3/4th of their length in sand, new plants form after a period of time.

Leaf bud cutting consists of leaf blade, petiole and a short piece of stem with the attached axiliary buds. Axillary bud gives rise to the shoot and the roots are initiated from the basal end of the stem piece (e.g. Tea, Camellia, Rhododendron).

Regeneration of Plants from Cuttings

Great differences exist among species and cultivars in the rooting ability of cuttings. The following are some of the important factors affecting the regeneration of cuttings:

1. **Food Supply:** The ratio of carbohydrate to nitrogen in the stems at the time of collection of cutting is directly related to rooting ability. High carbohydrate with low N level will aid easy and copious rooting.

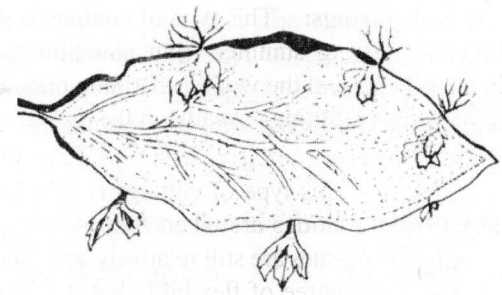

Fig. 15.4 Leaf cutting of Bryophyllum.

2. **Sex of the plant:** Cuttings taken from the male trees root better than the female trees (e.g., Red maple).

3. **Age of the stock plant:** Rate of formation of adventitious roots increases with increasing age of the cuttings (e.g., Apple, Pear, Eucalyptus). This is mainly due to the accumulation of rooting inhibitors as the plants grow older. The rooting ability in such old plants can be induced by severe pruning which tends to produce juvenile shoots.

4. **Time of year in which the cuttings are taken:** Broad leaved evergreen plants usually root most readily if the cuttings are taken after a flush of growth has been completed and this occurs during spring to late fall. An example is that in scented geranium the cuttings root readily during summer and autumn months but they root poorly during winter spring months.

5. **Presence of foliage on cuttings:** The amount of leaf area left on a cutting will determine the extent and amount of root production. The retention of a large photosynthetic leaf area will not only reduce losses of cuttings but produce more strongly rooted cuttings in a shorter period of time.

6. **Flowering or vegetative wood:** In most plants, cuttings could be made from shoots that are either in a flowering or vegetative condition. Cuttings will root easily and more in number when they are taken from the vegetative shoot than from flowering shoot (e.g., Jasmine).

7. **Type of cuttings:** In scented geranium, soft wood cuttings always root well than the hard wood or semi hard wood cuttings. This kind of variation in rooting has also been observed in many plant species.

8. **Treatment of cuttings with growth regulators and other materials:** Treatment of stem cuttings with synthetic auxin compounds like NAA, IBA and 2,4-D at appropriate concentration have resulted in improved rooting in many plant species. Addition of several nitrogen

compounds both organic and inorganic is beneficial in improving rooting in some plant species like Rhododendron. Boron stimulates root production in some plants by way of promoting its root growth.

9. **Wounding:** Root production on stem cuttings can be promoted by wounding the base of the cuttings (e.g., *Magnolia* and *Rhododendron*).

Environmental Conditions during Rooting

(a) **Water relation:** The soft wood and herbaceous cuttings which retain more number of leaves have to be kept without withering or wilting by reducing its transpiration rate. This is possible by maintaining the vapour pressure of water in the atmosphere surrounding the leaves equal to the water vapour pressure in the intercellular spaces within the leaf. High amount of rooting in the above kinds of cuttings is achieved in mist propagation by following the above principle. Such cuttings may not give adequate rooting or may fail to root under outside condition.

(b) **Temperature:** For most of the cuttings, it has been established that day air temperature of 21° to 27°C and night temperature of about 15°C are satisfactory for rooting. Any high temperature tends to promote bud development in advance of root development and increase the water loss from the leaves.

(c) **Light:** The basic principle of a cutting responding to light may be attributed to its effect of auxin production in plants. Cuttings under etiolation are found to have a higher level of endogenous auxin at the etiolated site during the period of root initiation. In rooting leafy cuttings products of photosynthesis are more important for root initiation growth, the synthesis of which is dependent on light.

(d) **Rooting media:** The kind of rooting media can affect the type of root system arising from the cuttings. Cuttings of some species when rooted in sand produce long unbranched coarse and brittle roots but they develop well branched roots when rooted in a mixture of peat moss and sand. The pH of the media is also important for rooting in some species. Tea cuttings tend to produce excess callusing (club roots) if pH of the rooting media is altered below 4.8 or above 5.0.

LAYERING

Layering is the development of roots on a stem while it is still attached to the parent plant. The layer is supported by the parent plant until it develops its own root system. In all forms of layering, the rooted branch is removed from the parent plant when enough roots have formed to enable it to lead an independent existence. The downward flow of metabolites and hormones in phloem is checked at the point of treatment causing accumulation of carbohydrates in the regions where roots are produced. This is a dependable method of inducing rooting as water and minerals are continuously supplied to the layered shoot.

Advantages of layering

1. It is easy to perform.
2. Any clone which cannot be easily rooted by stem cuttings may be made to root through layers.
3. Desired size of plants (big or small) can be obtained by layering.

Disadvantages

1. Layering is limited to certain types of plants which produce low-lying branches.
2. Number of new plants which can be produced by layering from mother stock plant is low as compared to cuttings.
3. Although the operation is simple, after care of the layers is essential which requires daily watering to maintain a very good growth.
4. Special platforms have to be erected and special provisions have been made for constant watering of the layered shoots in case where branches or shoots at high level are to be layered.

Techniques in Layering

The methods of layering can be broadly divided into two main groups *viz.*, ground layering and air layering.

Ground Layering

In this method the main principle is that the rooting of layers takes place in the ground media or in pots containing any rooting media. Different types of ground layering are adopted in many kinds of plants as described below:

(a) **Simple layering:** It consists of bending down a shoot and burying part of it in the soil so that the tip is above the ground level. The buried portion is sometimes wounded by a single stroke upwards on the underside of the shoot to be buried. The layered branches are held firmly in position by pegs or large stones. Roots are formed at the buried portion where the cut is made (Fig. 15.5). Low flexible branches of the plants can be bent to the ground (e.g. Rose, Jasmine, Guava).

Sometimes a single tongue-like cut is made in the shoots on the underside towards the growing point. At the cut portion, a pebble is placed to avoid fusion of tissues. This is completely covered with soil. After two months in the cut portion roots will develop.

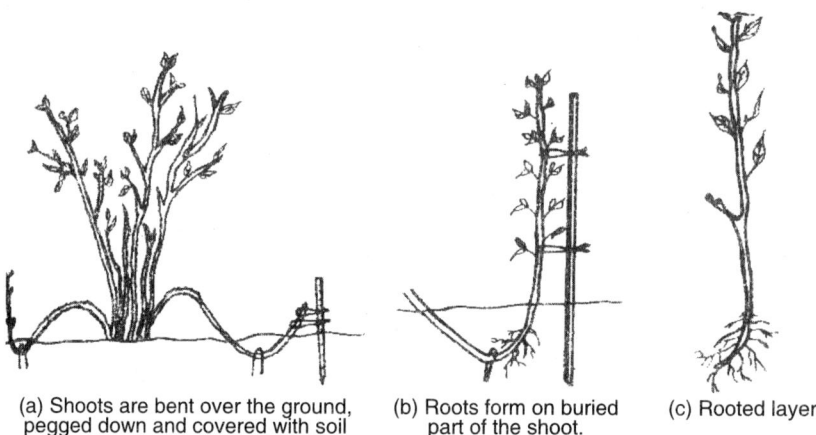

(a) Shoots are bent over the ground, pegged down and covered with soil

(b) Roots form on buried part of the shoot.

(c) Rooted layer

Fig. 15.5 Simple layering.

(b) **Compound or Serpentine layering:** This is again same like simple layering but here flexible branches may be covered with soil in several places that they are alternately covered and exposed over their entire length. Roots are formed usually at or in close proximity to the nodes that are covered and new shoots develop at exposed area. When a good root system has developed at each covered portion the new plants may be severed from the parent plants (e.g., American grapes, ornamental vines like Clematis and green house plant like Peperomia) (Fig. 15.6).

Fig. 15.6 Serpentine layer.

(c) **Trench layering or etiolation methods:** Etiolation refers to growing in darkness and in this method a branch is covered for its entire length to produce plants from all nodes or buds, wherein roots will strike. One year old plant is planted slantingly at an angle of 30° to 45° at a distance of 0.9 to 1.2 m apart in the ground. Once these materials get established these parent layers are bent over and laid flat on the bottom of a trench dug along the row about 5 cm deep. When buds start to swell a layer of 2.5 cm soil is placed on the branch. Buds don't grow, if covered deep. As the shoots grow, more soil is added until they are covered to a depth of about 12 to 15cm so that shoots are etiolated and roots are formed (e.g., Cherry, Plum and Apple rootstocks).

(d) **Tip layering:** In this method the apical ends or tips of current season's shoots are buried in the soil. The tip of the shoots grows downward into the ground (e.g. Blackberry and Raspberry) (Fig 15.7).

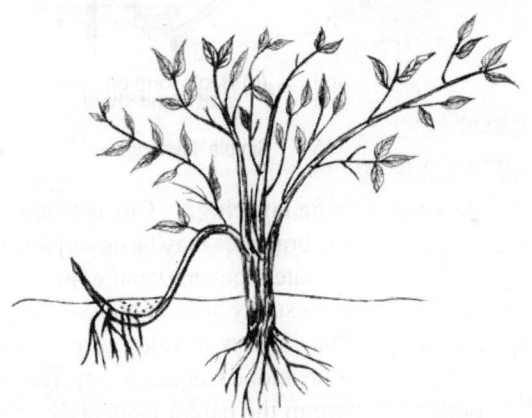

Fig. 15.7 Tip layering.

(e) **Mound layering or stooling:** It is a modification of etiolation method of layering. The parent plant which is already established is cut down to the ground and the resulting growths form the stud or stool is covered up with soil at intervals.

In this method stool beds are maintained by planting rooted layers in small trenches. During the winter season, the dormant shoot is cut back to 5 to 10 cm from the ground level. The soil is mounded half way up the new shoots. If the root development is good on the new shoots, the rooted shoots are severed from the parent plant and transplanted (Fig. 15.8) (e.g. Rootstocks of apple and pear).

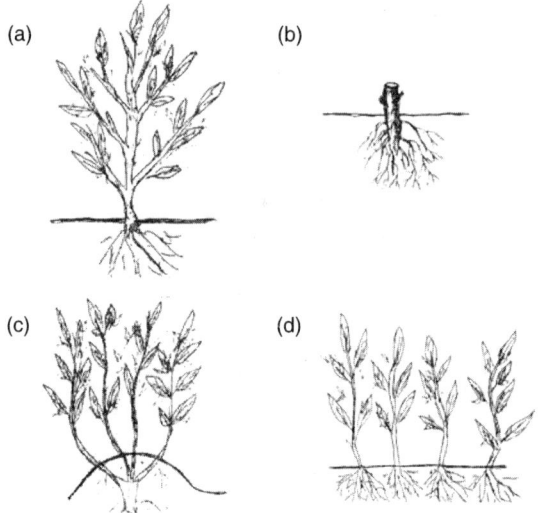

Fig. 15.8 Mound Layering

(a) – Parent Plant

(b) – Cutting it to ground level

(c) – Soil is mound to cover basal side of new growth

(d) – Rooted shoots are severed from the parent plant and transplanted

AIR LAYERING

It is also known as gootee or marcottage. In air layering roots form the aerial part of the plant where the stem has been girdled or slit at an upward angle. Removal of 2.5 to 3.0 cm girdling consists of bark at the nodal region. The injured portion is enclosed at the point of injury with rooting medium which is maintained continuously moist and is achieved by wrapping the moist sphagnum moss or vermiculite around the girdled portion by polythene sheets. These polythene sheets permit gaseous exchange but are impervious to water (Fig 15.9). The roots start emerging above the girdle or on the upper portion of the cut end of the ring. When the stem or shoot has produced a good root system, a first half cut is given just below the point of rooting and within 15 days next half cut is given in the same place to cut off the rooted portion. This separated portion is then transplanted in pots kept in the nursery for curing (e.g., *Ficus elastica, Guava* and *Crotons*).

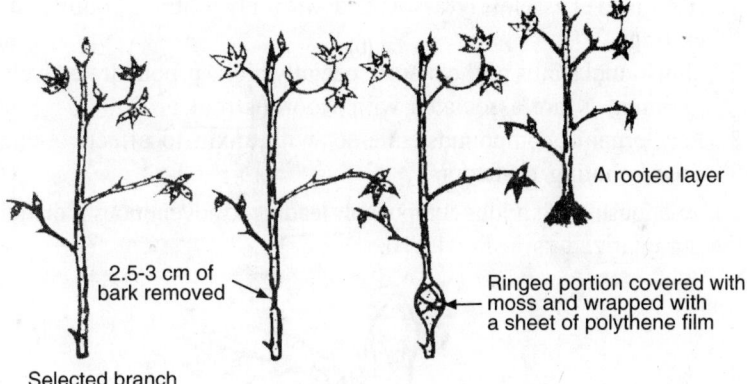

Selected branch

2.5-3 cm of bark removed

Ringed portion covered with moss and wrapped with a sheet of polythene film

A rooted layer

Fig. 15.9 Air layering.

Anatomical and Physiological Basis of Rooting

The formation of adventitious roots in cuttings or layering can be divided into two phases. One is initiation which is characterized by cell division and the differentiation of certain cells into root initials and then into recognizable root primordia. The second phase is the growth and emergence of the new roots, by a combination of cell division and cell elongation including rupturing of other stem tissues and formation of vascular connections with the conducting tissue of the cutting.

These root initials are formed adjacent to vascular tissue. In herbaceous plants which lack a cambium, the root initials are formed near the vascular bundle close to the phloem. In woody perennials, the adventitious roots in stem cuttings usually originate in the young, secondary problems although they may also arise from other tissues such as vascular layer, cambium or piths.

In some plants, adventitious root initials form during early stages of intact stem development and are already present at the time of preparation of cuttings. These are termed '*preformed*' or '*latent root initials*'. These generally lie dormant until the stems are made into cuttings and placed under environmental conditions favourable for further development and emergence of the primordia as adventitious roots. Willow, Hydrangea, Poplar, Jasmines, Citrons are some of the species which produce preformed root initials. The position of origin of these preformed root initials is same as that of other adventitious roots.

After elaborate studies with easy and difficult to root plants, some insight into the physiological basis of rooting has been established. The important aspects are summarized below as:

1. Auxin level is closely associated with adventitious rooting of stem cuttings
2. Nutritional status of the plant especially high carbohydrate levels with optimum N are associated with vigorous root growth
3. Few organic compounds, interact with auxin to affect rooting, are called rooting co-factors.

The relationship of various components leading to adventitious root initiation may be summarized as in Fig 15.10.

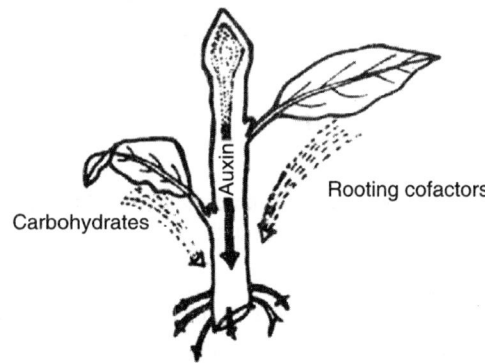

Fig. 15.10 Physiological basis of rooting.

Graftage

It is the process of operation of inserting a part of one plant into an or placing it upon, another in such a way that a union will be formed and the combination will continue to grow as one plant. The part of the graft combination which is to become the upper portion is termed as the scion (cion) and the part which is to become the lower portion or root is termed 'rootstock' or 'understock' or the 'stock' (Fig. 15.11). Root stocks are commonly grown from seeds, cuttings or layers. All methods of joining plants are popularly termed 'grafting' but when the scion part is only a small piece of bark (and sometimes wood) containing a single bud, the operation is termed 'budding'.

Reasons for Grafting and Budding

1. When other methods of asexual propagation are not successful in perpetuating a clone, e.g., mango and sapota can be successfully propagated in commercial scale only by grafting.
2. Plants propagated on their own roots may be weak, susceptible to pests and diseases, or to any adverse environmental condition or not

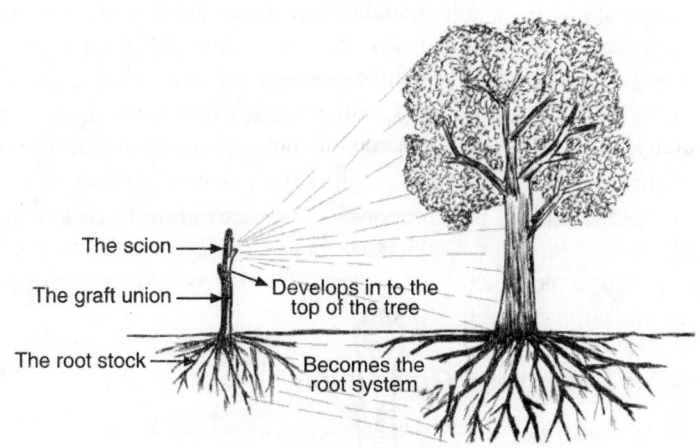

The scion
The graft union → Develops in to the top of the tree
The root stock → Becomes the root system

Fig. 15.11 Relationship between a grafted plant in the nursery and an established grafted tree in the field.

adaptable to a particular soil or climate. For many plant species, rootstocks are available which tolerate all the above cases and hence they may be exploited as a rootstock through grafting or budding.

3. For converting poor trees into more desirable ones by top-working.
4. For overcoming pollination problems; self-fertile varieties may be grafted over self-sterile trees.
5. For fancy purposes, different types of scion may be grafted in the same plant.
6. To modify the growth of the plant as dwarf one by employing suitable dwarfing rootstocks.
7. Occasionally the roots, trunk or large limbs of trees are severely damaged by winter injury, cultivation implements, certain diseases or rodent. But use of bridge grafting or inarching, such damage can be repaired and the tree saved.

Rootstocks

Rootstocks also influence the growth and productivity of scion. Rootstocks can be divided into two groups as follows:

1. **Seedling rootstocks:** Variation among seedlings can possibly make them undesirable as rootstocks. Variation in rootstock seedlings may cause variability in the growth and performance of the grafted trees. Seedlings which are weak should be avoided. Seedlings of 1½ to 2

years old with pencil thickness are considered optimum. Seedling rootstocks are employed for mango, plums and peaches.

2. **Clonal rootstock:** To avoid variation in rootstocks, thus to impart uniformity in the scion, often rootstocks are also propagated by cuttings or layers. Such rootstocks which are perpetuated asexually are termed clonal rootstocks. Nucellar seedling (polyembryony) in certain varieties of mango and all the species in citrus (excepting *C. grandis)* can be also considered as clonal rootstocks as they arise from the tissues other than the true sexual embryo. Clonal rootstocks are used in the propagation of apple and pear.

Factors for Successful Graft Union

1. Botanically the closer a rootstock and scion, the more will be the compatibility between these two.
2. Proper season of grafting is essential. For deciduous plants, grafting is done at the winter season or early spring season and for evergreen trees it should be done during its active growing season.
3. Any grafting or budding method should ensure intimate contact between the cambium of scion and rootstock.
4. Immediately after the grafting operation is completed all the cut surfaces must be carefully protected from desiccation.
5. Proper care should be given to the grafts for a period of time after grafting.

Formation of Graft Union

In graftage, freshly cut scion tissues capable of meristematic activity is brought into close, intimate contact with similar freshly cut stock tissue in such a manner that cambial regions of both are in close proximity. The healing of graft union takes place in a sequential step as indicated in Fig. 15.12.

1. Production of callus tissues (parenchyma cells) by the cambium regions.
2. Intermingling and interlocking of parenchyma cells of both graft components.
3. Differentiation of certain parenchyma cells of the callus into new cambium cells connecting with the original cambium in the stock and scion.
4. Production of new vascular tissues by the new cambium permitting passage of nutrients and water between the stock and scion.

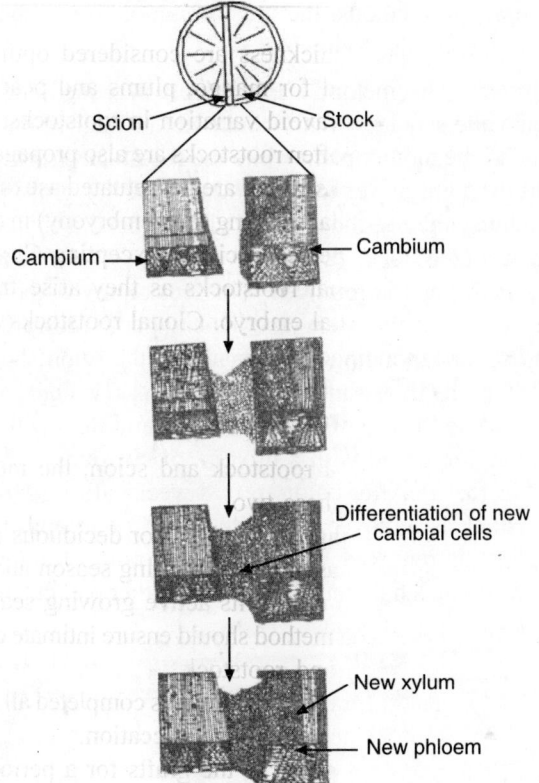

Fig. 15.12 Development sequence during the healing.

Method of Grafting

Grafting techniques may be employed for production of new plants, top working or for renovation and rejuvenation of the existing trees.

Methods for Production of New Plants

1. Inarching or approach grafting
2. Tongue inarching
3. Side grafting
4. Saddle grafting
5. Cleft grafting
6. Whip grafting
7. Whip and tongue grafting
8. Veneer grafting
9. Epicotyl grafting

Among the above methods, the most commonly employed methods in India for fruit crops are the following:

1. **Inarching:** This method is extensively practiced in the propagation of mango and sapota in Tamil Nadu. In this method the scion remains attached to the mother plant while the rootstocks grown separately on pots are brought closer to the mother trees for grafting. At the time of inarching, the seedlings stock should have a thickness of a lead pencil. From the stock, 22 cm above ground level, 5 to 8 cm long slice of bark and wood is removed. This cut removes about one third of the thickness of the stem and tapers gently towards the tip and bottom. A corresponding cut is made on the scion shoot so that two cuts fit in perfectly without leaving a chink. Two cuts are placed face to face and tied firmly with jute threads and then with twine over it. The union is covered with a mixture of cowdung and mud in equal parts (Fig.15.13). After union (6 to 8 weeks after grafting) the top of the rootstock is removed above the graft point and the base of the scion shoot is removed below the graft point. Removal is done in stages by giving a half cut first and an another half cut after an interval of 10 days.

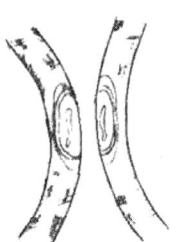 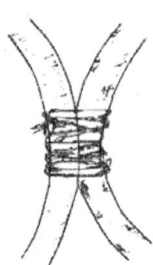 
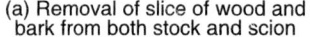

(a) Removal of slice of wood and bark from both stock and scion   (b) Positioning and tying both components   (c) Waxing the graft union

Fig. 15.13 Approach grafting.

If scion shoots are high up, special devices such as grafting pot stands, bamboo clefts, platforms have to be employed to accommodate rootstocks. In recent days, to avoid this expensive proposition mother trees are trained to produce low spreading branches in the initial stages itself so that the pots containing the rootstocks can be kept on the ground itself.

2. **Side grafting:** In this method, scion is prepared from the terminal of the past season's growth and used when they are not in active growth. Pre-curing of scion has to be done when it is still on the tree.

Pre-curing of scion consists of removal of all the leaves retaining their petiole intact excepting 10 cm at the top. Buds swell in axils subsequently and these pre-cured scions will be cut and used as scions after a week. Scion is inserted into the side of the rootstock which is larger than the scion and also the top of the rootstock is not removed. In the stock, a slanting cut of 2.5 cm is made at the base at an angle of 20° to 25°. After insertion of the scion, it is tied well. After a month buds in the scion begin to grow. When they grow to 7.5 to 9.0 cm long, the rootstock stem above the joint is removed. This method is useful in mango, sapota, fig and mangosteen.

3. **Whip or splice grafting:** This is simple and extensively employed in apples and pears. The scion shoot is given a slanting cut of 7.5 to 9.0 cm long at the basal end. A corresponding cut is made on the rootstock. The cut surfaces are placed together and secured tightly in position by tying or sometimes sealed with grafting wax (Fig 15.14). A tongue – shape cut is often given both in the cut surface of scion and rootstock which helps in holding the stock and scion more intimately together. This method is known as whip and tongue grafting (e.g. Apples and Pears). It heals quickly and makes a strong union because of close contact between the cambial regions (Fig. 15.15.).

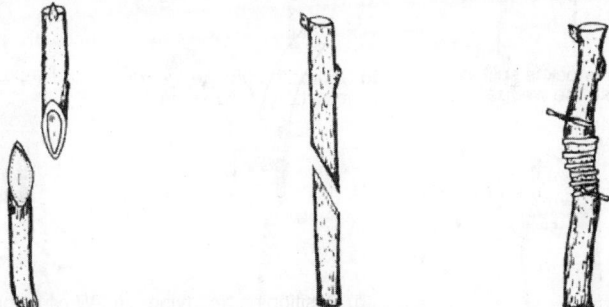

(a) Single oblique cuts made on   (b) Stock and scion placed   (c) The graft tied securely
      stock and scion   to match cambium   in place

Fig. 15.14   Splice grafting.

4. **Cleft grafting:** The base of the scion is prepared in the form of wedge. The rootstock is split in which the scion is inserted. For this method, the rootstock and scion shoots need not be of same thickness. Usually this is done on thick stocks of 2 to 8 cm in diameter. The stock is given a smooth cut and then it is split at the center and two scions are inserted at the ends in such a way that the cambial layers should be in contact (Fig. 15.16). Hence, scions are not inserted in the middle but done on either side. After the successful graft union, only the better developed one of these two will be allowed.

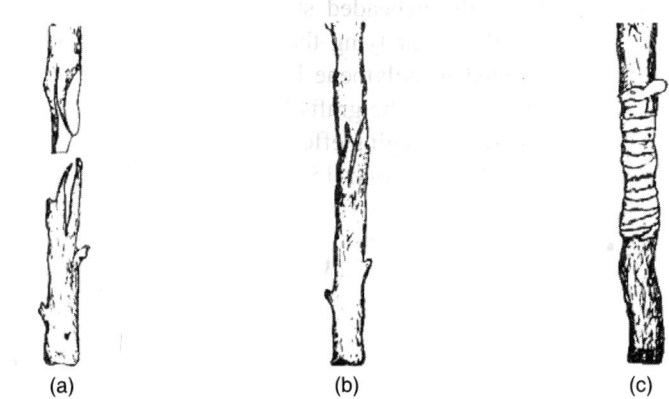

(a) The diagonal cuts on the stock and scion showing the position of the tongue
(b & c) Stock and scion inserted and complete graft wrapped with grafting taps

Fig. 15.15 Whip and Tongue grafting.

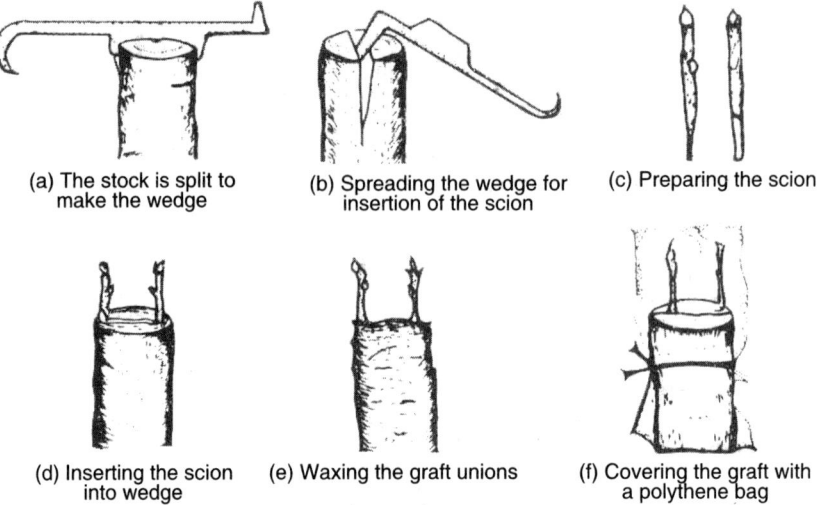

(a) The stock is split to make the wedge

(b) Spreading the wedge for insertion of the scion

(c) Preparing the scion

(d) Inserting the scion into wedge

(e) Waxing the graft unions

(f) Covering the graft with a polythene bag

Fig. 15.16 Cleft grafting.

5. **Epicotyl or stone grafting:** This type of grafting has been recently standardized in fruit crops like mango and cashew and this holds great promise for multiplying plants in large number and in lesser time. The seeds are placed on sand bed and covered with 5.0–7.5 cm thick layer of leaf mould for germination. Germinated seedlings of 8 to 15 days age are taken out and grafted indoor by beheading the seedlings about 5.0 cm above the seed and inserting the wedge shaped scion in the

vertical split of the beheaded stock. Polythene tape of 200 gauge thickness is utilized for tying the graft. Immediately thereafter, the grafts are planted in polythene bags filled with soil and farm yard manure mixture (1:1). The grafts are watered and kept in semi-shade condition to avoid damaging effects of sun and rain (Fig.15.17). If the age of the stock is beyond 15 days, the percentage of success is greatly reduced in mango. To overcome this problem, flush grafting and soft wood grafting have been developed. In flush grafting, the fresh growth of the second flush is used as rootstock while in softwood grafting; seedlings of one year old are used as rootstocks. The technique of grafting is same in all the three types of grafting.

6. **Top-working:** This is also known as 'top-grafting' or 'top-budding' if we adopt anyone of the grafting or budding techniques respectively. Top-working is aimed at changing the established plant, tree, shrub or vine with a desirable cultivar. For top-working, one has to select three to five well spaced scaffold branches which are not larger than about 10 cm in diameter and are conveniently cut close to ground. The selected branches are given smooth cut without tearing the bark from the trunk. Thus, these branches may be utilized as rootstocks and grafting may be done on them by one of the following grafting methods:

    1. Inarching
    2. Bark grafting
    3. Side grafting
    4. Veneer grafting
    5. Cleft grafting

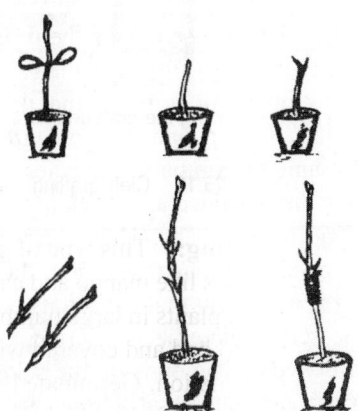

Fig. 15.17 Epicotyl or stone grafting.

In bark grafting, the bark is split and the scion is inserted as in graft. The scion is inserted between the bark and the wood and pushed carefully. The graft joint may be sealed with grafting wax. In veneer grafting, which is a modified method of side grafting, 2.5 to 3.0 cm cut is made downward. A small notch is made by removing a piece of wood by a diagonal cut at the base to accommodate wedge shaped scion (Fig 15.18).

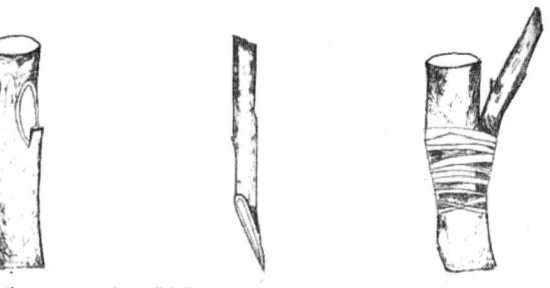

(a) Cut on the rootstock   (b) Base of the scion   (c) The completed graft
prepared for insertion

Fig. 15.18   Veneer grafting.

Trees just grafted should be amply supplied with water so that the tissues are in a high state of turgidity. This is necessary to have adequate callus production which is essential for the healing of the graft union.

FOR RENOVATION AND REJUVENATION

Occasionally the roots, trunks or large limbs are damaged by winter injury, cultivation implements, certain diseases or rodents. Such damaged trees can be repaired and saved by use of bridge and buttress grafting.

1. **Bridge grafting:**   This is used when the root system of the tree has not been damaged but there is injury to the bark of the trunk. It is done during the active growth of the tree so that the bark is slipping easily. The scion is selected from one year old growth, 0.5 to 1.5 cm in diameter of the same or compatible species. The first step in bridge grafting is to trim the wounded area back to healthy undamaged tissue by removing dead or torn bark. Then every 8 to 10 cm around the injured section, a scion is inserted at both the upper and lower ends into live bark. It is important that the scions be inserted right side up.

   After all the scions have been inserted, graft unions have to be thoroughly graft waxed. The buds on the scions will often push into growth, if the grafts are successful. These shoots are removed because no branches would be desired in this position. The scion will rapidly enlarge in size and completely heal over the wound in a few years.

2. **Buttress grafting:** This method is useful in supporting branches that may be in danger of breaking off or where there is a weak crotch. A small branch about a pencil size or little larger coming at about 30 cm or above the weak crotch is grafted into the adjacent branch to be supported.

Budding

Budding is the art of inserting a bud on the rootstock plant in such a manner that both will unite and continue to grow as a single individual plant. The special advantage of budding over grafting is the economy in the use of scion materials.

The shoots of mother tree from which buds are taken for use as scion materials for budding are called 'bud wood'. Bud wood should be collected from a selected tree and it should be of a well matured past season's growth with plumpy and well developed buds which should be dormant but ready to grow. Ten to fifteen days before the removal of the bud, the shoots should be defoliated so as to activate the buds. Buds can be removed with or without wood from the scion shoot. Budding should be done when the rootstock is still on active growing condition with a free flow of sap which will help in easy lifting of bark in budding operation and also for proper union.

Methods of Budding

1. **Shield budding or 'T' budding:** In this method a 'T' or inverted 'T' shaped incision is made on the rootstock stem. A transverse cut of 1.0 to 1.5 cm length is made first and then either below or above to this a vertical cut of 2.5 cm to 3.0 cm length is made and connected to the transverse cut. Then the bark is lifted by using the ivory edge of the budding knife. The scion bud is removed in the form of a shield with or without a piece of wood. The bark is raised and then the bud is inserted into the 'T' cut surface of the rootstocks. Then it is secured in position by proper bandaging (Fig. 15.19). This method of budding is practiced extensively in the propagation of sweet oranges, roses, plums and peaches.

2. **Patch budding:** A rectangular patch of bark of about 3 cm length and 1.5 cm width is removed from the stem of the rootstock. Similar patch of bud is removed from the bud wood with the bud in the center. Then it is placed in the rootstock and wrapped. This method is practiced in citrus, mango, rubber and annona.

3. **Flap or forket budding:** A transverse incision is made in the bark of the rootstock and then the bark is peeled off carefully to a length

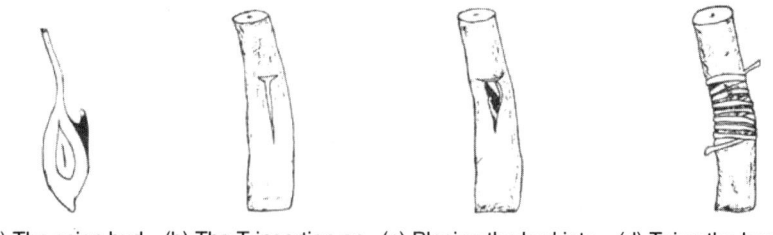

(a) The scion bud   (b) The T insertion on   (c) Placing the bud into   (d) Tying the bud union
                 the rootstock   the inscision   with budding tape

Fig. 15.19   T budding.

of 5.0 cm. The bud shield removed from the scion is pushed under the flap till the exposed edges of rootstocks meet. Flap is then cut to half and is brought to cover the bud shield partially and is then wrapped. If entire flap of the rootstock stem is retained instead of cutting the flap into half, the method is then known as modified flap or forket method.

4. **Ring budding:** The bud is prepared by taking a ring of bark 3.0 cm in length with the bud in the center. In the rootstock, two circular cuts 1.5 cm apart are made and these are connected with a vertical cut and the ring of bark is removed. The prepared scion bud with the ring of bark is fitted in the exposed portion of the rootstock and tied as usual. This is commonly practiced in cinchona.

5. **Flute budding:** This is exactly same as ring budding, the difference being that the ringed bark can be removed easily in the form of a flute. Here the stock plant is topped off at 25 cm height and at the top about 2.5 to 3.0 cm of bark is removed leaving the wood exposed. The bud in the form of a flute is then fitted in the rootstock. It is essential that the diameter of the rootstock and scion should exactly be same, otherwise the contact between them will not be proper (Fig 15.20).

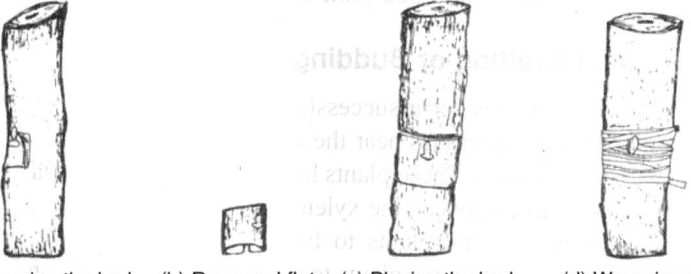

(a) Removing the bud   (b) Removed flute   (c) Placing the bud   (d) Wrapping with
along with a ring of tissue   with a bud   on the rootstock   waxed or polythene tape

Fig. 15.20   Flute budding.

6. **Chip budding:** In one-side of the rootstock, a slanting cut of 2.5 cm length going to a depth of half of the thickness of the stock is made. A notch is given at the bottom end of the cut. A similar cut is made in the scion shoot and a bud with a large piece of wood is removed. The scion bud is exactly fitted into the cut made in the rootstock and tied (Fig. 15.21) (e.g. Grapes).

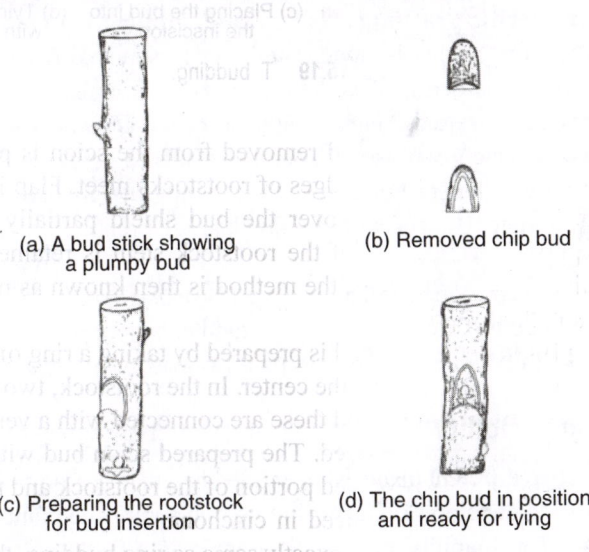

(a) A bud stick showing a plumpy bud

(b) Removed chip bud

(c) Preparing the rootstock for bud insertion

(d) The chip bud in position and ready for tying

Fig. 15.21 Chip budding.

In any of the above methods, the union will take place in 3 to 5 weeks. After the union is complete, the top of the rootstock is removed at gradual stages, 7 to 8 cm above the bud union. This has to be provided above the bud union so that drying of the tip of the rootstock may not extend upto bud joint and injure the bud.

Limitation of Grafting or Budding

One of the requirements for a successful graft union is the close matching of the callus-producing tissues near the cambial layers. Grafting is generally confined to dicotyledons. These plants have a vascular cambial layer existing as a continuous tissue between the xylem and phloem. For grafting, it should be borne in mind that the plants to be combined are capable of uniting. Generally, the more closely the plants to be grafted are related botanically, the more favourable is the chances of the graft union being successful.

1. **Intra-varietal grafting:** When a scion can be grafted back on same plant or a scion from a plant of a given clone can be grafted to any other plant of the same clone (e.g., Elberta peach on Elberta peach).

2. **Inter-varietal grafting:** When different varieties of a species employed as graft parents (e.g., mango).

3. **Inter-specific grafting:** In this case, grafting between two species of the same genus is done. But this is usually difficult but widely practiced between species in the genus Citrus. Japanese plum (*Prunus salicina*) is grafted commercially on peach (*Prunus persica*).

4. **Inter-generic grafting:** When the plants to be grafted together are in different genera but in the same family the chances of union are more remote. But successful union has been reported in the following cases:

   1. *Citrus* spp. on trifoliate orange (*Poncirus trifoliata*).
   2. Sathugudi (*Citrus sinensis*) on wood apples (*Ferronia elephantum*)
   3. Sapota (*Achras sapota*) on pala (*Manilkhara hexandra*).

Graft Incompatibility

The ability of two different plants when grafted together to produce a successful union and also to develop satisfactorily into one composite plant is termed as *compatibility*. The inability of two different plants to do so when grafted together is often defined as incompatibility or graft incompatibility. The distinction between a compatible and an incompatible graft union is not clear cut. On one hand, stocks and scions of closely related plants unite readily and grow as one plant. On the other hand, stocks and scions of unrelated plants when grafted together are likely to fail completely in union. Many graft-combinations lie between these extremes *viz.,* compatible to incompatible and therefore the characterization of incompatibility is not distinct. 'Partial incompatibility' where the stock outgrows the scion has been reported in certain fruit crops. For instance, mandarin when grafted onto trifoliate stocks outgrew the scions but the tree grew well and produced plenty of fruits of good quality. Incompatibility may be classified as 'translocated incompatibility' and 'localized incompatibility'. The former type refers those cases in which the incompatible condition cannot be overcome by the insertion of a mutually compatible interstock. This is due to apparently some labile influence moving across it. This type involves phloem degeneration and development of a brown line or necrotic area in the bark. Hale's Early Peach develops incompatibility when grafted on Myrobolan-B plum rootstock. But when a mutually compatible interstock 'Brompton Plum' is introduced, the incompatibility still persists indicating that the incompatibility is due to some factors translocated from the rootstock to the scion through the phloem causing phloem degeneration.

The second type *viz.*, 'localized incompatibility' includes a combination in which the incompatibility reaction apparently depends upon actual contact between stock and scion. Introduction of a mutually compatible interstock will normally overcome the incompatibility. Symptom of this kind of incompatibility is that the graft is often mechanically weak with discontinuity in cambium and vascular tissues. An example of this kind of incompatibility is that when Bartlett pear is grafted directly on quince stock, it is incompatible. When Old Home interstock is introduced in between these combinations, the three part combination is completely compatible and it grows satisfactorily. Another example is that when 'Eureka lemon' is grafted on trifoliate rootstocks, it proved to be incompatible, due to a toxic substance produced by the scion damaging the conducting tissues of the stock. When the interstock Valencia orange was introduced, the combination proved successful.

In some cases, the stock-scion combination grows in an apparently normal fashion for varying periods of times-perhaps for many years and then difficulties arise. This is called as 'delayed incompatibility'. A good example of the above phenomenon is the black line of walnut which occurs in certain Persian walnut orchards in California and France. When cultivars of *Juglens regia* are grafted on seedling rootstocks of *J. hindsii* or paradox rootstocks (*J. hindsii* × *J. regia.*), the trees grow satisfactorily for 15 to 20 years or even more years of age, thereafter the trouble starts. A thin-layer of cambium and phloem and the dead tissues develop at one point and gradually extend around the tree at the graft union until the trees become girdled. The vertical width of the dead area may reach 30 cm. Such girdling may kill the plants above the graft union but the stock remains alive and sprout. Another example is that sapota on *Bassia longifolia* stocks. Incompatibility is manifested by overgrowing of scion resulting in pronounced distortion at the bud joint and the graft dies prematurely. Delayed incompatibility has also been reported in many citrus species as indicated below:

Scion	*Rootstock*
Early Imperial Mandarin (*Citrus reticulata*)	Rough lemon (*Citrus jambhiri*)
Blood Red Orange (*Citrus sinensis*)	Karna (*Citrus karna*)
Sweet Orange (*Citrus sinensis*)	Sweet lime (*Citrus limettoides*)

Symptoms of Incompatibility

1. Failure to form a successful graft or bud union with a high percentage of success.
2. Yellowing of leaves in the latter part of the growing season followed by early defoliation accompanied by decline in vegetative appearance of shoot, die back and general ill health of the tree.
3. Premature death of the trees which may live only a year or two in the nursery.
4. Marked differences in the growth rate or vigor of scion and stock
5. Over growth at, above or below the graft union.

Causes of Graft Incompatibility

1. **Virus infection:** One component of the graft combination may carry a virus and be symptomless but the other component may be susceptible to it. For example, when Bartlett pear is grafted on *Pyrus pyrifolia* to tree declines due to virus infection of the susceptible rootstock while Bartlett on *P. communis* remains healthy, because *P. communis* is a virus resistant species.
2. **Growth differences:** In certain graft combination, the differences in the time of resumption of cambium activity of the stock and scion or differential growth characteristics of the stock and scion are reported to be the causes for graft incompatibility.
3. **Physiological causes:** Physiological incompatibility is due to the inability of the stock or the scion to supply the other component with necessary amount or quality of materials for normal functioning. There is some evidence that in certain graft combinations one component (scion or stock) produces chemicals that are toxic to the other, killing the entire plant, e.g. when pear is grafted 'onto quince rootstock, a cyanogenic glucoside, prunasin, normally found in quince is translocated into the phloem of the pear where it gets broken down in the region of the graft union into hydrocyanic acid. The presence of this acid leads to lack of cambial activity at the graft union, leading to graft Incompatibility.

Stock-scion Relationships

A grafted or budded plant can produce unusual growth patterns which may be different from what would have 'occurred if each component part of a graftage *viz.*, rootstock and scion was grown separately or when it is grafted or budded on other types of rootstocks. Some of these have major horticultural significance. This varying aspect of rootstocks in the performance of a scion cultivar or *vice versa* is known as 'stock-scion relationship'.

A. EFFECT OF STOCKS ON SCION CULTIVARS

1. **Size and growth habit:** In apple, rootstocks can be classified as dwarf, semi-dwarf, vigorous and very vigorous rootstocks based on their effect on a scion cultivar. If a scion is grafted on dwarf rootstocks (e.g. Malling IX), the scion grows less vigorously and remain dwarf only. On the other hand, if the same scion is grafted on a very vigorous rootstock (e.g. Malling II) the scion grows very vigorously. In citrus, trifoliate orange is considered to be the most dwarfing rootstock for grapefruit and sweet oranges. On the other hand, in mango, all plants of a same variety are known to have the same characteristic canopy shape of variety despite the rootstocks being of seedling origin. But seedling rootstocks of Kalapad, Olour and Bapakkai have been found to impart dwarfness in the scion cultivars of mango. Guava cultivars grafted on *Psidium pumilum* are found to be dwarf in stature.

2. **Precocity in flowering and fruiting:** The time taken from planting to fruiting i.e., precocity is influenced by rootstocks. Generally fruiting precocity is associated with dwarfing rootstocks and slowness to fruiting with vigorous rootstocks. Mandarin, when grafted on Jambhiri rootstocks, are precocious than those grafted on sweet orange or orange or acid lime rootstocks.

3. **Fruitset and yield:** The rootstocks directly influence on the production of flower and setting fruits in oriental Persimmon (*Diospyrous kaki Cv.* Hachiya). When it is grafted on *D. lotus,* it produces more flowers but few only mature; but when *D. kaki* is used as the rootstock, the fruitset is more. The influence of rootstock on the yield performance has been well documented in many fruit crops. Acid limes budded on rough lemon register nearly 70 per cent increased yield than those budded on troyer citrange, Rangpur lime or its own rootstock. Sweet orange var. Sathugudi budded on Kichili rootstock gave higher yield than on Jambhri or on its own seedling.

4. **Fruit size and quality:** Sathugudi sweet oranges grafted on Gajanimma rootstocks produced large but poor quality fruits while on its own rootstock-produced fruits with high juice content and quality. The physiological disorder 'granulation' in sweet orange is very low on Cleopatra mandarin seedlings, on the other hand, rough lemon seedling stocks induced maximum granulation. The physiological disorder 'black end' in Bartlett pear did not appear if *Pyrus communis* is used as the rootstock. When *P. pyrifolia* is used as the rootstock, the disorder appears affecting fruit quality.

5. **Nutrient status of scion:** Rootstocks do influence the nutrient status of scion also. Sathugudi orange trees have a better status, all nutrients in the leaves when it is budded on *C. volkarimariana* stock than on its own rootstock or on Cleopatra mandarin stocks.

6. **Winter hardiness:** Young grapefruit trees on Rangpur lime withstand winter injury better than on rough lemon or sour orange. Sweet oranges and mandarins on trifoliate stocks are cold hardy.

7. **Disease resistance:** In citrus considerable variability exists among the rootstocks in their response to diseases and nematodes. For instance, rough lemon rootstock is tolerant to tristesa, xyloporosis and exocortis but is susceptible to gummosis and nematode. On the other hand, troyer citrange is tolerant to gummosis but susceptible to exocortis virus disease. Similarly, guava varieties grafted on Chinese guava. (*Psidium friedrichsthalianum*) resist wilt diseases and nematodes.

8. **Ability to resist soil adverse conditions:** Among the citrus rootstocks, trifoliate orange exhibits poor ability, while sweet oranges, sour orange, Rangpur lime rootstocks exhibit moderate ability to resist excess salts in the soil. In Pome fruits similarly, variation exists among rootstocks to resist excess soil moisture or excess boron in the soil. Myrobolan plum rootstocks generally tolerate excess boron and moisture than Mariana plum root or other rootstocks' *viz.*, peach, apricot or almond.

B. EFFECT OF SCION ON ROOTSTOCK

1. **Vigour of the rootstocks:** In apple, it has been found that if apple seedlings were budded with the 'Red Astrochan' apple, the rootstock produced a very fibrous root system with few tap roots. On the other hand if scion 'Goldenburg' was budded on the seedlings, they produced two or three pronged deep roots without fibrous root system. In citrus, the scion cultivar is less vigorous than the rootstock cultivar the rate of growth and the ultimate size of the tree are more determined by the scion rather than the rootstocks.

2. **Cold hardiness of the rootstock:** Cold hardiness of citrus roots is affected by the scion cultivar. Sour orange seedlings budded to 'Eureka' suffered much more from winter injury than the unbudded seedlings.

3. **Precocity in flowering:** Young mango rootstock seedlings (6 months to one year old) were found to put forth inflorescence when the branches from old trees are inarched, which can be attributed to the influence of scion on the rootstock.

Factors influencing the healing of graft union

1. **Incompatibility:** Certain rootstocks and scions are incompatible, therefore, the graft union between these two will not normally take place.

2. **Kind of plant:** Some species like oaks are difficult to graft, but apple and pears are very easy in producing a successful graft union.
3. **Environmental factors during and following grafting:** There are certain environmental requirements which must be met for callus tissues to develop and heel the graft union.
    (a) *Temperature* has a pronounced effect on the production of callus tissues. An optimum temperature is essential for production of callus tissues. In most of the temperate fruit crops callus production is retarded after 42.5°C.
    (b) *Relative humidity* must be high in maintaining a film of water against the callusing surface to prevent these delicate thin walled parenchymatous cells from drying.
    (c) Presence of high *oxygen* content near this surface is essential.
4. **Growth activity of the stock plants:** Some propagation methods, such as 'T' budding and bark grafting depend upon the bark 'slipping' which means the cambial cells are actively dividing and producing young thin walled cells on the side of the cambium. These newly formed cells separate readily from one another as the bark slips.
5. **Propagation techniques:** Sometimes the techniques used in grafting are so poor that only a small portion of the cambial regions of the stock and scion are brought together. This may result in failure of the graft union.

PROPAGATION BY SPECIALIZED PLANT PARTS

Certain plants possess specialized vegetative structures whose functions are storage of food and vegetative reproduction. If such structures are naturally detachable for the propagation, this procedure is termed as '*separation*'. On the other hand, such structures are to be cut into sections for the purpose of propagation, then this process is called as '*division*'. The following specialized vegetative structures are used in propagation (Fig. 15.22).

1. Bulb

A bulb is a specialized underground organ consisting of a short, usually vertical stem axis bearing at its apex a growing point or flower primordium enclosed by thick fleshy scales. The outer bulb scales are generally fleshy and contain reserve food materials whereas the scales towards the inner contain relatively less food materials and are more leaf-like. Bulbs possessing dry and membranous outer scales are known as *tunicate bulbs* (e.g., onion) while bulbs which lack this cover is known as *non-tunicate bulbs* (e.g. lily). Dry outer scales in bulbs provide protection from drying and mechanical injury while non-tunicate bulbs are easily damaged.

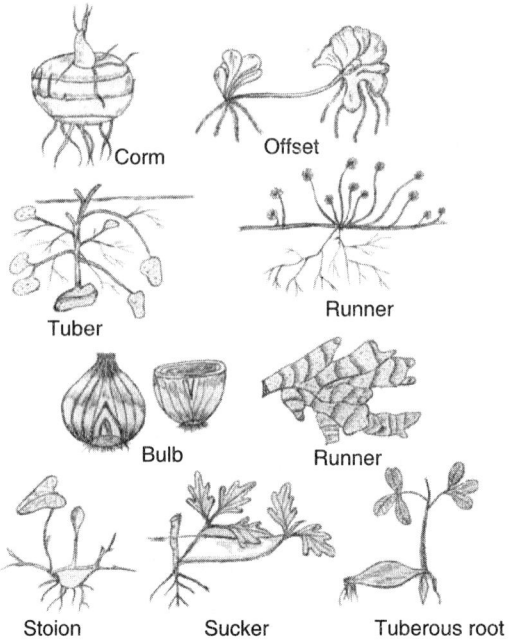

Fig. 15.22 Vegetative structures.

2. Corm

A corm is the swollen base of a stem axis enclosed by the dry scale-like leaves. It is a solid stem structure with distinct nodes and internodes e.g., gladiolus corm. The propagation of corm bearing plants is principally by the natural increase of new corms. The development of miniature corms between the old and the new corms is termed as cormels. Often large corms can be cut into sections, retaining a bud in each section and are used for planting. e.g., Elephant yam.

3. Tuber

A tuber is a modified stem structure which develops below ground as a result of the swelling of the subapical portion of a stolen and subsequent accumulation of reserve materials. A tuber has all the parts of typical stem. e.g., potato, Jerusalem Artichoke. These tubers are used for propagation either by planting whole tubers or by cutting them into sections each containing one or more buds. Certain plants produce aerial tubers in the axils of leaves which are known as tubercles (e.g., *Dioscorea bulbifera)*. They are also removed and used for propagation.

4. Tuberous roots and stem

In certain plants, the adventitious roots become thickened and they do have external and internal structures of roots but lack nodes and internodes. These are known as tuberous roots (e.g., Sweet potato, Dahlia). In other plants such as tuberous Begonia, Cyclamen or Gloxinia, they have thickened structures which have arisen from enlarged hypocotyl tissue. They have a vertical arrangement and may show features of stems. Propagation of plant with such tuberous roots or stems consists of division of such materials into sections but each section should have a section of the crown bearing a short bud.

5. Rhizome

It refers to a specialized stem structure in which the main axis of the plant grows horizontally at or just below the ground surface. A rhizome consists of nodes and internodes having leaf scars on the node (e.g., Ginger, Turmeric, many grass species and many fern plants). In determinate type of rhizomes each clump ends in a flowering stalk and growth continues only from lateral branches (e.g., cardamom). Indeterminate type of rhizomes does not produce a clump but spread extensively over an area and grow continuously from the terminal apex and lateral branch rhizomes. Propagation through rhizome is carried out by cutting it into sections; each piece has at least one lateral bud or eye.

6. Runner

Runner is a specialized stem which develops from the axis of a leaf at the crown of a plant and grows horizontally along the ground and forms a new plant at one of the nodes (e.g., Strawberry, *Mentha spp.*).

7. Offset (Offshoot)

It refers to a special type of lateral shoot or branch which develops from the main stem in certain plants and is characterized by shortened, thickened stem of rosette-like appearance. Offsets which produce sufficient roots can be removed by cutting them close to the main stem with a sharp knife and used for propagation (e.g., Datepalm, pineapple).

8. Sucker

A sucker is a shoot which arises on a plant from below ground from an adventitious bud on a root (e.g., Chrysanthemum). Suckers are further known as root suckers, ground suckers and shoot suckers if they arise respectively from root, near the ground and stem of the plant.

9. Crown

The term crown designates that part of a plant at the surface of the ground from which new shoots are produced. This kind of crown (also known as split) is observed in herbaceous perennials like strawberry, pyrethrum, Gerbera or African violet wherein the stem is a short and thickened structure from which the leaves are produced in a rosette-like arrangement. Division of the crown is the important method of propagation wherein each split unit should have few roots and a few leaves with a growing point.

Certain plants do have one or more of the above mentioned specialized structures useful for propagation. But particular structure is preferred for commercial propagation for obvious reasons. Strawberry can be propagated both by runners and splits from the crown. Splits give early yield but the establishment success is less when compared to runners which give the higher yield and establishment success. In the case of pineapple, slips are found to be the best planting material followed by side suckers and crowns.

Micropropagation

Micropropagation or *in vitro* propagation refers to the development of new plants in an artificial medium under aseptic conditions from very small pieces of plants such as embryos, seeds, stems, shoot tips, root tips, callus, single cells and pollen grains. This technique has been put into various applications in the discipline of agriculture, horticulture, forestry ever since the concept of 'totipotency' of plant cell was scientifically proved by scientists in late fifties. The various applications of micropropagation of plants are
1. Rapid rate of multiplication of a plant clonally
2. Production of disease-free and disease resistant plants
3. Induction of mutants and selection of mutants
4. Production of haploids through anther culture
5. Wide hybridization through excised embryo and ovule culture.
6. Somatic hybrids and cybrids through protoplast fusion
7. Transformation through uptake of foreign genome
8. Nitrogen fixation
9. Cryopreservation of germplasm types.

Requirements for Micropropagation

1. **Laminar air flow chamber:** This chamber is useful to perform all operations in aseptic culture. Sterilization is achieved by the ultra-violet (UV) germicidal lamp fitted in it and by the flow of filtered air towards the person doing the operations, which prevents dust particles which carry microorganisms from settling on the explant.

2. **Autoclave or pressure cooker:** It is used to sterilize the media, containers, petridishes and the various accessories required in the transfer operations. Normally sterilization is done for 15–20 minutes at 15 PSI pressure at 121–144°C.

3. Alcohol lamps, disinfectants and sterile water are also required.

4. **Culture medium:** A medium consists of mineral salts, carbon and energy source, vitamins, plant growth regulators and other organic components. The most commonly used medium is Murashige and Skoog medium (MS medium).

Procedure for Micropropagation

1. **Collection of explant:** The small piece of plant used to begin a culture is referred to as an explant. The size, age and type of explant decide the success of *in vitro* propagation. Explants may be excised from shoot-tip, root-tip, auxiliary bud, embryo, endosperm, hypocotyl, callus and anther. Explants are collected in containers having sterile water.

2. **Surface sterilization:** *Explants* so collected from field grown plants harbor numerous fungi and bacteria, which when inoculated into a nutrient medium contaminate the entire *in vitro* system. Hence, sterilization is resorted to prior inoculation of explants. The efficacy of the sterilants used are found to vary depending upon the type of chemical, concentration used, time of exposure etc. Sodium hypochlorite (0.5–2.0% *w/v*), Calcium hypochlorite (filtered 5–10% *w/v*) or 0.1% of Mercuric chloride are now most commonly used as surface sterilants. A few drops of teepol are also added to facilitate better contact between the explants and the sterilant. Then the explants are rinsed 3-4 times with sterile distilled water.

3. **Inoculation:** Transfer of the explant into the culture medium is known as inoculation. This must be done in an aseptic condition. This is achieved by surface sterilization of the working table of the laminar air flow chamber with absolute alcohol followed by UV light for 30 minutes. The forceps and scalpel used for inoculation besides Petri-dishes are also first steam sterilized in an auto-clave at 121°C for 15 minutes and later flame sterilized before each inoculation. The hand should be also cleaned with alcohol before use. The cultures are then incubated at 28 ± 2°C in an air conditioned culture room with 16 hours photoperiod (1000 lux) supplied by cool day fluorescent tubes.

4. **Sub-culturing:** After inoculation, the explant increases in volumes or it proliferates. At this stage, it is divided into different components or parts and transferred into a fresh medium-under above mentioned

aseptic sterile condition. This process is known as sub-culturing. This sub-cultured mass should produce a shoot and root system which is dependent upon the type of growth regulator and its concentration used in the medium. It is generally observed that if the concentration of cytokinins is high relative to auxin in a medium, shoots are induced and on the other hand, when the concentration of cytokinins is low to auxin, roots are induced and at intermediate concentration the tissue grows as undifferentiated callus. This basic approach is widely used to regenerate a wide variety of dicotyledonous and monocotyledonous plants and the ratio of cytokinins to auxin is specific to each kind to develop root or shoot induction. The most commonly used auxins are 2,4-dichloro phenoxy acetic acid (2,4-D), 3, indole butyric acid (IBA), alpha Naphthalene acetic acid (NAA), 3- indole acetic acid and 2,4,5-trichloro phenoxy acetic acid (2,4,5-T). Similarly, the commonly used cytokinins are 6-benzyl adenine (BA), 6-furfuryl amino purine (Kinetin) etc.

Various Methods of Culturing Plant Tissues and Organs

There are five classes of plant tissue culture

1. **Callus culture:** A piece of sterile plant tissue with living cells is transferred to a culture medium to induce callus proliferation. Sub culturing is then done onto a medium with or without altered growth regulator concentrations, ultimately resulting in the induction of adventitious organs or embryos. In the last stage, regenerated plants are removed from *in vitro* culture and slowly exposed to outer environment so that the plants can be fully autotrophic.

2. **Cell culture:** Cells are maintained in suspension cultures so as to produce free cells and are then subcultured to regenerate complete plant from single cells. This technique is now useful to induce variability in plant cells and to select desirable cell variants and regenerate complete plants from these variants.

3. **Meristem culture:** This technique involves aseptic culture of shoot meristems on nutrient medium so as to produce complete plants. Most important application of meristem culture is the production of virus free plant which is genetically identical.

4. **Embryo culture:** It involves aseptic excision of the embryo and its transfer to a suitable medium for development under optimum culture conditions. After the embryo has grown into a plantlet *in vitro,* it is transferred to sterile soil or vermiculite and grown to maturity in a green house. This technique is useful in the production of interspecific and intergeneric hybrids which could not be otherwise accomplished and also in overcoming embryo abortion.

5. **Protoplast culture:** From different sources, protoplasts, the plant without any rigid cellulose wall but with plasma membrane only, is allowed to fuse to form a somatic hybrid. These are cultured in suitable media to regenerate the cell wall and are again cultured in suitable medium for differentiation and morphogenesis.

Hardening

The plantlets developed in the culture tubes are acclaimatized to a specific environment having a high humidity, a low light level and a constant temperature. Besides, the roots developed *in vitro* are hairless and delicate, requiring care during transfer from culture medium. To have a better survival rate, the plantlets may be transferred to container in mist chambers where relative humidity is maintained at higher order. Once new growth is seen, the plants may be slowly transferred by exposing to increased light intensity in stages. Orchids, other ornamentals like Lilium, Gerbera, Cordyline, Begonia, Ficus, Potato etc. are some of the plants propagated by tissue culture on a commercial scale in Europe. In India, Banana, Cardamom, limited extent in ornamentals like gerbera and cordyline are propagated on a commercial scale through tissue culture.

Media for Propagating Nursery Plants

Several materials and their combinations are available as media for germinating seeds and rooting of cuttings. A good propagation medium should possess the following characters:

1. It must be firm and dense to hold the cuttings or seeds in place during rooting or germination.
2. It must possess sufficient moisture retaining capacity.
3. It must be sufficiently porous to permit excess water to drain away and to admit proper aeration.
4. It must be free from weed seeds, nematodes and pathogens.

The following are the commonly employed media for propagation of horticultural crops.

1. Soil

A soil is composed of materials in the solid, liquid and gaseous state. The materials must exist in the proper propoitions for satisfactory growth. Soils rich in humus are often used as media for forestry tree species in South Indian hills.

2. Sand

It is the most satisfactory medium for rooting of cuttings. It contains no mineral nutrients and hence mixed with organic materials for raising seeds.

3. Peat

It consists of the remains of aquatic marsh, bog or swamp vegetation which has been preserved under water in a partially decomposed state. When such peat is derived from sphagnum, hypnum or other mosses, it is known as peat moss. It is used in mixture after breaking them and moistened.

4. Sphagnum moss

Commercial sphagnum moss is the dehydrated young residue or living portion of acid-bog plants in the genus *Sphagnum* such as *S. papillosum, S. capillacem* and *S. palustre.* It is generally collected from the tree trunks of the forest species in South Indian hills above 1500 m. M.S.L. during rainy period. It is relatively sterile, light in weight and has a very high water-holding capacity. It is the commonly used medium in air layering.

5. Vermiculite

This is a micaceous mineral which expands markedly when heated. Chemically it is a hydrated magnesium aluminium-iron silicate. It is very light in weight and able to absorb large quantities of water. This can be used as a rooting medium for air layering and also in pots for raising certain plants.

6. Soil mixture

This is the most commonly employed medium for pot plants. It usually consists of red earth, horse or well decomposed cattle manure, leaf mould, river sand and also charcoal in some cases. In other countries, pot mixtures of better textures are made by mixing sand, soil and organic matter such as peat moss or saw dust or shredded bark.

The composition of soil mixtures which are used for growing particular class of plants are given in table 15.1.

Containers for Propagation and Growing Young Plants

1. **Seed pan and seed boxes:** Seed pans are shallow earthen pots about 10 cm high and 35 cm in diameter at the top. They have one large hole for drainage in the centre or 3 holes at equidistant from each other.

Table 15.1 Soil mixture for certain group of plants.

1. Fruit plants	
4 parts	Red earth
2 parts	Sand
1 part	Farm Yard manure

2. For potting rooted cuttings and young seedling	
1 or 2 parts	Sand
1 part	Loam soil
1 part	Leaf mould

3. Annuals	
3 parts	Horse manure
1 part	Red earth
1½ parts	Sand

4. Bulbous plants	
2 parts each of	Read earth, sand, leaf mould and manure
1 part	Loam soil
1/4 part	Charcoal

5. Palms	
4 parts	Leaf mould
3 parts	Red earth
3 parts	Sand
2 parts	Loam
2 parts	Horse manure

6. Ferns	
4 parts	Leaf mould
3 parts	Sand
2 parts	Loam
2 parts	Horse manure
1 part	Brick pieces
1 part	Lime rubbish

Seed boxes are made of wood, 40 cm wide and 60 cm long and 10 cm deep, with 6–8 properly spaced holes drilled in the bottom. Against each of the holes is placed a crock with its concave side down, some large pieces of crock are put over it and also by the side of crock, some coarse sand 2 or 3 handfuls is sprinkled on the crock pieces forming a thin layer to prevent fine soil from clogging the drainage. Over this, required soil mixture is added. Very delicate kinds of seeds like Cineraria, Begonia, Gloxinia etc. are best sown in these containers.

2. **Earthen pots:** They are made of burnt porous clay in various sizes to provide requisite amount of soil and root space to different kinds and sizes of plants. They have straight sides and are made wider at the top than at the bottom to hold the greatest bulk of compost where the feeding roots are and also to facilitate easy removal of soil, intact with roots (ball of earth) at the time of planting or repotting.

   In our country, pots of varying sizes *viz.*, tube pots, 1/4 size, 1/2 size, 3/4 size and 'thali' are used commonly. Tube pots are used to raise the rootstocks of mango and sapota for grafting purposes. 1/4 size pots are used for potting singly very small seedlings during first transplanting and also for layering in plants like West Indian Cherry and Guava. ½ size pots are extensively employed for growing well rooted cuttings of several kinds of plants and small plants of all kinds. 3/4 size pots are preferred for growing almost all kinds of indoor plants and annuals. Thali is preferred for growing Dahlia, Cannas, Palms, Shrubs, Roses etc.

3. **Polythene bags:** Small polythene bags with holes punched in the bottom for drainage and filled with a porous rooting medium are used for propagation of cuttings like jasmine, duranta, crotons etc., in the mist chamber. Sometimes, young seedlings which are raised in the nursery are subsequently transplanted in these polythene bags and kept there till they attain required growth for transplanting them to the main field (e.g. papaya, curry leaf etc.).

4. **Plastic pots:** Plastic pots, round and square are used to keep mostly indoor plants. They are reusable, light in weight, non-porous and they require only little storage space.

5. **Fibre pots:** These are available in small size varying from 5–10 cm width and are either round or square in shape. They are bio-non degradable and last longer periods with the soil and plants inside.

6. **Paraffined paper or styrofoam cups:** They look like ice-cream cups with drainage holes. They serve satisfactorily as temporary containers for growing and transferring young plants on a large seed bed. They are light, cheap and require little space. Recently, thermocool moulded pots have been gaining popularity as they are light weight and attractive.

7. **Plastic trays or pro-trays**: Generally plastic trays or pro-trays having different sizes of cells are used for raising seedlings. The trays may have cavities of 3.75 cm (1.5") in size or 2.5 cm (1.0') in size. These

plastic trays may be fixed over thermocool base trays having the same number and size of cavities on floor or firm base. The medium for filling these trays have three ingredients *viz.*, coco-peat, vermiculite and perlite, mixed in 3:1:1 (V/V) ratio before filling.

Seeds are usually planted/sown at a shallow depth after pressing the media with finger in a gentle way into the potting plugs or cells which are filled with artificial media. The actual depth of sowing depends on the crop and the size of the seeds. After sowing of seeds a thick layer of vermiculite is given to cover the seeds for better germination as this media is having water holding capacity. After seed sowing the pro-trays are kept in the germination room at the optimum required temperature for early and better germination. Usually one to two seeds are sown per cell and after the seeds germinate the seedlings are thinned to one per cell and the thinned out can be gap filled in those cells where seed germination could not take place.

Advantages

1. Seedlings can be raised under adverse climatic conditions where it is not possible under open field conditions.
2. Healthy seedlings can be raised in short period as compared to the time taken under open field nursery raising.
3. There is no chance of soil borne fungus or virus infection to the seedlings as the nursery is grown in soil-less sterilized media and insects cannot enter under the protected conditions.
4. Drastic reduction in the mortality in transplanting of the seedlings as compared to the traditional system of nursery raising due to perfect development of root system.
5. Early planting is accomplished by raising such nursery.
6. It is suitable for raising the nursery of sexually and asexually propagated vegetable and ornamental crops.
7. Management of insect/pests and diseases under greenhouse/protected conditions is quite easy particularly the infection of viruses.
8. Reduced seed rate (30–40 per cent) in comparison with the traditional nursery raising system, as individual seeds are sown in each cell which produces a very healthy seedling.
9. Easy for transportation after packing for long distances.
10. Farmers can get the nursery ready from such nursery greenhouse any time as per the requirement.

Part 3

POMOLOGY

16

Pomology

India is endowed with various climate, that it is possible to grow any kinds of fruits namely temperate, sub-tropical and tropical including humid and semi-arid or arid zone fruits in one or other parts of the country. India is the second largest producer of fruits in the world with 11 per cent production and has wide range of varieties in its baskets. Till eighties, fruits except mango were beyond the reach of common man, which were only to compliment the diet of sick people, have become a diet of common man now. There has been a tremendous change in production. Current production is 62.2 million tonnes from 5.7 million ha against 28.74 million tonnes during 1990–91. However, to meet the growing needs much more is required to be done having technology at driving seat.

The area under fruits in different states if analysed, it is obvious that Uttar Pradesh (23.89%) followed by Assam (10.22%) are topping first in having more area under fruits. Improved varieties and cultural practices including plant protection measures are being developed or standardized region wise for each fruit either by the Research wing of the various Agricultural Universities in India or by Research Institutes of the Indian Council of Agricultural Research (I.C.A.R).

In this Chapter, important fruits are dealt with under the broad categories of (a) Tropical and sub-tropical fruits (b) Humid fruits (c) Arid/semi fruits and (d) Temperate fruits.

(A) TROPICAL AND SUB-TROPICAL FRUITS

Tropical fruits generally prefer high temperature and they cannot stand even slight frost and sub-tropical fruits can withstand low temperature or occasional frost during winter and also high temperature during summer months. A strict classification between these categories can not be made since most of the fruits coming under this category can be grown in both tropical and sub-tropical zones. Important fruits under these categories are dealt here.

1. Mango *(Mangifera indica)* Family: (Anacardiaceae)

Mango undoubtedly deserves to be the national fruit of India. It is a favorite fruit of all parts of India and has been repeatedly acclaimed as the 'King of fruits'. It is one of the most ancient fruits of India and mention has been made in many ancient old literatures about its cultivation. It is found in several countries also but in no other country does it enjoy the status as it does in India. India is the only country where truly wild forests of mango are found in Assam and its adjoining Chittagong Hills. This genus is reported to have originated from South-East Asia with a natural spread in the Indo-Malayan region. About 40 species are recognized in this genus *Mangifera* and some important ones are *M. altissima, M. sylvatica* and *M. zeylanica* which closely resemble to *M. indica.*

Mango is an outstanding source of vitamin A and a good source of vitamin C, apart from the usual content of minerals and other vitamins. Good mango varieties contain over 20 per cent of total soluble solids (sugars) in which the non-reducing sugars are more than the reducing sugars. The acid content of ripe fruit varies from 0.2 to 0.5 per cent and the protein content is about one per cent.

In India, it is grown in almost all states of India with a total estimated area of 1.02 million hectares with an annual production of 8.33 million tonnes. India's contribution to the global production is the maximum (nearly 65%), followed by Pakistan, Philippines and Indonesia. Among the states in India, Uttar Pradesh, Bihar and Andhra Pradesh are the leading states in area and production wise. In Tamil Nadu, it is grown in about 41,000 ha almost in all districts excepting Nilgris. However the districts like Dharmapuri, Salem, North Arcot and Kanyakumari have maximum area.

Soli and Climate

Mango is not very exacting in its soil requirements. It grows successfully almost in all states under different soil types. It is a large tree with deep tap-root system. The nature of the sub-soil is therefore important. The water-table should be below 180 cm. A well drained deep loamy soil is generally conducive for successful mango culture. A pH range of 5.5 to 7.5 has been found good. In alkaline soils, young plants are susceptible to injury and they exhibit symptoms of burning. Mango is a tropical as well as semi-tropical plant. It grows upto an altitude of 1200 m, but the fruiting is poor above 600 m. It thrives equally well from Kanyakumari in Southern India to sub-mountainous region in the north. Annual mean temperature at which mango thrives best is around 26°C. But at low temperature (below 1°C), the plants are severely affected by frost. The prevailing temperature also affects the sex expression in mango. Higher temperature tends to produce more percent of perfect flowers. In India, mango grows equally well in area receiving both low and heavy rainfall (75 cm annually). However, with annual rainfall of 75 cm and above, it can be grown with little or no irrigation.

Propagation

A number of vegetative propagation techniques such as grafting, budding, layering and cuttings have been successfully practiced. Inarching, veneer grafting and epicotyl grafting are commercially adopted in India. Yet, inarching is the only technique in vogue in commercial nurseries. For inarching, one year old seedlings are used as rootstocks. About 90 per cent success is obtained in this method. From the time of sowing of seeds, the grafts take one and a half to two years to be ready for planting in the field.

Recently, propagation by veneer grafting has been perfected. It can be done any time from spring to early monsoon. Success upto 80 per cent can easily be obtained. The grafts make rapid growth and are cheaper to be produced. The grafting can be done on rootstocks raised in pots or in the field (*in situ*). Besides epicotyl grafting has been standardized recently. Germinating seeds of about 8 to 15 days old are used as the rootstocks. The scions are prepared by prior defoliation of shoots of comparative thickness. Splice and wedge methods are used for grafting. The percentage of success in splice and wedge methods is 50.0 and 33.0 respectively.

There is no commercial practice of propagating mango either through cuttings or layering although success was indicated by various workers. Similarly propagation through budding is nowhere practiced as a commercial method of propagation.

Varieties

There are hundreds of varieties in mango, out of which only a few happen to be of commercial importance. Some of the important commercial mango varieties in India are described in table 16.1. With the object of evolving a regular bearing variety with desirable characters, breeding work was initiated in many research institutes and as a result more hybrid varieties are now available. They are described in table 16.2. The hybrid varieties No. 6–8 were developed at IARI, New Delhi, 9–12 were developed at IIHR, Bangalore and 13 and 14 were developed at Tamil Nadu Agricultural University, Periyakulam.

Polyembryonic Varieties

The phenomenon of polyembryony is known to occur in number of mango varieties. Seedlings arising from the adventitious embryos nucellar origin are highly uniform. These can therefore be used for vegetative multiplication of a polyembryonic variety. If found suitable, they can also be utilized as standard rootstocks for some of the monoembryonic varieties. For example, experiments conducted at Horticultural Research Institute, Periyakulam showed that Bappakkai (Polyembryonic rootstocks) were found to be better rootstocks for 'Neelum' in respect of growth, yield and quality of fruits. In India, almost all the commercial varieties are monoembryonic. Few, that are polyembryonic, are comparitively of little economic value and are mostly conifined to Kerala. Some of them are Bappakkai, Olour, Kurukkan, Chandrakaran, Goa and Bellary.

Planting and Aftercare

Under Tamil Nadu conditions, planting season varies from June to September. In North India, planting can be done in February–March if irrigation is available.

The planting is done in previously dug, exposed and filled pits of the size of 0.9 m × 0.9 m × 0.9 m. Generally, a distance of 7 to 10 m is satisfactory. Recently at the advent of many hybrid cultivars, a closer spacing of 2.5 × 2.5 to 5 × 5 m is recommended.

Variety	Spacing	No. of plants/ha
Arka Aruna, Sindhu	5 × 5 m	400
Amrapali	2.5 × 2.5 m	1000

Training and Pruning

Rootstock sprouts and low-lying branches have to be removed in the initial years of establishment. Besides, training is done to provide frame work for the future so that branches are spaced properly and do not break with the crop load at the bearing stage. Thereafter, overlapping, intercrossed, diseased, dried and weak branches are to be removed after their harvest season is over. In old and irregular bearing trees, Tamil Nadu Agricultural University recommends a kind of severe pruning to make it a regular bearer. In this method, few internal branches are removed only once to have open centre.

Table 16.1 Important mango varieties.

S. No	Variety	Brief description
1.	Neelum Syn: Kasaladdu	Appreciably a regular bearer, quality fruits, precocious in bearing, each fruit weighs 200-250 g, taste is good with slightly acidic.
2.	Bangalora Syn: Kallamai or Kilimooku or Totapuri	Regular bearer, very good yield, early yielder in South India, large fruit weighing 400-450 g, shape is typical oblong, with a prominent beak, pulp is excellent for processing.
3.	Rumani	High yield, good keeping quality, fruit quality is only moderate, it remains sour if harvested slightly early.
4.	Alphonso Syn: Gundu, Khader, Badami	Fruits with excellent flavour, shape, size and colour, good keeping quality, each fruit weighs 250–300 g, very good variety for canning, flavour stable, suitable for export, limited adaptability, biennial in habit, flesh develops spongy tissue.
5.	Dashehari	Most popular variety in North India, excellent fruit quality, good keeping quality, each fruit weighs 200-250 g, biennial habit, external appearance not very attractive. Susceptible to mango malformation.
6.	Mulgoa	A late season, shy bearing variety, fruits are large, each fruit weighs 450-500 g, taste is excellent with good keeping quality.
7.	Langra	Most popular variety in North India, wide adaptability, fruits are medium to large, each weighing 250–300 g, excellent in quality, biennial in habit, poor keeping quality.
8.	Paiyur-1	A clonal selection from Neelum, developed at Regional Research Station, Paiyur (T. Nadu). Trees dwarf, suited for high density planting, fruits are medium sized, each fruit weighing 150–200 g.
9.	Niranjan	A clonal selection developed at Marathwada Agricultural University, Parbhani. An off-season bearing variety, it flowers during June–July and matures in October, yields 600 fruits/tree, each fruit weighs 165 g.

Table 16.2 Important Mango Hybrid Varieties.

S. No.	Name	Parentage	Brief description
1.	Neeludin	Neelum × Himayuddin	Regular bearer, yields about 500 – 700 fruits/tree, fruits are medium sized (200 g) and intermediate in shape between pareznts, skin is smooth, flesh firm, juicy, fibreless, very sweet comes to harvest during end of May.
2.	Neelgoa	Neelum × Yerra Mulgoa	Regular bearer, yields about 1000 fruits/tree, fruit weighs 270 g, smooth skinned, flesh firm, melting, fibreless, moderately juicy, very sweet with a delightful flavour, matures in May.
3.	Neeleshan	Neelum × Baneshan	Each tree bears about 1200 fruits, each weighing about 300 g, with thick smooth skin, flesh firm, very sweet, fibreless, moderately juicy with a pleasant flavour, suitable for caning, comes to harvest in May.
4.	Swarna Jehangir	Chinna Swarnarekha × Jehangir	Fruits resemble Jehangir in shape and skin characters but slightly less in size weiging 230 g, moderetaly firm fleshed, fibreless, abundantly juicy, very sweet with pleasant flavour, come to harvest in June and stands transport well.
5.	Aurumani	Rumani × Mulgoa	Fruits resemble female parcnt in shape, beak and other characters of the skin but size is as big as male parent (300 – 500 g), excellent quality-flesh melting, absolutely fibreless, very sweet with excellent flavour abundantly juicy, comes to harvest in June and it stands transport well.
6.	Amrapali	Dashehari × Neelum	Hybrid developed at IARI, New Delhi, dwarf, regular bearer with good fruit quality.
7.	Mallika	Neelum × Dashehari	Regular bearer, mid season variety, fruits are medium sized, cadmium coloured with good quality including good shelf life.
8.	Manjira	Rumani × Neelum	Semi-vigorous, regular bearing variety, fruits medium sized, skin colour is light yellow, flesh firm, sweet and fibreless.

(Contd.)

Table 16.2 Important Mango Hybrid Varieties.

S. No.	Name	Parentage	Brief description
9.	Arka Neelkiran	Alphonso × Neelum	Regular bearer, fruits are medium in size (260 g), develops good red blush upon ripening, free from fibre and spongy tissues, suitable for export.
10.	Arka Aruna	Banganapalli × Alphonso	A distinctly dwarf tree, precocious and regular but medium in bearing, matures in June (Late Variety). Fruits are large (500 – 700 g), skin smooth, firm flesh, fibreless with a TSS of 20. Brix. Pulp percentage is 80% free from spongly tissue, suitable for processing.
11.	Arka Puneet	Alphonso × Banganapalli	Regular bearer, heavy yielder, mid season (May – June), fruits-medium (230 – 250 g), skin medium thick, flesh firm, fibreless, TSS 210 Brix, resembles Alphonso in shape, colour and quality and can be used as alternate for export and canning industry.
12.	Arka Anmol	Alphonso × Janardhan Prasad	Regular bearer, medium yielder, mid season (May – June), Fruits medium in size (270 – 300 g), skin medium thick, flesh firm, fibreless, good flavour, high pulp to seed ratio (76.0%), free from spongy tissues, suitable for export.
13.	PKM. 1	Chinna Swarnarekha × Neelum	Regular and cluster bearer, each fruit weighs 250 – 300 g, fruits are long, narrow with prominent beak.
14.	PKM. 2	Neelum × Mulgoa	Fruits are large, each weighs 650 – 700 g, taste very good, with good keeping quality.
15.	Ratna	Ralnagiri Alphonso × Neelum	Regular bearer, fruits are with excellent taste, flavour and attractive shape, colour and size without development of spongy tissues, good keeping quality and early maturing type.
16.	Sindhu	Ratna × Alphonso	Produces quality mangoes abundantly, the fruits are free from spongy tissues, medium sized (215 g) with appealing colour and better keeping quality. The fruits are seedless since it has a very thin (0.30 cm) non viable stone weighing 6.76 g compared to Ratna (37 g) and Alphonso (25 g). The pulp recovery is 83.13%.

and the terminal whorl of shoots are thinned out annually during August, September so as to retain only one or two healthy shoots (Fig 16.1). This has resulted in greater yield than in previous years in Cv. Mulgoa. Recently, another method of pruning is recommended. It involves heading back of flowering shoots after harvest below a node. This helps to produce 3–5 new shoots below the pruned node which will bear the fruit buds during the next season by Nov–December. This pruning should be followed by immediate irrigation and thereafter also at regular intervals to get desired effect.

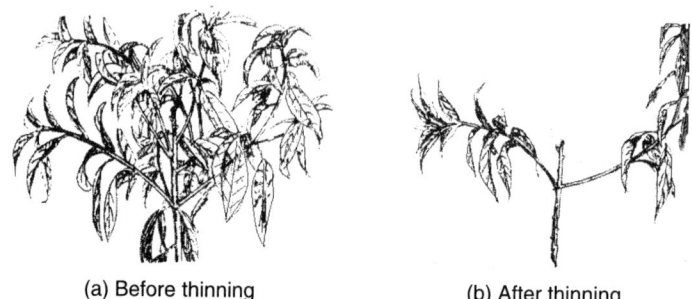

(a) Before thinning   (b) After thinning

Fig 16.1   Cluster thinning of shoots in mango.

Manures and Fertilizers

The following quantities of manures and fertilizers are applied

Manures and Fertilizers (Kg/tree)	1st Year	Annual increase	6th year onwards
FYM	10.00	10.00	60.00
N	0.20	0.20	1.00
P	0.20	0.20	1.00
K	0.25	0.20	1.50

For the bearing trees, half the quantity of NPK is applied after harvest and the remaining half of NPK besides full quantity of FYM are applied during September–October. Application of manures and fertlisers is usually done in a ring around the tree. Such rings are dug around the base of the tree 2.0 meters away from the main trunk with a depth upto 15 to 20 cm.

Intercropping

The intercrops chosen for mango should be separately irrigated or fertilized so as not to compete or clash with those of mango. Vegetable crops like onion, tomato, radish, beans and fruit crops like papaya, guava can be grown in the initial years and with the increasing age of mango plants, area under intercrops should be progressively decreased to minimize competition.

Irrigation

Mango needs 125 to 170 cm of precipitation annually; but on account of deep tap root system, it requires little irrigation after the first few years. When the planting is done in spring,the young plants require irrigation fourth or fifth day during the first summer till the rains break. During rains, irrigation should be given only if the soil dries up. The bearing plants are often given no irrigation but irrigation during flowering reduces fruit drop and gives larger fruits. Irrigation if withheld during the preflowering phase increases flowering, otherwise, irrigation during this period is likely to promote vegetative growth. Trees receiving heading back pruning needs to be provided 2–3 irrigation at that time to induce new shoot formation.

Cropping

Fruits buds are borne on past season shoots terminally in mango and is a pure bud. The inflorescence is a large terminal panicle which carries both male and bisexual flowers. Each panicle carries 300 to 4000 flowers in which the percentage of perfect flowers vary from 2.0 to 65% depending upon cultivars. Nectar is produced to attract insects. Fruit is a fleshy drupe. Fruit bud differentiation occurs between October to December and flowering takes place as early as November–December in A.P, December–January in Tamil Nadu, February–March in Northern India. The duration of flowering is very short, usually of 2 to 3 weeks. The fruitset varies with cultivars and generally in a shy bearing cultivar, 0.1% is considered as optimum. Many chemicals are recently recommended to induce flowering and fruit set. Potassium nitrate (KNO_3) at 1% concenetration sprayed at the time of flowering/fruit set stage improved the earlier flowering and fruiting.

Due to the various causes, fruitdrop occurs in mango rather at a higher rate even upto about 99 percent in various stages of growth, more during the initial four weeks. The extent of fruits drop can be reduced significantly by (a) Regular irrigation during the fruit development period, (b) Timely and effective control measures against major pests and diseases, and (c) Through the application of growth regulators like NAA (50 ppm) and 2,4-D (20 ppm) during off years about six weeks after fruitset.

Grafted mango trees start bearing from the age of five years and commercial crops can be obtained in about seven years. Seedlings take nearly 10 years. A year of good crop may follow one or more years of poor or no crop. The year in which good crop is obtained is referred to 'on year' and the year in which poor or no crop is obtained is referred to 'off year'. This phenomenon is called 'irregular bearing' or 'biennial bearing'. Various factors are reported to be responsible for this phenomenon. They are:

1. **Climatological factors:** High humidity, heavy wind and low temperature during the 'on' year convert into an 'off' year directly or indirectly by promoting the incidence of diseases like powdery mildew or anthracnose.
2. **Age and size of shoots:** It is generally found that the flowering shoots should have attained certain amount of physiological maturity atleast 8–10 months old. But this hypothesis does not hold true in many cases.
3. **Carbon/Nitrogen ratio:** It is generally indicated that higher starch reserves, total carbohydrates and C/N ratio favour flower bud formation and anything to alter this or to favour higher N content may tend favoring vegetative growth.
4. **Hormonal balance:** Physiological studies on flowering in mango established that higher level of auxin-like substances and an inhibitor (ABA) and lower level of gibberellin like substances are vital for a floriferous shoots in mango. This situation is often altered by the development process as there is an inhibiting-influence of developing fruit on vegetative growth.

Control of Biennial Bearing

1. Proper care and maintenance of trees in a healthy condition may help to reduce erratic or irregular bearing but it may not induce regular bearing in biennial cultivars.
2. Deblossoming during the 'ON' year so as to obtain some crop every year has been recommended but the response to this practice depends on the cultivar, stage of the panicle etc.,
3. Use of growth retardants: spraying of growth regulators like ethrel during 'OFF' years to induce flowering is recommended which results in some success in few cases not in all the cases. Similarly, application of Paclobutrazol, a anti-gibberellin synthesis, when drenched around the trees @3–5ml/tree during August/September induces the trees to flower within 100–120 days of application. Drenching process involves dissolving the growth retardant in 6 litres of water and pouring it around the trees in drip circle in previously made 5–6 holes using the crowbars. This operation should follow regular irrigation. Use of paclobutrazol is recommended wherever the trees are in healthy stage with regular manuring and irrigation.
4. Growing of regular bearing cultivars like Neelum, Bangalora, Ratna, Sindhu etc.

Harvesting

The mango fruit ripens in about 90 days after flowering. All fruits on a tree is generally harvested when a few fruits ripen naturally and fall from the trees. Another criterion for harvesting is a slight change of colour in certain varieties. The fruit is harvested with a harvesting pole.

The harvested fruit takes about five days under tropical condition for ripening and becomes over-ripe after seven or eight days. Usually, merchants ripen the fruits in closed but well ventilated store-houses. The fruits are placed in a single layer over paddy or wheat straw spread 6 to 8 cm thick on the floor of the house. The fruits can be ripened quickly in about two days if a small piece of calcium carbide is placed in a covered heap of mangoes. The acetylene gas emitted by the calcium carbide gives the fruit good colour but the rating quality remains poor. This technique of ripening is now banned on health ground, instead, artificial ripening using ethrel is recommended. Ethrel at 2500 ppm kept in a beaker in the ripening room with a pinch of Sodium or Potassium Hydoxide pellets evolve the ripening gas, ethylene, which induces ripening in 24 hours with good colour and quality. Fruits ripened in the trees are unsuitable for storage due to their high susceptibility to low temperature injury. Most varieties are best stored under cold storage at 5.6 to 7.2°C with 85 to 90 per cent relative humidity. The cold storage fruits once brought outside at 15.5 to 21°C will ripen in about six days. The yield of mango varies greatly. The average yield in our country is about 7.5 to 10 tonnes per hectare. The number of fruits per tree during its steady bearing age (i.e., after 20 years) generally varies from 1000 to 2000 fruits.

Physiological disorders

1. **Spongy tissue:** This disorder is common in the fruits of Alphonso wherein a non-edible, sour, yellowish sponge-like patch with or without air pockets develop in the mesocarp of the affected fruits during ripening. Externally, the affected fruits present a healthy appearance. The incidence is reported to be more if fruits are harvested at full maturity than at 3/4th maturity. Higher temperature coupled with radiation is believed to induce more of this problem. No other exact cause for the malady is known as on today. Mulching using organic materials along with irrigation can reduce this incidence.

2. **Mango malformation:** This is a serious one in North India than in the South. It is of two types *viz.*, vegetative and floral. The former is more common in the nursery seedlings and young plants, wherein malformed vegetative shoots are characterized by the appearance of small leaves in whorls on thick; stunted shoots which may crown the

apex, giving a bunchy-top like appearance. In the case of floral malformation, the floral parts are transformed into compact masses of sterile flowers appearing in the form of a bunch, causing total loss to the crop. Various causes such as cultural practices, nutritional practices, mites, virus, fungi etc., have been attributed, but the precise causal organism/substance is yet to be established. Malformed trees can be improved by a single spray of NAA (200 ppm) in the first week of October and deblossoming and at the budburst stage.

3. **Black tip:** It refers to the development of a small etiolated area at the distal end of the fruit which gradually spreads and turns black and covers the tips completely. Such black tipped fruits fetch very low market price. This disorder is prevalent in Punjab, U.P., Bihar and West Bengal if orchards are located near the brick kilns (within a distance of 1–1.5 km). Gases like carbon-di-oxide, sulphur-di-oxide and acetylene which constitute the brick kiln fumes are reported to cause the above symptoms. Spraying borax (0.6%) and caustic soda (0.8%) thrice, *i.e.*, before flowering, during flowering and at the fruitset stage are reported to prevent the disorder as these chemicals are basic in nature, they neutralize the acidic fumes from the brick kilns. (Plant protection: Table 16.3).

2. Banana (*Musa sp.*) Family: Musaceae

Banana is one of the oldest and most popular fruit. The Indo-Malayan region is believed to be the place of origin. It is widely used as a fresh fruit. The central core of the pseudostem is used as a vegetable. The banana pseudostem is also used for manufacturing paper and boards.

India ranks first in terms of area and production, growing in about 4,90,700 ha with an annual production of 168,13,500 mt, sharing about 17% of global production. Among the various states in India, Tamil Nadu, Karnataka and Maharashtra account for major share in area and production.

In Tamil Nadu banana is grown in almost all the districts with more area under the districts like Trichy, Tutucorin, Coimbatore and Kanyakumari. Banana is a tall herb producing pseudostem which is the aerial stem made up of a number of leaf sheaths completely encircling the axis of the stem. Inflorescence is a terminal complex spike, each spike being subtended with big bracts. The flowers in the basal side of the peduncle behave as pistillate one while in the terminal ones as staminate one and those in the intermediate as transitional stage but are functionally male. The axis beyond the female phase is generally bare, but in some cultivars, flowers and bracts are retained. Fruit is a berry and in edible bananas they develop by vegetative parthenocarpically.

Table 16.3 Important pests and diseases of mango.

Pests/diseases	Symptoms/damages	Control measure
1. Mango hopper (*Amritodus atkinsoni, Idioscopus clypealis, I. niveosparsus*)	Adults and nymphs suck the sap from tendershoots and panicles during the flowering season, panicles wither away and fruitset is adversely affected.	Spray two rounds of acephate 75 SP @ 1g/lit or phosalone 35 EC @ 1.5 ml/lit or carbaryl 50 WP @ 2 g/lit or phosphamidon 40 SL 2 ml/lit of water, first at the time of panicle emergence and the second two weeks after first spray. Wettable sulphur @ 2 g/lit may be sprayed after spraying carbaryl to avoid mite resurgence. Neem oil 5 ml/lit of water can be mixed with any insecticides for the control of hopper and shoot webber.
2. Nut weevil (*Sternochetus mangiferae*)	The grubs travel through the pulp and enter the seeds where these pupate and the adults come out piercing through the stone and the pulp.	General cleanliness in the orchard, destroying the adults in the bark crevices and holes and spraying with Fenthion 100 EC 1 ml/lit spray during marble stage and second spray 15 days after the first spray.
3. Stem borer (*Batocera rufomaculata*)	Tunnels through the main trunk or its branches, weakens the plant and in extreme cases the plant may die. Its presence can be identified by dry hard balls of excreta emerging from the tunneled portion.	1. Padding with monocrotophos 36 WSC 10 ml in 2.5 cm patch per tree by soaking in absorbant cotton when the trees are not in bearings stage. 2. Application of carbofuron 3G @ 5 g per bore hole and plugging with mud
4. Shoot borer (*Chlumetia traversa*)	Caterpillars enter the young shoots from the terminal end and bore down, causing the shoots to wilt and dry up.	Spray carbaryl 50% @ 2 g/litre of water.

(Contd.)

Table 16.3 Important pests and diseases of mango.

Pests/diseases	Symptoms/damages	Control measure
5. Fruitfly (Dacus spp.)	Fruitflies lay eggs in clusters of 150-200 under the skin of the fruit just before ripening and the affected fruits begin to rot and drop down.	1. Plough the inter spaces to expose pupae. 2. Monitor with methyl eugenol traps. 3. Prepare bait with methyl eugenol 1% solution mixed with malathion 0.1%. Take 10 ml of this mixture per trap and keep them in 25 different places in one hectare between 6 a.m. and 8 a.m. 4. Remove the fallen fruits now and then and bury them deep into soil.
6. Powdery mildew (Oidium mangiferae)	Greyish white powdery bloom on the flower buds and fruitlets, causing the panicles dried and black, resulting in total failure of the crop.	Apply sulphur dust (350 mesh) in the early morning to protect new flush or spray wettable sulphur 0.2% or Tridemorph 0.05%.
7. Anthracnose and stalkend rot (Colletotrichum gloeosporioides)	Black necrotic areas appear on the affected parts and the affected young shoots show 'die back' symptoms or diseased young fruits drop or the matured affected fruits show more symptoms in storage.	Spray Mancozeb 0.2% or Carbendazim 0.1% as preharvest spray, 3 times at 15 days interval.
8. Sooty mould (Capnodium sp)	Common in areas where hopper incidence is more; the fungus develops on the honey dew secreted by the hoppers in the leaves twigs and inflorescence.	Spray Dimecron 0.03%+ Maida 5% (1 kg maida or starch boiled with five litre of water and diluted to 20 litre).

Climate and Soil

Banana is essentially a humid tropical plant, coming up well in regions with a temperature range of 10°C to 40°C with an average of 23°C. In cooler climate, the duration is extended, sucker production is affected and bunches are smaller. Low temperatures (less than 10°C) are unsuitable since they lead to a condition called 'choke' or impeded inflorescence and bunch development. Banana comes up well from sea level upto an altitude of 1800 m above sea level. It enjoys an annual rainfall ranging from 100 to 325 cm. When wind velocities are more than 80 km they cause havoc to plantations. Young plants break and older ones, especially those bearing bunches are uprooted. In such areas, the dwarf types such as Dwarf Cavendish are preferred. It can be grown in a variety of soils, i.e., sandy, alluvial, lateritic etc. in India. However, deep well-drained soil with abundant organic matter is highly suitable for its cultivation. A soil pH of 5.5 to 8.5 is found to be optimum.

Varieties

The present day edible cultivars of banana are believed to have been derived from two wild species *viz., Musa acuminata* and *M. balbisiana*. These cultivars have a basic chromosome of 11 and they are of either diploids (2n = 22), triploids (3n = 33) and tetraploids (4n = 44). Each cultivar is hence nomenclatured based on genomic contribution. For example, the popular cultivar Poovan has to be referred as *Musa* sp. Cv. Poovan (AAB) which means that it is a triploid, having derived from two genomes AA (2n) from *M. acuminata* and one genome B (n) from *M. balbisiana*.

Simmonds and Shepherd, the world renowed scientists in banana derived a method of indicating the relative contributions of these two species to the development of any cultivar through scoring technique of fifteen distinguishing morphological characters and applying the derived information for distinguishing the *M. acuminata* types from those of *M. balbsiana* (Table 16.4) (Fig. 16.2). If a character of cultivar agrees with *M. acuminata*, the score of 1 is given and if it agrees with that of *M. balbisiana* a score of 5 is given and the intermediate expression of the characters are assigned with a score of 2, 3 or 4 according to their intensity. According to this scoring technique, the score may range from 15 for *M. acuminata* and 75 for *M. balbisiana*. Hence, depending upon their contribution of these parents to the contribution of the progeny and combining their chromosomal status; the naturally occurring edible banana can be classified as in Table 16.5.

To apply the above scoring technique, a knowledge on judging its ploidy level based on certain morphological characters is essential. Usually it can be done based on the variation in the orientation of leaf characters between diploids, triploids and tetraploids (Fig. 16.3).

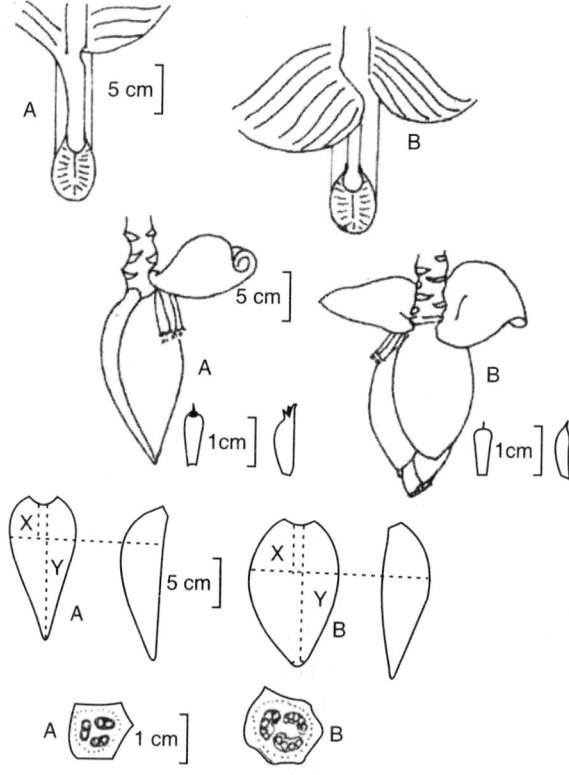

M. acuminata (A) on left, M. balbisiana (B) on right. From top to bottom; petioles, male buds (to show shape, bract curling and bract insertion), free tepals of male flowers, male bracts (ratio = x/y) and ovule arrangement (note also the different ovule sizes). Redrawn (Some-what modified and expanded) from Simmonds and Shepherd, 1955.

Fig 16.2 Distinguishing morphological traits between *M. acuminata* and *M. balbisiana*.

There are more than 100 varieties available in different banana growing areas of the world. Commercially grown varieties in India are alone described in Table 16.6.

TNAU, Coimbatore has developed a multiple hybrid variety *viz.*, CO.1 banana which is akin to hill banana in respect of flavour and taste.

Ladan (AAB) × *M. balbisiana* (BB)

↓

F_1 (AB) × Kadali (AA)

↓

CO.1 (AAB)

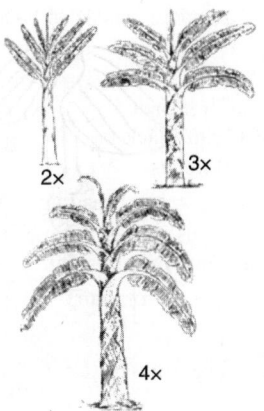

Fig 16.3 Distinguishing morphological characters among diploid, triploid and tetraploid banana.

The bunch weighs about 10.0 kg, having 7 hands with a total number of 80–85 fruits, crop duration is 14–15 months. Similarly, KAU, Trichur has released two hybrids *viz.*, BRS–1 (Agniswar x Pisang lilin) and BRS–2 (Vannan × Pisang lilin). BRS–1(AAB) is 100 days earlier than Rasthali with significant differences in bunch weight. It has been released for homestead cultivation of Kerala as it is resistant to Sigatoka leaf spot. BSR–2 (AAB) is a medium statured hybrid, tolerant to leaf sport and Panama disease, rhizome weevil and nematodes. The average bunch weight is 14 kg with 8 hands and 118 fruits in crop duration of 314 days. FIHA– Hybrids 1,3,17 and 20 are tetraplods, developed by Fundacion Hondurena de Investigacion Agricola, Honduras possess good horticultural traits and quality fruits besides possessing resistance to fusarium wilt and tolerance to nematodes. These hybrids however are not preferred by Indian farmers.

Propagation

Banana is vegetatively propagated by means of suckers, whole or bits of the rhizomes of the parent plant or daughter suckers with atleast one sprouting bud. There are two types of suckers *viz.*, sword sucker and water sucker. Sword sucker has a well developed base and pointed tip with narrow shaped half blades at the newly formed stage. The sword sucker weighing 1.5 to 2.0 kg should be selected for planting. Such sword suckers at the time of flowering of the parent plant is the best planting material. Water sucker can be identified by the presence of broad leaves on small, undersized corms. These are not recommended for planting as it has a tendency of slow growth and poor yield. Healthy clumps do not produce water suckers but they normally

Table 16.4 Taxonomic scoring of banana cultivars.

Characters	M. acuminata	M. balbisiana
1. Pseudostem colour	More or less heavily marked with black or brown blotches.	Blotches slight or absent
2. Petiolar canal	Margin erect or spreading with scarious wings below, not clasping pseudostem.	Margins enclosed not winged below, clasping pseudostem
3. Peduncle	Usually downy or hairy	Glabrous
4. Pedicel	Short	Long
5. Ovules	Two regular rows in each locule	Four irregular rows in each locule
6. Bract shoulder ratio	Usually high (ratio < 0.28)	Usually low (ratio > 0.30)
7. Bract curling	Bracts roll after opening	Bracts lift but do not roll
8. Bract shape	Lanceolate or narrowly ovate, or tapering sharply from the shoulder	Broadly ovate not tapering sharply
9. Bract apex	Acute	Obtuse
10. Bract colour	Red dull purple or yellow outside; pink dull purple or yellow inside	Distinctive brownish purple outside, bright crimson inside
11. Colour fading	Inside bract colour fades to yellow towards base	Inside bract colour continues to base
12. Bract scars	Prominent	Scarcely prominent
13. Free tepal of male flower	Variably corrugated below	Rarely corrugated
14. Male flower colour	Creamy white	Variably flushed with pink
15. Stigma colour	Orange or rich yellow	Cream, pale yellow or pale pink

develop in old, poorly maintained shady plantations. The suckers before planting have to be trimmed so as to remove the roots and decayed portion of the corm. The top of pseudostem should be cut leaving 20 cm from the corm.

Recently, tissue culture derived planting materials are extensivey used as they are found to be vigourous, early and high yielders besides possessing uniformity. Suckers derived from tissue culture planted clumps can be used to raise first and second ratoon crops, thereafter it is not advisable for ratooning as the yield will be less.

Table 16.6 Important commercial cultivars of banana.

Name of the cultivar with synonyms	Special attributes
1. Poovan (AAB) Chempa (W. Bengal) Lalvelchi (Maharashtra) Palayangodan (Kerala)	Foremost commerical variety in Tamil Nadu, A.P. and West Bengal states 11 months duration, average bunch weight 15 kg. Fruits medium size, held firmly in the bunch and have a distinct mammillary tip, fruit rind thin, pulp cream coloured. Sub-acid taste, ripe rind golden yellow and keeps well, resistant to Panama wilt.
2. Monthan (ABB) Bontha (A.P.) Kanch Kela (W.S.) Ponthan (Kerala)	A leading commercial culinary banana of India, plant is hardy, drought resistant, duration is 12-14 months, each bunch weighs 10 kg, fruits are long with good girth. In the middle, plump, angulate, slightly curved with blunt or knobbed apex, thick green rind, in some parts of Tamil Nadu ripe fruits are eaten, but mostly for culinary purpose.
3. Dwarf Cavendish (AAA), Basrai (Maharashtra) Kadali (W. Bengal). Pachavazhai, Mauritius, Kuzhi vazhal (Tamil Nadu)	Dwarf type, cavendish group, widely cultivated throughout the banana growing tract of the world, duration is 10 months, each bunch weighs 15 kg, fruits large, curved, thick, greenish, flesh soft and sweet, retains green colour to some extent even after ripening, keeping quality poor, susceptible to bunchy top but resistant to Panama wilt.
4. Harichal (AAA), Robusta (Tamil Nadu) Bombay Green (Maharashtra)	A semi-tall sport of Dwarf Cavendish, duration is 12 months, bunch weighs 25-30 kg, fruits large, skin thick, greenish to dull yellow, sweet and delicious better keeping quality.
5. Rasthali (AAB) Martaman (W.B) Mutheli (Maharashtra) Amruthnapanl (A.P.)	Choice table variety, duration 15-16 months, each bunch weighs 12 kg, fruits are medium sized similar to Poovan in appearance, skin thin, ivory yellow in colour, pulp firm sweet with a pleasant flavour. Ayiranka Rasthali is a sport of this variety which has only pistillate flowers upto the length of the axis.

(Contd.)

Table 16.6 Important commercial cultivars of banana.

Name of the cultivar with synonyms	Special attributes
6. Hill Banana (AAB), Virupakshli, Vellavazhai Sirumalai	This variety is a speciality of Tamil Nadu, grown in lower Pulney hills from 1.000-1500 m as perennial shade crop for Coffee. The duration is 12 months, each bunch weighing 12 kg with 80 – 90 fruits per bunch. Two main types viz., Sirumalai and Virupakshi are known to exist. The fruits of former one are having a thick rind which comes off clean while in the latter threads of inner cushiony substance sticks to the pulp. Besides, fruits of Sirumalai are tastier with more mellowy pulp with fine flavour than those of Virupakshi which are slightly subacid.
7. Nendran (AAB) Ethankai (Kerala) Rajali (Maharashtra) Plantain (Trinidad)	Fruits are always cooked to make it more palatable and hence known as Plantain, commercially grown in Kerala and parts of Tamil Nadu. 11-12 months in duration bearing 12-15 Kg of bunch, each bunch has 6 hands. 8-15 fingers per hand, fruits are relatively longer and thicker than most banana fruits, rind is thick and buff yellow when ripe flesh is firm yellowish with a characteristic good keeping quality (15 days even after ripening).
8. Sevazhai (AAA) (T.N), Red Banana (Trinidad)	This cultivar is grown largely in Kerala, lower Punlneys etc. The characteristic feauture of this variety is that pseudosterm, midrib and fruit rind is purplish red, hence the name red banana.Fruit is of good size, slightly curved with a blunt apex, thick red rind and on ripening develop a characteristic strong flavour.
9. Kunnan (AB)	A popular dessart variety of Kerala, fruits have thin rind but with firm flesh.
10. Ney Poovan (AB), Rasa Kadali (Kerala)	A popular cultivar in South India, duration 11-12 months, bunch weighs 10-15 kg, fruits are firm and sweet in taste.

Table 16.5 Ploidy and Classification.

Ploidy	Score	Contribution	Example of cultivars
2 ×	16–23	A A	Matti, Anai Komban, Sanne Chenkadali, Kadali, Namarai, Tongat, Pisang Lilin
3 ×	15–21	A A A	Robusta, Dwarf Cavendish, Thenkadali, Red Banana, Gros michel.
4 ×	15–20	A A A A	Bodles Altaford (Not existing in nature but is a synthetic hybrid)
2 ×	46–48	A B	NeyPoovan, Kunnan
3 ×	26–46	A A B	Poovan, Rasthali, Pachanadan, Virupakshi, Nendran
3 ×	59–69	A B B	Monthan, Peyan, Karpuravalli
4 ×	63–69	A B B B	Klue Teparod

Preparatory Cultivation and Planting

The preparatory cultivation for banana varies with the system of cultivation. The common system of cultivation for banana are (1) Garden land (2) Wet land (3) Perennial (Padugai) and (4) Hill banana cultivation. The preparatory cultivation practices followed and the spacing recommended for various systems of cultivation are furnished in Table 16.7. Recently, high density planting (HDP) involving planting of 2–3 suckers per pit has been found to be beneficial in getting higher yield per unit area. The drawbacks associated with this system are extra duration (3–4 months) and slightly reduction in individual bunch size and quality.

Pits of 45 cm are dug in garden land, padugai and hill systems of cultivation and the pits are refilled with top soil, mixed with 10 kg of FYM, 50 kg of BHC 10% to each pit. In the case of HDP, pit size of 0.6 m × 0.6 m × 0.45 m is preferred.

Irrigation

Banana is normally grown as a rainfed crop in west coast and hills of south India. In garden land, regular irrigation is a must for banana. Normally it is irrigated immediately after planting, life irrigation is given after 4 days and subsequent irrigation at 10–15 days interval. In the case of banana fields, water is allowed in channels which seep through the blocks of banana by capillary action. Under wet land condition, irrigation channels are formed for every two rows of banana and cross channels of the same depth with 4–6 plants in each bed. Drip irrigation is found to be highly beneficial in improving the yield. Banana requires 10–25 litres per plant through drip depending upon the stage of the crop.

Intercropping

At the early stages of growth, intercrops like leguminous vegetables, beetroot and radish can be grown as intercrops. Cucurbitaceous vegetables are not recommended the incidence of CMV affecting the crops. Banana itself is grown as intercrops and mixed crops along with plantation crops such as coffee, coconut and arecanut plantations.

Manures and Fertilizers

Banana requires higher fertilization due to its rapid and vigorous growth and high fruit yield. The nutrient uptake studies also reveal that the uptake per unit area is more than any other crop. The nutrient uptake pattern analysis conducted in different countries showed that a crop of 40–60 t of yield/ha removes nearly 250–300 kg N, 25–40 kg P, 800–1200 kg K, 150–180 kg Ca, 40–60 kg Mg and 14–20 kg S per hectare. This reveals that the fertilizer applied should contain more of Nitrogen and potash in the ratio of 1:3. The current fertilizer recommendation for garden land, wet land banana are given below:

Particulars	N	P	K
	(g/plant/year)		
Garden land			
Nendran	150	90	300
Other varieties	110	35	330
Wet land			
Nendran	210	35	450
Rasthali	210	50	390
Poovan, Robusta	160	50	390
Hill Banana	375 g of 40:30:40 NPK/Plant besides 130 g of Muriate of Potash per clump per application during April, July and November.		

*For tissue culture banana apply 50% extra fertilizers at 2nd, 4th, 6th and 8th month after planting.

Time of application is also very important in banana. Generally, it is better that nitrogen is applied in 3–5 split doses at 30–40 days interval and K in two split doses, one half at the time of planting and the other at the time of initiation of flowering. The latest recommendation is that the doses are divided into three splits and applied at 3rd, 5th, 7th month stages. N should be applied as neem coated urea to minimise the waste of N. Similarly *Azospirillum* @ 3 g/plant on 2nd and 4th month is also recommended now for all the varieties.

Recently, fertigation has been found to be highly beneficial in improving the yield and quality of banana. Fertigation at weekly intervals starting from 2nd month after planting with 75% of the recommended fertiliser doses as per the following schedule is found to be good. Water soluble fertilizers if used the yield enhancement is tremendous as against conventional fertilisers.

Weeks after planting	N (%)	$P_2 O_5$ (%)	$K_2 O$ (%)
9–18 (10 weeks)	30	100	20
19–30 (12 weeks)	50	–	40
31–42 (12 weeks)	20	–	32
43–45 (3 weeks)	–	–	8
Total	100	100	100

Sulphate of potash (SOP) when given as foliar spray @ 1.5–2.0% immediately after shooting and followed by one more spray a month after improves the bunch weight, bunch grade and quality of banana fruits.

After Cultivation

Desuckering: It refers to the removal of unwanted suckers. They are periodically removed otherwise they compete with the mother plants for nutrients, resulting in lower bunch weight *vis à vis* yield. Desuckering is accomplished by cutting the suckers at ground level or pushing a chisel shaped crowbar between the parent and the sucker and giving a twisting motion. Under garden land and wetland conditions, desuckering is done at monthly intervals. In hill banana, two bearing plants and two followers per clump are considered optimum and the rest are desuckered. Normally desuckering is done at 3 times in a year.

Weeding: Depending upon the type of cultivation, 2–3 diggings are given to keep the field free of weeds. Chemical weed killers are also recommended. Paraquat @ 3 ml per litre is recommended as a post emergence spray to kill broad leaved weeds.

Propping: The pseudostem requires the support from propping bamboo or casuarina poles at the time of bunch emergence. The pole is fixed and tied around on, one side of the plant or two poles are tied at the tip in such a way, they form 'x' shaped and are kept against the pseudostem to bear the weight of the bunches.

Other Special Operations

(a) **Removal of withered styles and perianth:** In some varieties the styles are persistent and hence they remain in the bunch till maturity. They can be easily removed by a light brushing movement of the hand a few days after flowering and if it is delayed, it is difficult to remove

Table 16.7 Systems of cultivation in banana.

System of cultivation	Preparatory cultivation	Variety	Season of planting	Spacing
Garden land	2-3 ploughings, then taking pits and forming channels and basins.	Rubusta Nendran, Dwarf Cavendish, Grand Naine	Jan – Feb and Nov – Dec	1.8 × 1.8 m 1.5 × 1.5 m
Wet land	No elaborate preparation is required since banana follows paddy. Planting is done in small pits, sufficient to accomodate the corms.	Poovan, Rasthali, Monthan, Neyvannan	Feb – April	2.1 m × 2.1 m
Perennial (Padugai)	One deep mummutti digging, annual deepening of the water channels and spreading of the silt in the bunds.	Poovan, Monthan	Jan – Feb Aug – Sept	3.3 m × 3.3 m
Hill banana	Scrub jungles are removed, contour walls are constructed.	Virupakshi, Sirumalai	Apr – May Jun – Aug	3.6 × 3.6 m

and later they become brown and shriveled. This operation ensures removal of infection by saprophyte fungi especially to avoid the fingertip disease.

(b) **Removal of male bud:**   Removal of the male bud along with the withered styles 8–11 days after bunch emergence is recommended for better fruit development.

(c) In certain pockets of Trichirapalli, the banana Cv. Poovan is with a peculiar development disorder *viz.*, 'kottaivazhai' which is charecterized by the presence of distinctly conical and ill filled fruits with a prominent central core having many under developed non-viable seed like structures rendering the fruits inedible. This disorder can be overcome by spraying 2,4-D 20 ppm when the last hand of the bunch is opened. The same chemical at the same dose and stage results in increased bunch weight and uniform grade especially Cvs. like Nendran and Monthan.

(d) In Cv. Rasthali, hard mass or hard lump, characterized by pinkish brown, firm pulp than the usual soft pulp occurs, tastes like immature or unripe fruits. Spraying the bunches uniformly with 2,4-D at 1000 ppm or dipping the cut end of peduncle of the bunches for a period of 5 minutes appears to favour the reduction of lumps and improve the size.

(e) The peduncle of bunches may be covered with flag leaf to prevent 'main stalk rot' and also the bunches with banana leaves to avoid sunscald.

Harvesting and Yield

The time taken for maturity and thus to harvest depends upon the cultivars. Poovan, Monthan, Rasthali and Dwarf Cavendish are ready for harvest in 11–12 months from the date of planting. Red Banana takes 18 months from planting. A bunch usually takes 90–150 days from shooting (flowering) to maturity.

The bunches are harvested when the ridges on the surface of the skin change from angular to round i.e. after attainment of 3/4th full stage. After harvesting, the pseudostem should be cut leaving a stump of about 0.6 m height. This practice is called, 'mattacking'. The food material stored in the left out stump continues to nourish the daughter suckers till it withers and dries up. Bananas are not usually allowed to ripe on the plant as it takes time. Besides, the fruit peel splits; fruit ripens unevenly and fails to develop good colour and aroma. Hence banana is made to ripen artificially.

Smoke treatment is the commonest method to induce ripening in Tamil Nadu and Maharashtra. Smoking is done with straw, leaves and cowdung in a enclosed chamber with bunches arranged in a heap for 18–24 hours in summer and 48 hours in winter. After taking the bunches from the chambers, they are placed in a well ventilated room for development of colour. Bunches take 3 days for ripening in this method. Ethrel treatment (2500 ppm) also causes ripening within 2 days. In this method, the bunches are arranged in a closed chamber and ethrel is taken in a beaker and added with few pellets of sodium or potassium hydroxide. This facilitates the release of ethylene gas which aids in ripening the fruits.

The yield of banana depends upon a number of factors such as cultivar, density, management practices etc. The probable yield from important cultivars is:

Cultivar	Yield
Poovan	40–50 t/ha
Monthan	30–40 t/ha
Rasthali	40–50 t/ha
Robusta	50–60 t/ha
Dwarf Cavendish	50–60 t/ha

Plant Protection: Please see table 16.8

3. Citrus *(Citrus spp)* Family: Rutaceae

Citrus fruit is the world's leading tree fruit crop and in India, it is the third important fruit crop next to mango and banana. This group of fruits includes sweet oranges, mandarins, limes, lemons, grapefruits, pummelo, citron, kumquat etc. It is believed that most of the species under the genus *Citrus* are native to tropical and subtropical regions of South-East Asia, particularly India, China and in the regions between these two countries. Now citrus is grown in more than 100 countries of the world and the major citrus producing countries are USA, Spain, India, Italy, Israel etc. USA is the largest producer of Citrus fruits covering about 35–40 per cent of the world production. In India citrus occupies about 1,90,000 ha with an annual production of about 15,00,000 tonnes. Among them, the mandarins occupy the largest area, followed by sweet oranges and lime and lemons. The major mandarin growing areas in India are Coorg (Karnataka state), Nagpur belt (Maharashtra), Darjeeling District (West Bengal) etc. In Tamil Nadu, mandarins are grown in Pulney, Nilgris and Shervaroy hills.

Citrus plants are perennial, small trees. Its leaf is uniformly compounded. In most species, the petioles are winged exception being lemon and citron. Grapefruit and pummelo have large wings while sweet oranges and acid lime have small wings. Flowers are produced on current season growth in cymes both axillary and terminally. Two types of flowers, perfect and imperfect

Table 16.8 Important pests and diseases of banana.

Name of the pest/Disease	Symptoms/Damages	Control measures
1. Banana stem Borer (Rhizome Weevil) *Cosmopolites sordidus*	The grubs tunnel the corms and affect the growing point. Leaves turn yellow, wither and dry.	Apply BHC 10% @ 20 g/plant or Carbaryl 10-20 g/ plant in the soil around the stem.
2. Banana aphid (*Pentalonia nigronervosa*)	It is the vector of bunchy top virus disease.	1. Spray phosphamidon 0.1%. or Methyl demeton 0.2% or monocrotophos 0.1% on the crown and pseudostem base upto ground level at 21 days interval atleast thrice. 2. Injection of monocrotophos 36 WSC 1 ml/plant (1 ml diluted in 4 ml of water) at 45 days interval from the 3rd month till flowering is very effective. Avoid injection of monocrotophos after flowering.
3. Nematodes – Burrowing nemetode – (*Radopholus similis*), lesion nematode (*Pratylenchus coffeae*) and root knot-nematode (*Meloidogyne incognita* and *M. javanica*)	Leaves become pale, plant can be easily toppled down due to poor anchorage of roots, black lesions with purple margins in the roots and corms.	Pralinage (the corms are dipped in a slurry solution containing 4 parts of clay + 5 parts of water and the dipped corms are sprinkled with carbofuran (at the time of planting) with of Carbofuran 3 G 40 g granules per sucker. Pare and dip the corms in 0.5% monocrotophos solution for 15 minutes, shade dry and plant. Pre-treat the suckers with 40 g of Carbofuran 3 G. If pre-treatment is not done, apply 40 g of Carbofuran around each plant one month after planting.

(Contd.)

Table 16.8 Important pests and disease of banana.

Name of the pest/Disease	Symptoms/Damages	Control measures
		or pare and dip the corm into 0.75% Monocrotophos solution; shade dry and plant. Then grow Sunhemp after 45 day and incorporate one month later. Press mud application @ 15 t per ha one month after planting and neem cake 1.5 t per ha one month after planting.
4. Sigatoka leaf spot (*Mycosphaerela muscicola*)	Pale or light green small lesions on the leaves, these spots eventually dry and become light grey with a dark brown or black border. Severe spotting results in drying of entire leaves thus affecting the growth of the fruits	Remove the affected leaves and burn Spray any one of the chemicals after planting at monthly interval from 3rd month onwards— Carbendazim (0.1%), Mancozeb (0.2%), Copper oxychloride (0.2%). Add any wetting agent.
5. Panama disease (*Fusarium oxysporum f.sp. cubense*)	Buckling of green leaves, splitting at the base of the pseudostem, brownish streaks in the roots and corms	Uproot and destroy severely affected plants. Apply lime 1-2 kg in the pits after the removal of the affected plants. In the affected field, corm injection method may be followed i.e. a small portion of the root is removed to expose the upper portion of the corm and an oblique hole at 45° angle is made to a depth of 10 cm and immediately 3 ml of 2% carbendazim solution is injected into the hole with the help of corm injector on 2nd, 4th and 6th month after planting.

(Contd.)

Table 16.8 Important pests and diseases of banana.

Name of the pest/Disease	Symptoms/Damages	Control measures
6. Bunchy Top	Affected leaves show dark green streaks and dots on the secondary veins on the under surface of the lamina and on the midrib and petiole. Leaves also show marginal chlorosis and curling, leaves become brittle in texture, smaller in size and eventually the crown of the plant becomes crowded with stunted leaves.	Use virus free suckers, paring and pralinage, destroy the virus affected plants, use the suggested methods to control the vector, insert a gelatin capsule by capsule applicator containing 200 mg Fernoxone (2,4-D) into the corm 7 cm deep or inject 5 ml Fernoxone solution (125 g/ litre of water) into the pseudostem using the injection gun. The plant collapses and topples in 3-5 days.

(with rudimentary pistils) are found. The fruits of citrus are of a special type of berry known as 'hesperidium'. The pericarp is divided into three layers.

(a) Exocarp (Flavedo)—Outer most layer of the fruit having many oil glands.

(b) Mesocarp or albedo—is a white spongy portion adhering to the inner side of the exocarp.

(c) Endocarp—is the edible portion containing many carpel segments. The segments are separated by carpel walls. Within the carpels are many juicy vesicles developed from hair-like papillae on the segment membrane and some seeds. The juicy vesicles are attached to the membrane with the thread like stalk.

The principal constituent of the edible portions of the Citrus fruits of sweet group are sugars and acids while the fruits of acid group contains primarily the acids in the fruit juice. They also contain vitamin C ranging from 25 to 85 mg per 100 ml of the juice. The TSS in most of the groups may vary from 8–12 percent.

Major Citrus Species of Horticultural Importance are

1. **Sweet orange (*C. sinensis*):** A highly polyembryonic species, trees medium-large, fruits sub-globose to oval in shape, orange colour skinned, segments 10–12, orange coloured, pulp juicy. Important cultivars are Mosambi, Malta Blood Red, Sathugudi, Pineapple, Valencia Late etc.

2. **Mandarin Orange (*C. reticulata*):** Also known as 'Tangerine' or loose skinned Oranges. Fruits medium sized, globose, sweet in taste, segments easily separable. Core open at maturity, loose skinned, orange in colour, rind thin, rind and segments easily separable, usually 10–14 segments in each fruit. Important cultivars are Coorg Orange, Kodai Orange, Nagpur Santra. Khasi Mandarin etc.   .

3. **Limes (*C. aurantifolia*):** A highly polyembryonic species, flowers are white, fruits are small, round to oval, greenish yellow and thin skinned, core solid at maturity. e.g. kagzilime, PKM-l-developed at HCRI, Periyakulam; a selection from seedling progenies of Kadayam type, fruits are big, each weighs about 50 g rich in juice (52%). Each tree bears on an average 934 fruits weighing 37 kg.

4. **Lemon (*C. limon*):** The petals are pink coloured and the fruits are oval to elliptic with prominent nipple. Fruit surface smooth, light yellow, core solid and juice abundant and acidic. Important varieties are Sevelle, Malta, Rajamundhry, Mayor, Eureka etc.

5. **Pummelo (*C. grandis*):**   A monoembryonic species with large sized fruits, sub-globose to pyriform in shape, with thick and spongy rind, white fleshed and pink or red fleshed types are available.

6. **Grapefruit (*C. paradisi*):**   A polyembryonic species, fruits large, sub-globose in shape, fruit rind and flesh are yellowish in colour, fruits highly juicy, sweet with bitter taste, central axis or core opens at full maturity.

Besides the above six citrus kinds, there are some more kinds which are cultivated in isolated areas. They are

1. Sour orange   *C. aurantium*
2. Rough lemon   *C. jambhiri*
3. Sweet lime   *C. limettoides* and
4. Citron   *C. medica*

Recently; many artificial citrus hybrids have been developed as a result of breeding and many of them indicated below are popular.

Common Name	Parentage
1. Citrange	*Poncirus trifoliate* * × *C. sinensis*
2. Citrangequat	Citrange × *(Fortunella marginata)* **
3. Tangelo	*C. reticulata* × *C. paradisi*
4. Tangors	*C. reticulata.* × *C. sinensis*
5. Lemonime	*C. limon* × *C. sinensis*
6. Lemonnage	*C. limon* × *C. sinensis*
7. Lemandarin	*C. limon* × *C. reticulata*

* Trifoliate orange, principally used as a dwarfing rootstock for orange trees.
** An ornamental species, called kumquat, fruits are used for candying.

Soil and Climate

Citrus can grow well in a wide range of soils but it prefers deep, well drained loamy soils. The best pH is 5.5 to 7.5, good drainage is an important prerequisite of soils and the water table should be always below 2.0 meters.

Citrus thrives well in frost-free subtropical to semi-tropical climate. However, each citrus kind needs a specific climate for its better performance. Sweet oranges prefer dry arid conditions coupled with well-defined summer and winter seasons having low rainfall i.e. 20–40 cm. Places receiving high rainfall and high humidity are unsuitable. On the other hand, mandarins perform well in sub-mountain tracts like Coorg, Wynad. Nilgiris, Pulney and Shevroys hills ranging from 600–1000 m and the rainfall ranging from 75 to 250 cm. They do equally perform well in Nagpur area with an elevation of 300–600 m, the rainfall of 110 cm, relative humidity of 53 percent besides extremity in temperature i.e. maximum temperature being 47.0°C and the minimum temperature being 6.0°C. Lime and lemon on the other hand prefer warm regions, moderately moist and free from strong winds from sea level upto 1000 m. MSL. Pummelo and grapefruit can thrive both under humid and dry tracts. They can grow well upto an altitude of l00 m.

Propagation

Citrus fruits are propagated both by vegetative and seedling means. Since most of the species exhibit polyembryony, the nucellar seedlings obtained through seed propagation carry the true characters of the mother plants. Budding is also common in sweet oranges and mandarins. Selection of a suitable rootstock for each citrus kind is an important factor as a large number of graft compatible species and cultivars are available, however, each one impart varying effect on the vigour, precocity, productivity, longevity, quality of the fruits besides resistance to pests and diseases.

A good rootstock for citrus should ensure an economic orchard life of about 30–50 years along with its tolerance to or resistance to problems in soil and virus diseases. In addition to it, it must be polyembryonic, easily propagatable with adequate and cheap sources of seeds. Some of the commonly used rootstocks are sour oranges, rough lemon, trifoliate orange, citrange, Rangpur lime, mandarins etc. (Table 16.9).

Based on the experiments conducted at India, the best root-stocks recommended for important citrus crops are:

Scion	Rootstocks
1. Sweet Oranges	Jatti Khatti (C. jambhiri), Troyer citrange
2. Mandarins	Rough Lemon Rangpur lime

Acid lime is propagated through seedlings while lemon is propagated either through layers, cuttings or seedlings. Grapefruit and pummelo are mostly propagated through seeds or by budding.

Recently, a new propagation technique viz., 'Microbudding' has been standardized for citrus at NRC for Citrus, Nagpur for fast and year round multiplication of disease-free planting materials and also for commercial propagation for getting the marketable budded seedlings earlier by atleast 6 months to one year depending upon the rootstock. This technique involves taking tender scion bud that can be grafted on 6 month old rootstock in humidity–controlled screen–house/low cost shade net structure where a success of 70–80% will be normally obtained. The scion is collected from disease free planting material of Nagpur mandarin produced through in-vitro shoot tip grafting for early and timely release to growers.

Table 16.9 Commonly used rootstocks in citrus.

1.	Sour oranges	Impart medium sized tree growth, good yield and better quality fruits, susceptible to tristez a virus and nematodes.
2.	Rough lemon	High yield and with good fruit size but fruit quality is poor, Tolerant to tristeza, saline and calcarious soils.
3.	Trifoliate orange	A dwarfing rootstock, good yield and fruit quality, susceptible to exocortis virus, but tolerant to most other virus diseases.
4.	Rangpur lime	A vigorous, hardy rootstock, tolerant to tristeza, susceptible to exocortis, xyloporosis and foot rot.
5.	Cleopatra Mandarin	Tolerant to tristeza, exocortis and xyloporosis, root rot including saline soil, however yield potential is less.
6.	Citranges	Tolerant to tristeza and xyloporosis, gummosis but susceptible to exocortis and nematodes, trees are very productive. Most widely used rootstocks now.

Planting

The spacing for citrus varies with species. The following spacings are generally recommended:

Species	Spacing in metres
Sweet oranges	8 × 8 or 7 × 7
Mandarins	6 × 6
Lime	5 × 5 or 6 × 6
Lemon	6 × 6
Grapefruit	6 × 6
Pummelo	7 × 7

The pits of 75 × 75 × 75 cm are dug out and the planting season is normally during rainy months, avoiding heavy rainy months.

Irrigation

The frequency and quantum of irrigation depends upon many factors such as species, stage of the crop, soil type etc. Sweet oranges are to be irrigated at weekly intervals, limes at 7–10 days interval. Mandarins are mostly grown under rainfed conditions, however, irrigation during dry months is beneficial. During the periods of rapid growth, flowering and fruitset, the citrus trees are sensitive to moisture stress. Moisture stress during fruit development phases affects the size of the fruit besides causing excessive fruit drop. They are equally sensitive to excessive moisture and water logging conditions. Care should be taken to avoid the irrigation water directly contacting the tree trunk.

Manures and Fertilizers

Citrus crops demand major and minor nutrients for proper growth and development. In general, citrus crops of medium yielding capacity remove more than 100 kg N, 30–40 kg P_2O_5, 130–150 kg K_2O and 200 kg CaO from one hectare of land and hence there is a need to maintain proper nutrient balance in citrus orchards. The fertilizer recommendations available for Citrus crops under Indian conditions are given in Table 16.10. Recently, integrated nutrient management is extensively recommended which encompasses the use of organic manures, organic cakes, biofertlisers which result in increased yield, quality fruits etc.

For the non-bearing trees, the fertilizer application may be done in area more than the drip circle, leaving 15–30 cm radius around the tree trunk. In the bearing trees, since the entire area is covered with roots, fertilizer may be applied by broadcasting from trunk to trunk leaving about 30cm around the trunk. Manures and fertilizers are to be applied in two split doses once during March–April and the second dose during October/November.

Citrus crops experience various symptoms of micro nutrients deficiency especially in neglected and old orchards. Hence, micro nutrients are usually supplied through foliar sprays. Better absorption and effects of foliar sprays are observed when the new spring flush leaves are half expanded. Spray solution should contain sulphates of Zinc (0.5%), Manganese (0.5%), Iron (0.25%) and Magnesium (0.5%) besides Borax or Boric acid (0.1%) and Sodium or Ammonium Molybdate (0.003%). Micro nutrient spray is normally given once in three months.

After Cultivation

Water shoots, rootstock sprouts dead and diseased shoots are to be removed regularly. In the young budded plants or seedlings, the laterals should be removed upto 45 cm from the ground level. In the young orchards, ploughing may be done to keep the field free from weeds and hoeing with spade in older orchards.

In the early stages of establishment of a citrus orchard till bearing, the inter-space can be utilized with short duration crops (leguminous types). Wherever soil erosion is a problem, cover crops may be grown.

Cropping

Seedling trees of mandarin orange, sweet orange, grapefruit and most other citrus fruits come to bearing within 6–8 years after planting. In these citrus crops, though the budded plants are precocious, they may be allowed to bear after 4–5 years. Lime and lemon start bearing within three years of planting. In acid group like limes and lemons, flowering is seen almost throught the year but 2–3 peak waves of flowering can be noticed. In other group, flowering normally takes place during early spring (Feb–March) under North Indian condition, while in South India, where there no well defined winter with very low temperature, the flowering season is longer and not very distinct. Under rainfed condition, mandarins normally flower during March/April. It is common to get two crops, occasionally three also in many citrus types.

In most Citrus species, poor fruitset is often observed due to many reasons. Growth regulators have been put into use to increase the fruitset. In acid lime and mandarins, 2,4-D spray @ 20 ppm is recommended during spring to increase the fruits set. Not all the fruits set are carried to maturity. High rate of fruit drop is also a serious problem in most of the orchards. Many growth regulators have been recommended to reduce the fruit drop.

Most of the citrus species take 6–8 months from flowering to fruit ripe stage. Mandarin should be harvested as soon as it is ripe, when it begins change its colour from green to orange. Sweet oranges are harvested as they

Table 16.10 Fertilizer recommendations for citrus.

Manures and Fertilizers	Years					
	I	II	III	IV	V	VIth and above
			(Kg/Plant)			
(1) Pulney hills						
FYM	10.00	15.00	20.00	25.00	25.00	30.00
N	0.100	0.200	0.300	0.400	0.500	0.600
P	0.040	0.080	0.120	0.160	0.160	0.200
K	0.050	0.100	0.200	0.300	0.300	0.400
(2) Shervarayan hills						
FYM						30.00
N		*Same as Pulney hills from I^{st} to V^{th} year*				0.700
P						0.375
K						0.600
VAM						1.00
(3) Coorg region						
FYM	10.00	15.0	20.0	25.0	25.0	30.0
N	-	0.160	0.325	0.500	0.600	0.700
P	-	0.165	0.310	0.380	0.400	0.500
K	-	0.140	0.255	0.500	0.520	0.600

develop a yellow or pale colour. Lime and lemon are harvested mature but still green, so that their acidity is of the highest level. Oranges may be kept at room temperature for about a month after harvesting while mandarin oranges can be kept for a week. Ripe fruits are stored in a cold storage for a period of 3–4 months.

Yield

Species	Fruit No./Tree	Yield/ha in (MT)
Sweet oranges	300	30
Mandarins	1200–1500	15–20
Acid lime	600–1500	25
Lemons	100–250	20

Viral Diseases

Many viruses are known to affect citrus plants causing severe losses to the growers. They are tristeza, greening, psorosis, xyloporosis and exocortis. Since the symptoms caused by these virus diseases resemble those nutritional imbalances and certain trees are symptomless carriers of viruses, it is extremely difficult to say by visual examination that a particular tree is infected with a virus or a complex of viruses. Hence, to ensure whether a particular virus is present or absent in a citrus plant, bud wood from that plant can be indexed in an 'indicator' plant. Each virus has a different indicator plant. For example sour lime seedling is an indicator plant for tristeza.

The indexing of old citrus trees for freedom from viruses is an elaborate process involving several steps. They are

1. Identification of apparently healthy citrus trees *i.e.,* these 'candidate trees' must be true to type, vigorous, high yield potential, free symptoms of virus like diseases.
2. Bud woods from these 'Candidate trees' are then indexed for different viruses in different varieties of indicator plant in an insect proof house.
3. Indexing is done in two stages (a) Short-term indexing for exocortis, greening and psorosis and (b) Long term indexing for xyloporosis.
4. Simultaneously, few buds from these selected 'Candidates' are budded onto rough lemon seedlings which later on serve as the 'primary increase trees' for further propagation if these candidates are declared as' virus free' after the indexing.
5. The bud woods from these primary increase trees are used for virus free 'Foundation or Mother trees' that are maintained in insect-proof screen houses or isolated places.

Table 16.11 Plant protection.

S. No.	Name	Symptoms/damages	Control measure
1.	Leaf miner (*Phyllocnistis citrella*)	Larvae usually mine the underside of leaves making serpentine mines which are silver coloured	Spray Fenthion 10 EC @ 1 ml/litre or Monocrotophos 36 WSC @ 1.5 ml/litre or 5% Neem seed kernel extract during flushing period. Spray dichlorvos 76 WSC @ 1 ml/lit or dimethoate 30 EC @ 2 ml/lit or fenthion 100 EC @ 1 ml/lit or monocrotophos 36 WSC @ 1.5 ml/lit or neem seed kernel extract (NSKE) 50 g/lit or neem cake extract or neem oil 3%.
2.	Sucking pests (whiteflies, Aphids and Rust mite)	Crinkling, curling and yellowing of leaves Cause weakening of infested twigs and withering of leaves.	Spray Monocrotophos 36 WSC @ 1.5 ml/lit **Whitefly:** Spray quinalphos 25 EC @ 2 ml/lit **Aphid:** Spray Methyl demeton 25 EC or monocrotophos @ 2 ml/lit or neem oil 3% or Fish oil rosin soap 1.25 g/lit. or endosulfan 35 EC @ 2 ml/lit. **Rust mite:** Spray dicofol 18.5 EC @ 2.5 ml/lit or wettable sulphur 50 WP @ 2 g/lit.

(Contd.)

Table 16.11 Plant protection.

S. No.	Name	Symptoms/damages	Control measure
3.	Fruit sucking moth *Othreis materna*, *O. fullonica* *O. ancilla*	The insects pierce through ripe fruits rind and expose it for secondary infection of diseases and infestation to flies, such infested fruits fall off.	(i) Bait with fermented molasses + Malathion 50 EC @ 1ml/litre (ii) Apply smoke and set up light trap or food lure (pieces of citrus fruits). (iii) Destroy *Tinospora* weed host. (iv) Bag the fruits with polythene bags punctured at the bottom.
4.	Stem borer *Chelidonium cinctus*, *Chloridolum alcmene*	The grubs bore into the trunk near the tree base and make tunnels. The trees	Prune the branches containing grubs, inject trunk with monocrotophos 36 WSC @ 10 ml per bore hole and plug with wet clay. Plug the fresh holes with cotton soaked in monocrotophos solution mixed @ 5 ml/20 ml of water.
5.	Fruit fly (*Dacus dorsalis*)	The female flies insert 2-15 eggs at a time by making punctures in the rind of semi-ripe fruits, the maggots develop in the fruit pulp and such fruits usually	(i) Spray Malathion 50 EC @ 1 ml/litre or fenthion 100 EC @ 1 ml/litre with 1% crude sugar (10 g/lit). (ii) Set up bait with methyl eugenol 0.1% solution mixed

(Contd.)

Table 16.11 Plant protection.

S. No.	Name	Symptoms/damages	Control measure
		drop prematurely. Secondary infections of fungus and bacteria in the punctured fruits result in fruit rottening as well.	1., with malathion 50 EC 0.05% between 6 a.m. and 8 a.m. 2. Use polythene bags fish meal trap with 5 g of wet fish meal + 1 ml dichlorvas in cotton. 50 traps are required/ha, fish meal + dichlorvos soaked cotton are to be renewed once in 20 and 7 days, respectively.
6.	Root nematodes (*Tylenchulus semipenetrans*): Burrowing nematode (*Rodopholus similis*) and root lesion nematode (*Pratylenchus sp.*)	Reduction in terminal growth and vigour, yellowing, mottling and drying of the twigs from tip downwards causing slow decline	(i) Application of Carbofuran 3 G @ 75 g/tree to control citrus nematodes in severe infestations. Application of 20 g *Pseudomonas fluorescens* formulation per tree at a depth of 15 cm and 50 cm away from the trunk once in four months. Soil application of phorate @ 2 g followed by drenching with metalaxyl + mancozeb 72 WP @ 1% 50 ml/ cutting/ poly bag/ kg of nursery soil for citrus decline. (ii) Use of resistant rootstocks like *Poncirus trifoliate*

(Contd.)

Table 16.11 Plant protection.

S. No.	Name	Symptoms/damages	Control measure
7.	Phytophthora rot (Foot rot/Collar rot/Crown rot) (*Phytophthora parasitica, P. citrophthora, P. palmivora*)	Rotting of rootlets, girdling of the tree trunk and defoliation. The bark cracks, shreds in length wise, gums start exuding	(i) Avoid water stagnation around the tree trunk. (ii) Provide good drainage (iii) Paint Bordeaux paste on the trunk upto 50-60 cm. (iv) Use tolerant rootstocks.
8.	Powdery mildew (*Oidium sp*)	Whitish powdery growth develops on the young leaves (both sides) and twigs. Affected leaves get distorted in shape and reduced in size, causing leaf fall, pre-mature fruit drop etc.,	Apply suplhur dust (350 mesh) in the early morning to protect new flush or spray wettable sulphur 0.5%
9.	Citrus canker (*Xanthomonas citri*)	This bacteria affect leaves, twigs, fruits and thorns, infected fruits fetch less market value	(i) Prune and burn the diseased twigs before monsoon, (ii) Spray streptomycin 100 ppm + 0.3% copper oxychloride.

6. The bud wood from these foundation trees is then distributed as 'standard primary sources of virus free propagative material'

Control of Viral Diseases

1. Use of virus free propagating material
2. Control of insect vectors
3. Use of tolerant rootstocks (e.g. Rough lemon is tolerant to Tristeza and exocortis)
4. Certain virus like exocortis is also transmitted mechanically, tools should be disinfected by dipping them in sodium hypochlorite solutions.
5. Use of cross protection seedlings. Cross protection involves plants that are systematically infected with one strain of virus, are commonly protected from infection by other related strains of the same virus. This technique is now adopted in acid lime against tristeza. The mild strain will be maintained on acid lime indicator plants. A piece of bark taken from this plant is grafted to healthy one year old lime seedlings after removing a similar piece of bark in the stem preferably 22 to 30 cm above the soil level. The bark is held in position and tightly fastened by means of polythene strips. Greenish bark patch indicates the union after 15 days ie mild strain invasion into the system of the plant. Such cross protected plants tolerate the viral infections.

Physiological Disorders

Citrus species manifest many physiological symptoms. They are

(a) **Granulation:** It refers to drying of the juice vesicles in fruits, resulting decrease in soluble solids,acids and accumulation of polysaccharides like starch, cellulose and pectin. Sweet oranges exhibit more granulation. Several factors such as high relative humidity, high temperature during spring, time of picking and tree vigour influence the granulation in Citrus. Application of micronutrients sprays (Zn + Cu + K each at 0.25%) reduces the granulation percentage.

(b) **Splitting:** It refers to the splitting of fruits in oranges and certain lemon varieties. Dry period followed by heavy rain and high relative humidity is mostly responsible for this. Abundant K-supply probably reduces this disorder.

(c) **Degreening:** In certain varieties of oranges and lemon, the fruits though attain full maturity, their colour remains green and as such the fruits do not fetch remunerative price in the market. The fruits are to be degreened artificially by spraying Ethrel @ 2000 to 4000 ppm at pre harvest stage.

Citrus Decline

In recent years, there have been numerous complaints from all over the country that citrus plantations decline following a period of satisfactory performance lasting several years. The trees make excellent growth for first 5 to 6 years, but with increase in cropping, growth slows down and in some cases trees begin to decline. Citrus decline signifies a continuous drying of twigs. Apparently, many causal factors such as cultural practices, fungi, viruses, nematodes etc. are responsible though exact causal factors are not still fully understood.

1. **Varieties:** Certian varieties are more prone to citrus decline than others. The problem of decline may, therefore, be partly solved by planting varieties which have proven tolerant to decline. e.g. Blood Red, Musambi varieties of orange show higher indidence of decline.

2. **Rootstocks:** Rootstocks are known to have a profound effect on the vigour, precocity, productivity, internal quality and longevity of scion varieties grafted on them. They are also known to influence the susceptibility of trees to various diseases and insects. It is also possible that citrus decline in India may be partly due to scionic incompatibilities.

3. **Soil factors:** Much of the citrus decline may result from improper selection of soil involving pH, salt status, structure, and excessive free lime, presence of hard pan, defective drainage and soil erosion effects.

   (a) Soil reaction—An ideal pH for citrus is considered to be between 5.5 to 7.5 because the soils in this pH range are more fertile and trouble free than those with higher and lower pH.

   (b) Excessive salts—Citrus trees are extremely susceptible to salt injury and should never be planted on land that has even a slight accumulation of saline material. Application of gypsum to sodic soils will decrease changes of sodium hazard.

   (c) Defective drainage—Citrus is sensitive to over-moist soil condition within its root zone and thrives best on soils that are well drained.

   (d) Presence of hard pan in soils—The main cause of citrus decline at Fruit Research Station. Ananthrajupet, was the presence of hard pan caused by calcium carbonate. The main reason for the lack of congeniality in the soils of the Punjab was also believed to be the occurrence of a lime layer upto 75 cm of the soil surface or within the feeding zone of plants. Presence of these pans affects both the permeability and aeration of soils.

   (e) Excessive free lime—Survey of citurs orchard soils in Punjab had shown that declining trees were generally associated with soils containing more than 5 per cent lime in any of the horizons upto 180 cm in depth.

4. **Nutrition:** It has often been reported that malnutrition is one of several factors responsible for citrus decline. A study of the decline orange trees in Kerala showed that poor nutrient status of the soils was the greatest single factor responsible for decline. Further deterioration occurred due to neglect and lack of manuring. The decline was negligible in the orchards that were regularly and adequately manured. It is, therefore, important that growers be educated on the role of nutrition in the success of citriculture.

5. **Irrigation:** Improper irrigation can sometimes lead to decline of trees. The basic problem of water control in citrus lies in the avoidance of (i) periodic moisture deficits, especially during critical periods of fruit setting and growth (ii) water excess, (iii) salinity build-up in the soil and (iv) contact of water with the trunks of citrus trees.

6. **Intercropping:** One of the causes of citrus decline in Punjab State was the excessive and indiscriminate intercropping of citrus orchards, primarily due to fertility depletion, shading effect, incompatible irrigation, pests and diseases affecting the intercrops.

7. **Nematodes:** The citrus nematode, *Tylenchulus semipenetrans*, is wide spread, its population is rather high at most places and it is one of the important factors responsible for decline of citrus trees in India. Since this nematode spreads from nurseries, it is very important that nurserymen assure nematode-free plants by periodical fumigation. The burrowing nematode, *Radopholus similis* is also reported to cause problems in area where banana is grown as intercrop.

8. **Insects:** It is clear that although a large number of insects infect citrus and reduce yields in our citrus orchards, there are only a few which appear to play an important role in causing citrus decline in India. Important among these are the Citrus psylla, various aphids and the leafminers. The Citrus psylla is probably the most serious because in addition to sucking the plant sap in large quantities, it has been incriminated as the vector of greening virus. In addition, it is suspected that its saliva contains a toxin that cause young shoots upon which they feed in large numbers are invariably killed. The citrus leaf-miner causes very serious damage in the nursery and is also serious on growing as well as on bearing trees, especially in northern India. Its attack increases the incidence of citrus canker, a serious bacterial disease.

9. **Diseases:** Prominent among the various agencies causing citrus decline are deseases produced by viruses, bacteria and fungi. Virus diseases particularly psorosis, xyloporosis, exocortis and tristeza cause decline. Among the various fungi and bacterial diseases, responsible for decline, Phytophthora, pink diseases, citrus canker are more important.

Grapes *(Vitis vinifera)* Family: Vitaceae

The grape is one of the important fruit crops grown in India. It is a native of Armenia (USSR) and was introduced into India during 11th century and in Tamil Nadu it was introduced in Madurai and Salem districts in 1832 by French Missionaries. The area under grapes in India has been increased substantially in the last four decades and the present area is about 0.25 lakh ha with an annual production of 4.08 lakh tonnes. The principal centres of grape cultivation are Theni in Tamil Nadu, Nasik and Poona of Maharashtra, Hyderabad and Aurangabad in Andhra Pradesh, Punjab, Haryana etc. In Tamil Nadu until recently grapes were grown only in two districts *viz.*, Madurai and Dharmapuri. Now it has been extended to other districts like Coimbatore, Salem, Tiruchirapalli and Tirunelveli.

Climate and Soil

The grapevine is a subtropical plant and in temperate regions the vines are deciduous while in the tropics the vine is evergreen, growing continuously. But it originated as a temperate plant only but subsequently acclaimatized to subtropical and tropical conditions also. For best development, the vinifera grapes requires long warm to dry summers and cool winters in temperate regions. The grapevine in Tamil Nadu grows luxuriously without an unproductive dormant-bud, characteristic of most vineyards in the temperate and subtropical regions. The grapes growing districts receive moderate rainfall distributed in such a manner that it does not harm to any phase of the crop. The rainy seasons are broken by frequent spell of dry weather so that the vine receives heat and sunshine sufficient to mature two crops satisfactorily. The significant climatic feature of Tamil Nadu which in contrast to northern grape belts enables to harvest even five crops in two years against only one crop in other parts of India. Further, hail storms, gales and floods are practically unknown in the grape districts of Tamil Nadu.

The weather conditions at the time of pruning and following pruning upto berry development should be free of heavy rain as continuous cloudy and rainy weather, besides interfering with flower and fruit set, encourages pests and diseases. In Tamil Nadu, pruning is done in such a manner that vines are not in flowering during October and November, the North-East monsoon seasons.

Grapes can be cultivated in a wide range of soil from low to high fertility. An ideal soil for grapevine should be well drained, rich loamy one with low water table. The soil depth should be atleast 0.6 m.

Varieties

Grapes are richer sources of carbohydrates (15–25%) particularly glucose and fructose but traces of minerals are also available (0.2 to 0.6%). The fruit is used in many ways as table grapes and for making juices, canning and wine making.

There are more than 10,000 cultivars in the world. There may be number of synonyms for the same cultivar in different regions. The existing cultivars can be broadly grouped under the following categories:

(a) **Table grapes:** Utilized either as a fresh fruit or for decorative purpose.They must have an attractive appearance, good eating quality, shipping and storage qualities. E.g. Thompson Seedless, Beauty Seedless and Bangalore Blue.

(b) **Raisin grapes:** Grapes which produce acceptable dried products are called raisin grapes. Seedless cultivars with high sugar content producing yellowish or greenish raisins are preferred. E.g. Black Corinth, Thompson Seedless.

(c) **Juice grapes:** The juice of some cultivars with muscat flavour produces an acceptable unfermented beverage when it is preserved by pasteurization or other means. E.g. Bangalore Blue, Beauty Seedless.

(d) **Wine grapes:** Cultivars which have high sugar content and low acid which can produce satisfactory wine is called wine-grapes. E.g. Muscat Canelli, Black Champa.

(e) **Canning grapes:** Seedless cultivars with white grapes with larger berries are preferred for canning.

Important varieties are described in Table 16.12.

Propagation

Grapes are commonly propagated by rooted stem cuttings; 20–30 cm long cuttings are made from ripened wood of the previous season. The lower cut is made just below a node. The cuttings prepared should have atleast 3 to 4 buds. The cuttings are planted in the nursery with one or two internodes exposed. Well matured healthy cuttings form the roots within 3 to 4 weeks and strike roots to the extent of 90%. After a period of three months, the rooted cuttings could be potted in the tube pots. Grapes are also propagated by budding (chip budding) and grafting (whip, cleft; side graftings) so as to establish vines on a rootstock resistant to soil borne pests, diseases, nematodes or alkaline or saline conditions.

Table 16.12 Description of important varieties of grapes.

S. No.	Name	Brief descriptions
		(a) Seedless cultivars
1.	Beauty Seedless	Medium vigour, bunches-medium to large, long, compact, bluish black, spherical, medium sized berries. TSS. 18-21%, acidity 0.6 to 1.0%, suitable for Punjab, Haryana, Delhi and Uttar Pradesh.
2.	Thompson Seedless (Australian Seedless. Kishmish, Sultania. Bedana)	Weak vines, bunches are medium to long, with spherical, small, light green berries. TSS. 20-22%, acidity 0.5 to 0.7%, suitable for Tamil Nadu also.
3.	Perlette	Bunches - Medium large, long conical, berries - medium in size, whitish green and spherical, TSS. 16-18%, acidity 0.5-1.0%.
4.	Pusa Seedless	Bunches large, long berries, greenish white, ellipsoidal, TSS. 20-22%, acidity 0.5%, keeping quality good.
5.	Arka Vati (Black Champa × Thompson Seedless)	Vigorous, medium bunches, long conical to cylindrical, medium berries, TSS 22-25%, acidity 0.6 to 0.7%.
6.	Arka Neelamani (Black Champa × Thompson Seedless)	Moderately vigorous, bunches medium in size, well filled, berries slightly elongated, darkred to black with TSS. 22 to 24%.
7.	Sarath Seedless	A selection from Beauty Seedless made by grapevine growers.
8.	Sonaka Seedless	A budsport of Thompson seedless.
9.	Tas - a - Ganesh	A budsport of Thompson seedless.
		(b) Seeded cultivars
1.	Arka Kanchan (Anab-e-Shahi × Queen of the Vineyard)	Very vigorous, bunches medium to large, large berries, golden yellow colour, TSS 19-22%, acidity 0.5 to 0.6%.
2.	Arka Shyam (Bangalore Blue × Black Champa)	Vigorous, medium bunches, compact medium large berries, TSS 22-25%, acidity 0.5 to 0.6% suitable for wine and juice making.

(Contd.)

Table 16.12 Description of important varieties of grapes.

S. No.	Name	Brief descriptions
		(b) Seeded cultivars
3.	Arka Hans (Bangalore Blue × Anab -e-Shahi)	Vigorous, medium bunches, well filled, medium large berries, yellowish green, spherical to ellipsoidal, TSS 18-21%, acidity 0.5%.
4.	Anab-e-Shahi	Vigorous vine, large cluster, large berries, amber-coloured, TSS-16-18%, a popular cultivar throughout India.
5.	Pachadraksha (Bhokri/Green grapes)	Vigorous vine, large bunches, medium berries, light green, TSS 18%, acidity high (0.70 to 1.00%), poor keeping quality, best adapted to the tropical climate of Tamil Nadu.
6.	Muscat (Muscat Hamburg, Gulabi, Karachi Gulabi, Black Prince, Panneer)	Medium vigour, small bunches, berries small to medium, ripening uneven, TSS 16-17%, acidity 0.7 to 0.8% grown widely in Tamil Nadu.
7.	Bangalore Blue	Medium vigour, berries small to medium, dark purplish in colour suitable for juice and wine-making.

Rootstocks

Rootstocks have so far not been employed in commercial viticulture in India. However, with the increasing problem of soil salinity, drought, nematodes and poor fruitfulness in varieties, the need for the rootstocks has been felt in Indian viticulture also during the past few years.

Dogridge: It is a variety of *Vitis champini*. This stock imparts very high vigour to its scions. Because of this, it should not be used for vigorous varieties and environments such as mild tropics and subtropics where excessive vigour is determined. Dogridge is only recommended for use in the lighter and less fertile sandy soils.

Saltcreek (Ramsay): It is a variety of *Vitis champini*. This stock also imparts great vigour to scions, but not as much as Dogridge does. It has been performed well with wine and raisin varieties of California in light sandy soils of low fertility. Rooting of the cuttings of this stock is also poor but the bud/graft take is high as in Dogridge. It has less suckering habit as compared to Dogridge.

1613: It is a hybrid between *Vitis solonis* and Othello. This stock is compatible with all wine, raisin and table grape varieties of California. Suited for all except very light soils. Cuttings of this stock root easily. Bud/graft take is very high with this stock. It has poor suckering habit.

Nematode resistance rootstocks: Rootstocks 1613, Dogridge and salt creek were highly resistant but 1616 and Teleki 5–A were moderately resistant.

Salinity tolerant rootstocks: The rootstocks Salt Creek, 1613 and Dogridge exclude the sodium and chloride even under saline conditions.

Planting

In Tamil Nadu, they are usually planted from June to January. The rooted cuttings are planted in pits of 1 m × 1 m × 1 m size dug at required distances recommended for different varieties.

Variety	Spacing
Muscat	3 × 5 m
Pachadraksha	5 × 5 m
Anab-e-Shahi	6 × 6 m
Thompson Seedless	5 × 5 m

Well decomposed farm yard manure or, compost, red earth and sand should be mixed in equal proportion and are used to fill up the pits. Rooted cuttings are to be planted in the centre of the pit.

Training

There are numerous systems for training grape vines such as head, kniffin, trellis and bower system of training. The choice of training systems depends on many factors such as apical dominance, vine-vigour, cultivars, bearing zone and climate. In Tamil Nadu, the grapevine is trained over pandal of 2m height. In some areas, live stakes of *Commiphera* (Killuvai) are used for erecting the bowers, in the middle of which the vine is set out. The height of the bower is however low and the root's effect of the stakes is an objection. Recently stone and cement concrete pillars or even G.I. tubes are being used by the growers.

Pruning

The vines are pruned at the end of 11 to 12 months from planting for the pandal trained vines. At the end of 18 months, the vine is ready for pruning for fruit. When canes (twigs of past season) are cut back, the terminal and penultimate buds generally sprout giving rise to shoots which carry the flower cluster and develop into bunches, not all the buds which sprout carry bunches, some turn out to be unfruitful leafy shoots. Therefore, it is important that canes have to be cut to a certain specific length to yield good results. The recommended bud level of pruning for Muscat and Pachadraksha is 5 buds while it is 4 to 5 buds for Anab-e-Shahi and 8 buds for Thompson Seedless. Weak and immature canes should be pruned to one or two buds to induce vegetative growth.

The pruning consists of mainly thinning out and heading back. All mature shoots (canes), whether those have given fruit or only vegetative growth are to be headed back to a certain level which largely depends upon fruit bud bearing zone of the cultivar. Some of the canes are to be severely headed back to retain one or two buds to serve as renewable spurs (thinning out). Generally in a medium to vigorous cultivar, 50% canes are to be headed back to fruit bud level and 50% are to be thinned out. Since every season the canes are pruned at fruit bud level, the fruiting spurs are moved quite away from the centre after some years resulting in an unproductive zone in and around the centre of the vine. This is the effect of 'Forward pruning'. To make these zones also productive, the secondary or tertiary branches are pruned severely so as to encourage more shoots. 50% of these shoots are then pruned to fruit bud level in the subsequent season and rest are to one to two bud level to encourage vegetative growth only. This is called 'backward pruning'.

In Tamil Nadu, the vines are pruned 'twice a year'; pruning in December–January and harvesting in April –May (summer crop) and pruning May–June and harvesting in August–September (monsoon crop). In some areas, it is staggered throughour the year in order to have succession crops for every five months. However, pruning is done only once in North India during the month of February after the winter, while in S. India in the states of Andra Pradesh, Karnataka and Maharashtra; twice a year, once in summer again in winter. Summer pruning is done during March–April and canes are cut back to 1–2 bud level for vegetative growth. It is also called 'back pruning' or 'growth pruning'. Winter pruning is done during last week of September–October at fruit bud level.

Irrigation

The vines should be irrigated soon after pruning and fertilizer application. Until the berries attain pea size, water is given at intervals of 5 to 7 days and thereafter until the commencement of maturity at 10 days interval. Irrigation is usually withheld until harvest which will help to improve the quality of berries, irrigation prior to harvest increases the weight of the bunches at the expense of quality and should be avoided. On an average, about 15–20 irrigations may be necessary per season. Nowadays, drip irrigation is mostly followed in most of the vineyard.

Manures and Fertilizers

A grapevine removes an appreciable quantity of nutrients from soil both in terms of wood and fruits. It has been estimated that an average crop of grape removes from the soil 40–60 Kg N, 10–15 Kg P and 50–70 kg K per hectare. Hence, to maintain the soil fertility for consistent yield, it becomes necessary to replenish these nutrients through proper manures and fertilizers. The following manures and fertilizers are recommended for grapes in Tamil Nadu.

The manures and fertilizers are applied to the soil soon following pruning, removing the soil to a depth of 15 cm leaving a space of 30 to 60 cm around the trunk of the vine. Muriate of potash is applied in two split doses as it will improve the qualities of the berries. Half the dose is applied after pruning and the other half 60 days later. Under North Indian conditions, each vine is applied with 0.4 kg N, 0.4 kg P_2O_5 and 0.8 kg K_2O.

Particulars	I Year (per plant)	Yearly increase	Full grown vine
Farm yard manure	50 kg	25 kg	200 kg
Green leaves	50 kg	20 kg	200 kg
Ammonium sulphate	500 g	250 g	1 kg
Super phosphate	500 g	250 g	1 kg
Muriate of potash	500 g	250 g	1 kg
Bone meal	–	500 g	2 kg

Nowadays, fertigation is extensively followed in grapes in India following the schedule indicated below:

Stage	Unit value			Total units		
	N	P	K	N	P	K
Days after back pruning						
0–30	3	2	0	90	60	O
31–60	1	3	1	30	90	30
61–90	0	2	3	0	60	90
91–120	0	0	1	0	0	30
Days after forward pruning						
0–40	3	1	1	120	40	40
41–70	2	2	2	60	60	60
71–110	1	1	3	40	40	120
111–140	0	0	2	0	0	60
Harvest to pruning	1	1	1	15	15	15
Total units				355	365	445

Cropping

Grape has two kinds of functional buds *viz.*, (1) leaf buds and (ii) fruit buds. The leaf bud is a rudimentary sterile shoot and when it opens it will not bear fruit. Fruit bud contains a shoot having both rudimentary leaves and flower clusters. The fruit buds take 47 to 70 days from bud break to blooming. The position of the inflorescence is terminal and further extension growth takes place by means of the axillary buds of the terminal leaf. During flowering the flower does not open from the tip and proceed downward unlike the other flowers. Instead, the petals open from base but does not complete as they are firmly united at their tips. The cap like calyptra becomes detached at the base and is shed as a tissue cap at the time of flowering. As soon as calyptra is thrown off, pollination and fertilization take place, resulting in fruitset.

Normally the bud immediate below the pruning point in a fruiting spur starts sprouting in about 15 days. As it is a mixed bud, it produces a leafy shoots and produces flower clusters in about 3–4 weeks after bud burst. In this current season shoot, the formation of the embryonic cluster in the fruitbud for the next season crop takes place within 30–45 days after pruning and this primordia grow to a certain extent in the current season itself and cease its activity and again resume their activity after next pruning.

Bunches of most of the cultivars are ready for harvest between 100 to 120 days after pruning. It has to be harvested at proper stage for quality. Generally the berries are to be harvested at ripening stage, characterised by showing colour and softening of berries. The sugar content increases at this stage. Grapes do not ripen after they are harvested so must be left on vine until they are fully ripe. While harvesting, the clusters should be handled only by the stem so that the natural appearance is not impaired due to rubbing of the cluster. The cluster should be removed from the vine by cutting with a sharp knife near its attachment to the cane.

Many types of containers are used for packing. Now, the grapes harvested in baskets and brought to packing sheds. After trimming, they are finally packed into market containers *viz.*, wooden boxes and corrugated fibre board cartons. Table grapes meant for short distance local markets are placed in the field in bamboo basket or deal wood boxes, while those for long distance and overseas markets are packed in CFB cartons in the shed. The CFB boxes are superior to all other containers. Bruises to berries will be least. Their handling and staking in the carriage is easy and convenient due to their uniform dimensions and flat surfaces. They have greater stability in movement and can with stand jerks better. They also absorb moisture released by the grapes and provide insulation against minor fluctuations in temperature during cold storage.

Before packaging and precooling the clusters are exposed to fumigation with sulphur-di-oxide especially the consignment is meant for overseas market. This fumigation retards the activity of decay causing microorganisms on fresh grapes. Fumigation is normally done for 20 minutes and afterwards the containers are brought to normal conditions. This fumigation is not practiced in India.

Yield

Although grapevines yield in the 2nd or 3rd year of planting, good crops can be expected only after the third year. Yields of grapes are highly variable, as much as 12.5 to 21 tonnes as early as 10th month from planting followed by equally heavy crops every six months thereafter until about 10 Years. Pachadraksha yields 35 to 37 tonnes in about 18 months followed by 20 tonnes within another 6 months. This continues over a number of years. In the case of Anab-e-Shahi, the cropping is unpredictable ranging from few bunches to 25 tonnes per hectare. Yield of seedless ranges from to 2.0 to 7.5 tonnes per hectare.

Special Techniques to Improve Fruit Quality

1. After fruitset, tipping of the shoots and the clusters are to be done.
2. The tendrils in the shoots are to be removed.
3. Nipping the growing shoots of axillary buds and terminal buds at 12–15 node stage.
4. Thinning the bunches-excessive bearing of vines can be reduced by cluster or berry thinning. This helps to increase the size of berries and

Table 16.13 Plant protection in grapes.

Pest/Disease	Symptoms/Damage	Control measures
1. Downy mildew. (Plasmopora viticola)	Foliage, shoots and fruits are affected, light yellow translucent spots appear on the upper leaf surface, then white mould patches develop on the lower surface of the leaf, severely affected leaves dry and drop off.	Spray 1% Bordeaux mixture or any copper fungicides at 6-7 days interval.
2. Powdery mildew (Uncinula necator)	Small, indistinct white patches, powdery growth appears on the leaves, later the leaves fall off, berries if affected, fail to mature	Karathane 1 ml/litre of water or wettable sulphur 2 g/litre of water.
3. Thrips (Scirtothrips dorsolis)	They cause white patches on the leaves, curling of leaves, and yellowing of blossoms, fruitset is also affected.	Spraying of any systemic insecticides like methyl demeton 25 EC or dimethoate 30 EC @ 2 ml/lit of water.
4. Mealy bugs (Ferrisia virgata)	Cause shrivelling of the fruits which become unfit for consumption.	Malathion @ 0.1% will control the pest. Apply quinalphos or methyl parathion dust in the soil @ 20 kg/ha to kill the phoretic ants. Spray methyl demeton 25 EC or monocrotophos 36 WSC @ 2 ml lit of water or spray dichlorvos 76 WSC @ 1 ml/lit with fish oil rosin soap @ 25 g/lit. Release Coccinellid beetle, Cryptolaemus montrouzieri @ 10 per vine.

(Contd.)

Table 16.13 Plant protection in grapes.

Pest/Disease	Symptoms/Damage	Control measures
5. Nematode *(Meloidogyne sp.)* *(Rotylenchus sp)*	Damage the roots, growth is severely affected leading to dropping of leaves and slow dying of the vines.	Apply 60 g of Carbofuran 3 G or 20 g Phorate 10 G per vine a week before pruning and the plots are irrigated profusely. The soil should not be disturbed for atleast 15 days. Thereafter normal manuring may be done. Application of neemcake 200 g vine also controls nematodes. Alternatively, application of *Pseudomonas fluorescens* formulation in talc containing 15×10^8 colony forming units/g, 30 cm away from base of the vine atleast 15 cm depth at the time of pruning.

also the quality. Thinning should be done when the berries are of pea-size in varieties producing compact bunches. The degree of thinning is limited to 20 per cent of the berries in the cluster.

5. Use of growth regulators-application of gibberellic acid at 50 ppm at the just opened flowers in the clusters (calyptra stage) increases berry size. Besides, ethrel is used to advance ripening by a week and improve the quality including colour improvement and uniform ripening. It is normally applied at 250 to 500 ppm concentration at berry start ripening stage. Plant protection: See table 16.13.

Sapota *(Achras sapota)* Family: Sapotaceae

Sapota is a native of tropical America. It is commonly known as chikku, Sapodilla and Sapota. It is mainly cultivated in coastal regions of peninsular India. Recently, it is also grown on larger scale in dry, arid region of India. Maharashtra, Andhra Pradesh, Tamil Nadu, Karnataka, Kerala, Uttar Pradesh and West Bengal are the leading states in sopota production accounting for about 5000 ha in India. Sapota fruit, when fully ripe is delicious and is eaten as dessert fruit. The pulp is a good source of sugar (12–14%), Sapota is grown not only for its fruits, but in other countries the milky latex from the bark of the tree yields an important commercial product, gutta-perdha, which forms the base of the manufacture of chewing gum.

It is a slow growing evergreen tree. Flowers are perfect and the fruits are borne in the axils of the leaves of current season growth. The fruit is a berry, the epicarp (skin) is thin and rusty brown in colour, mesocarp (flesh) is yellowish brown and the seeds vary in number 0–12.

Soil and Climate

Sapota can be grown in almost all types of soil but with good drainage. It comes up well in alluvial soils of the river banks, sandy loams near coastal areas, red laterite soils of the heavy rainfall area and medium black soil. The unsuitable soils are those underlaid with hard rock or lime content. Sapota is a tropical fruit and it comes up well from sea level to upto elevation of 1000 m in the hills. It prefers a warm and moist weather and can be grown in both dry and humid areas. Coastal climate is also best suited. Optimum temperature is between 15–35°C and at higher temperature above 43°C during summer, the flower and fruitlets may drop. Areas with an annual rainfall of 125–250 cm are highly suitable.

Varieties

Many varieties are popular among the growers and they are given in Table. 16.14. Recently, many hybrids have been released. They are furnished in Table 16.15.

Propagation

The sapota is not usually propagated from seed as the plant raised from seed grows very slowly, takes longer time to fruit and does not inherit the true characteristics of the parent tree. It is therefore, commercially propagated by vegetative methods such as air-layering and inarch-grafting. Air-layering is a popular method of propagation in West India. In other parts of India, inarch-grafting is generally practiced.

The best time for air layering is the beginning of rainy seasons. Application of 10,000 ppm of IBA is recommended to increase the success of rooting coupled with better quality of roots. This method is most common in Maharashtra, Gujarat and Karnataka. Though it is possible to get a whole plant in a short time, mortality is high and root system is shallow, making it unsuitable for wind prone area.

In Tamil Nadu and certain other parts of India, approach grafting has been practiced as a commercial method. The rootstock commonly used is the pala or khirnee *(Manilkhara hexandra or Mimusops hexandra)* as it is found to be vigorous, strong, healthy and compatible for many years. In nature self sown seedlings of *M. hexandra* occur in plenty in Ratnagiri district of Maharashtra, Surat district in Gujarat and West Bengal in the forest areas and are collected and raised in pots and trained to single stem for grafting. Besides, many other rootstocks have been tried but the results are not encouraging as seen below:

Name of the rootstock	Demerits
1. Mahua (*Madhuca latifolia*)	Rootstock tends to overgrow scion, causes it to die-back, fruits are also poor in quality
2. Mee tree (*Bassia longifolia*)	Exhibits incompatibility by over growing the scion, pronounced distortion at the graft union and the grafts die prematurely
3. Sapota seedlings	Grafts are slow in growth
4. *Manilkhara elengi*	Grafts not successful

Training and Pruning

All the growths that appear on the rootstock below the graft or bud joint must be removed. After 3–4 years from planting, the lower most branches upto a height of 60–90 cm may be removed as with an advance in age, lowermost branches, if left undisturbed, may weigh down to ground and become unfruitful. Similarly,

Table 16.14 Popular varieties in sapota.

Name of the varieties	Brief description
1. Cricket Ball (Syn: Kolkata Large)	Fruits are large (300-350 g) and are round in shape, pulp is gritty and granular and moderately sweet (TSS. 16-20%), 2-12 seeded.
2. Baramasi	Fruits are medium sized, round in shape, medium sweet (TSS 20.23%), 2-8 seeded, does not bear throughout the year.
3. Thagarampudi	Fruits are medium sized, round or oval shaped, soft flesh, melting, very sweet (TSS. 22-23%, 1-2 seeded).
4. Badami	Fruits are small, soft gritty medium sweet, (TSS. 18-21%), 1-4 seeded.
5. Dwarampudi	Fruits are large, round, soft flesh but gritty, mild flavour, very sweet (TSS 23-27%), 1-6 seeded.
6. Oval	Fruits are small to medium sized egg shaped, slightly gritty, medium taste (TSS 18-22%), 2-6 seeded.
7. Kirtibarti	Fruits are small to medium sized, oval or egg shaped, soft, gritty, medium taste (TSS 21-22%), good keeping quality
8. Pala	Fruits are small to medium sized, oval or egg shaped, melting slightly gritty, good in taste (TSS 18-21%).
9. Guthi	Fruits are small, elliptic in shape, soft flesh gritty good in taste (TSS: 20-26%)
10. Kolkata Round	Fruits are large, flesh is gritty, moderate quality

overshaded and crowded branches are also removed periodically to permit adequate sunlight and air circulation. Recently, it has been established that centre opening of the grown up trees above 3–4 tiers of scaffold branches is beneficial to get higher yield with quality fruits. Similarly, pruning of side branches by retaining 1/2 to 2/3 length encourages new growth, flowering and yield.

Manuring

Sapota is rarely fertilised but the following fertlisers are recommended for maximizing the yield.

Age	N	P_2O_5 Kg/ha	K_2O	Farm Yard Manure (kg/tree)
Annual increase	30	30	50	10
I year	30	30	50	10
Adult	150	150	250	50

Table 16.15 Hybrid varieties in sapota.

Name of the Improved Cultivar	Parentage	Brief description
Co-1 developed at Horticultural College. Coimbatore.	F$_1$ hybrid between Cricket Ball X Oval	Fruits 125 g, the flesh is granular in texture, TSS 18%.
Co-2 developed at Horticultural College. Coimbatore	A clonal selection from Baramasi	Fruits are medium (140 g), obovate to round in shape, flesh soft grity, sweet in taste (20–26%), 2-3 seeded.
Co-3 developed at Horticultural College. Coimbatore	F$_1$ hybrid between Cricket Ball X Vavilavalasa	Its yield potential is 150 kg/tree/year, fruits are good in size (133 g), TSS (24.2°C Brix), trees grow upright with compact canopy, hence suitable for high density planting.
PKM-1 developed at Horticultural College, Periyakulam	A clonal selection from Guthi	Tree is dwarf. Fruits are round or oval, skin is very thin, TSS 24 unit, medium sized fruits (120 g)
PKM-2 developed at Horticultural College, Periyakulam	A hybrid between Guthi X Kirth Barthi	Each fruit weighs 90-95 g, fruits are oblong to oval in shape, TSS 25–27%
PKM-3 developed at Horticultural College, Periyakulam	Guthi X Cricket Ball	Tree has a vertical growth habit and hence lends itself for high density planting. Trees bear big sized fruits with oval shape and have cluster bearing habit.
PKM-4 developed at Horticultural College, Periyakulam	A clonal selection from Open pollinated seedling of PKM-1	Spindle shaped fruits, pulp attractive with light pinkish honey brown colour, crisp and sweet flesh (TSS 2° brix)
PKM-5 developed at Horticultural College, Periyakulam	Clonal Selection from Virudhunagar	An off season bearer, putforth flowers during May and Crops during September
DHS-1 developed at Agricultural College, Dharwad, Karnataka	Kalipatti × Cricket Ball	Tree vigorous, bearing round to slightly oblong fruits, each weighs 150 g, TSS 26%
DHS-2 developed at Agricultural College, Dharwad, Karnataka	Kalipatti × Cricket Ball	Tree vigorous, bearing round fruits. sweet, TSS 23% each fruit weighs 180 g.

Irrigation

Irrigation may not be necessary except during early stages of establishment and growth. However, during hot weather period, irrigation results in good crop, besides helping in the development of fruits.

Intercropping

Sapota has long pre-bearing age and till the trees cover the entire area with their canopy, intercrops such as legumes and short duration vegetables and fruit crops like papaya and banana can be grown.

Cropping

Sapota starts bearing small crops from the second or third year of planting but economical yield can be obtained from sixth or seventh year onwards. Under South Indian conditions, flowers appear almost throughout the year in several flushes at shorter intervals out of which two distinct profuse flowering *viz.*, July to November and again February–March is generally observed. Fruitsetting is a problem in an isolated tree as fruitfulness and protogynous condition exist and the pollination is effected by wind. Mixed planting of different cultivars is recommended to have good fruitset. Wherever fruitsetting is a problem, use of growth regulator such as NAA (100–300 ppm) during flowering and again 15 days interval increases the fruitset and the yield. It takes seven months from fruitset to maturity. Maturity can be judged by certain external symptoms such as:

1. Fully matured fruits have a dull orange or potato colour.
2. A matured fruit when scratched lightly show a yellow streak while immature fruit shows a green streak.
3. Brown scale-like materials on the surface disappears at full maturity.
4. The dried spine at the tip of the fruit falls or drops off easily if touched.
5. Milky latex content is reduced as the fruit matures.

The peak harvest periods are during February–June and September–October. Fully matured fruits are harvested with the stalk intact indirectly giving a twist by and collected without bruising.

Fruits are kept in gunny bags for ripening them and they may take five days. To hasten ripening within 48 hours, ethrel 5000 ppm along with sodium hydroxide pellets are kept in a beaker, surrounding which sapota fruits are placed and immediately closed with air-tight containers. Fruits get ripened well and ready for eating within 48 hours. Ripe fruits keep well for 7–8 days from picking and can be kept for six weeks if stored at 2–3° C and 85–95% relative humidity.

Yield

It depends upon several factors. On an average, 12–25 tonnes of fruits/ha/year can be harvested. On individual tree basis, the following is the average number of fruits which can be harvested per tree.

Year from planting	No. of fruits/tree
3	100
5	250
7, 8	700–800
10–12	1000–1500
15	2000
30	2500 to 3000

Plant Protection

Table 16.16   Important pests and diseases in sapota.

Pest/diseases	Damages/symptoms	Control measures
1. Leaf webber *Nephopteryx eugraphella*	Feeds on buds, leaf and young fruits	Spray phosalone 35 EC @ 2 ml/lit
2. Hairy caterpillars	Flowers and flower buds are damaged	Spray choloropyriphos 20 EC or Endosulfan 35 EC @ 2 ml/lt
3. Sooty mould *(Capnodium sp.)*	Photosynthetic functions of leaves get affected and fruits get disfigured	Spray starch or maida (1 kg boiled in 5 litres of water and diluted to 20 litres)
4. Leaf spot *(Phavophleospora indica)*	Pinkish to reddish brown spots are seen, resulting in defoliation	Spray mancozeb @ 0.2%

6. Papaya (*Carica papaya*) Family: Caricaceae

The papaya is native of tropical America and was introduced to India in 16th century. It is now grown in almost all tropical and subtropical countries in the world. In India, it is largely grown in Bihar, Assam. Maharashtra, Madhya Pradesh and Andhra Pradesh with a total area of 34,000 ha, producing about 3,50,000 tonnes.

Papaya is a wholesome fruit and is rich in Vit–A (2000 1U/100 g). Products such as jam, jelly and nectar can be prepared from the fruits. Papaya yields a valuable proteolytic enzyme, papain, which has several and varied uses in medicine and industry. Papain is used to correct certain digestible ailments, for tenderizing meat in the manufacture of leather and in clarifying beer. The other uses of papain are in the treatment of ulcer, diphtheria, pre-shrinking of wool, manufacture of chewing gum, degumming natural silk and rayon, in cosmetics, dental paste preparation etc. The raw fruits are cooked as vegetables and consumed.

Papaya is usually dioecious but hermophrodite type and gynodioecious types are also recognized. In dioecious type, both male and female plants are separate. The male flowers are found on long pendulous panicle. Female flowers are solitary and are much larger than male. The ovary is usually large in female flowers. In the case of hermophrodite flowers, two kinds are often observed *viz.*, one with long corolla type and 10 stamens and another type with a short corolla and 5 functional stamens. Fruit is a large hollow berry elongated or globular in shape. In gynodioecious the female and bisexual flowers are borne one on the same plant. The fruits develop from female flowers are globular in shape while fruits develop from bisexual flowers are elongated in shape.

The edible fruits are found only in *Carica papaya* while the other species are now included in the genus, *Visconella. V. candamarcensis,* known as 'mountain papaya' thrives well at an elevation between 1500 to 2000 m in western ghats. *V. monica* is found growing wild in Amazon basin.

Soil and Climate

It does well in varied soil types, the best performance is observed on loams of uniform texture upto 1.8 m in depth. The most important requirement is that the soil should have good drainage. Even two to three cm of water stagnation around the tree for a few hours is likely to damage them due to the collar-rot disease occurrence.

Papaya performs well in tropical climates where summer temperature ranges from 35.0°C to 38.0°C. At higher elevations, the fruit quality is usually lower. It cannot tolerate very hot summer or frost, this limits the cultivation in Northern India. It cannot tolerate very hot summer or frost, a dry warm climate tends to increase the sweetness of the fruits. In wind prone areas, wind breaks have to be provided to save the trees from wind damage. Tamil Nadu is an ideal home for growing papaya of the mild temperatures and freedom from mosaic and leaf curl diseases. These features help all the year round cultivation of papaya.

Varieties

Papaya is a highly cross-pollinated crop. Seeds taken from a fruit would rarely breed true to type. If a variety is to be maintained pure, controlled pollination between selected female and male progenies of the same parent i.e., crossing of sister and brother, called sibmating, has to be done. This consists of collection of pollen from the male parent and applying it on the previously bagged female flower. Seeds from such sibmated fruit should be used for further multiplication. Failure to observe this precaution leads to the deterioration of the variety resulting in the progeny being a mixture of all kinds of types within a few years. Many varieties have been bred and released for cultivation in India. They are described here under Table. 16.17.

Propagation

The most common method of propagation of papaya is from seeds. Seeds are collected from well mature, ripe and large fruits borne on female plants or hermaphrodite plants as the case may be. The fruits are cut open and seeds are carefully extracted in trays. They are washed and dried in the sun or shade and are stored in bottles. Fresh seeds may be mixed with fine cold wood-ash which absorbs the slimy coating on them and helps to keep the seeds separate on drying. About 500 g seed is required for raising in one hectare. Seedlings can be raised in the raised nursery beds or in polythene bags; however the seedlings from the latter one are good. Two seeds in gynodioecious type or 5 to 6 seeds in dioecious type should be sown in each polybag. The papaya plant can also be propagated from cuttings and grafts. Propagation from seeds is, however, preferred, because the vegetative methods of propagation are not economical.

Planting

Pits of 45 cm × 45 cm × 45 cm size are dug at about 1.8 m apart either way. This would accommodate 3000 plants per hectare. Due to sex variations about 40 to 60 per cent of the plants may turn to be male in the case of dioecious varieties. Therefore, in such case 2 to 3 seedlings per hole at 30 cm apart in the pit should be planted, so that when they reach the flowering phase, the unproductive male trees can be removed to keep the population ratio of one male tree for every 15 to 20 female trees. In the case of bisexual varieties, such contingency may not arise; one good seedling per pit may be allowed to establish. The best time for planting papaya is the beginning of the South-West monsoon in most parts of India. In south India, June to October and January to March are suitable for planting as the other months are either too hot or rainy. Sowing the seeds during cold months may be avoided as it may affect germination.

Manures and Fertilizers

The nutrition of papaya is different from other crops because of its quick growing, continuous and heavy fruiting nature. Nutrient uptake studies conducted at TNAU showed that the uptake of N, P, K is more between

Table 16.17 Important cultivars of papaya.

Name	Parentage	Brief description
Co-1	A selection from Ranchi typ1e	Dwarf plant, first fruit within a reach of 60-75 cm from ground level, fruit medium sized (1200-1500 g) spherical in shape, ridged at the apex, flesh orange yellow in colour, soft firm, TSS 12.5 to 13.5° Brix, lacking the objectionable papain odour.
Co-2	A pure line selection from local type	Fruits are medium sized (1000-1500 g), obovate, ridged at the apex, flesh orange in colour, soft to firm, TSS (11.5 -12.5° Brix), suitable for extraction of papain yielding 4-6 g of dried papain per fruit.
Co-3	A hybrid derivative of Co-2 × Sunrise Solo	Hermaphrodite type giving 100-120 fruits in its span of life, fruits are medium sized (800 g), sweet (13.8 to 14.6° Brix), reddish orange flesh, suitable for home gardens.
Co-4	A hybrid derivative of Co-1 × Washington	Fruits are medium sized (1300-1500 g), round shaped, flesh is yellow with purple tinge, sweet (13.2-13.5° Brix).
Co-5	A selection from Washington type	Fruits are exclusively suitable for papain extraction yielding 14.15 g dry papain per fruit, each tree produces 75-80 fruits in two years of its span of life.
Co-6	A selection from Giant	Dioecious type, fruits are large in size, each weighing 1.5–2.0 kg, yields 80-100 fruits/tree, suitable for table purpose and latex extraction.
Co-7	A hybrid derivative of CP75 (Pusa Delecious × Co-3) × Coorg Honey Dew	A gynodioecious type, with each tree produces about 78 fruits, each fruit weighs 1.15 kg, flesh red in colour very sweet (TSS 16.4° Brix).
Pusa Delicious		A gynodioecious type, fruits are medium sized (1000 g) with deep orange flesh, colour having excellent flavour and taste (13.0° brix).

(Contd.)

Table 16.17 Important cultivars of papaya.

Name	Parentage	Brief description
Pusa Majesty		A Gynodioeclous type, fruits are medium sized (950 g) round in shape, flesh yellow colour, with good transportation quality
Pusa Giant		A dioecious cultivar, fruits are big (1900 g), mostly suitable for vegetable canning industries, TSS low (7° Brix).
Pusa Dwarf		A dioecious cultivar, dwarf in stature, start bearing from 30 cm above the ground level, fruit size medium (1000 g) and oval in shape, flesh is orange red colour with a low TSS of 6.5° Brix.
Coorg Honey Dew		Hermophrodite type, fruits are medium to big (1350 g), shape long to oval, flesh orange in colour, central cavity big, TSS 9.8° Brix.
Washington		Fruits medium and ovate, pulp sweet and agreeable flavour, the stem and petiole have characteristic purple colour.
Solo		Small sized fruits, developed and grown in Hawaii are called Solo varieties, some of the important Solo strains
Varieties		are Sunrise Solo, Line 5 Solo, Line 8 Solo, Line 10 Solo, Kapaho Solo and Waimanalo. The fruits are usually pear shaped with star shaped cavity.
Surya		F_1 Hybrid between Sunrise Solo x Pink Flesh Sweet, released from IIHR, Bangalore. Gynodioecious nature, produces 75-80 fruits of medium size weighing 600-800 g, flesh red in colour, firm, sweet in taste with a TSS of 14° Brix.

Recently, Sunup and Rainbow (the transgenic papaya from Hawaii), Eksotica, Frangi from Malaysia and other exotic cultivars like Red Lady, Zinta are also popular among the growers. Since the PRSV strains available in India is different from other strains, most of these introduced varities are also showing more susceptibility to PRSV in farmers' field.

flowering and harvesting stage, its peak requirement being between fruit development and harvesting. As three stages i.e. flowering, fruit development and harvesting concurrently occur in papaya plant, regular fertilizer application is recommended 10 kg of FYM/plant as basal besides 50 g each of N, P and K per plant at bimonthly interval is recommended by TNAU. Application of these quantities of fertilizers at monthly interval @ 25 g each of N, P and K per plant is also highly beneficial. At Indian Institute of Horticulture Research, Bangalore, a dose of 250 g each of N, P_2O_5 and 500 g K_2O per plant per year in six split application is recommended to get higher yield.Balanced fertilization with 300 g of K along with 300 g each of N and P per plant per year also helps to increase the papain yield with higher TSS and enzyme activity.

Irrigation

Papaya responds well to copious irrigation in well drained soils. Regular irrigation helps fruit development and induces the tree to bear larger sized fruits. However, stress irrigation at preflowering phase tends to keep the plants dwarf and produce fruits at the basal portion in the trunk. Water stagnation should be avoided. In most parts of India, papaya are irrigated once in 8 or 10 days. Drip irrigation if followed 5–10 litres of water is enough for each plant depending on the stage of the crop.

After Cultivation

It is not possible to identify the sex at the early stage until they putforth flowers which may take 4–5 months from planting. At this stage, male trees should be removed maintaining one male tree for every 20 female trees for proper pollination and fruitset. In each pit only vigourously growing female/hermophrodite tree should be retained and other plants are removed. During the pre-bearing age, short duration vegetables like cabbage, cauliflower, onion, chillies, radish, etc. can be grown as intercrops. Weeding should be done regularly to keep the field weed free in the young plantation and weeding may not be required in the grown up field as the interspace remain well covered with the top canopy cover which checks the weeds.

Sex Expression

Many sex forms such as dioecious, hermophrodite, gynodioecious etc., have been reported in papaya. There are no distinct or definite methods to ascertain the sex of the plants at the early stage itself. Besides, many factors have been reported to influence the sex expression.

1. **Environment:**   Low temperature tends to produce perfect flowers on the male tree and female flower production is increased in cool weather and short days. Season of planting also affects the sex expression. Planting during February shows more male plants while planting in March/April produces an equal number of staminate and pistillate plants.
2. **Growth regulators:**   GA (50 ppm), ethrel (200 ppm), SADH (250 ppm) and phosphon-D (2500 ppm) increase the femeleness in dioecious types.

Harvesting and Yield

The first crop of fruits becomes available in 8–9 months from the time of planting; the cropping is practically continuous during the life of the tree. In the plains of North India fruits continue to mature through the spring and summer, but in the cooler places in the hills only 3 to 4 months from February to May. Fruits should be harvested when the colour changes from green to yellowish green. It should be harvested individually with hand, taking care to avoid injuries on the fruits. Yield varies considerably producing 50 to 100 fruits/ plant. The yield may also vary according to the number of female and hermophrodite trees in the orchard. TNAU bred varieties yield 100–160 t/ha. Papaya gives economic crop upto 2 year and thereafter it declines drastically. Certain growers at this stage cut the main trunk leaving about 1.0 m length of stumps which upon regular management produces the fruits for next one year. Fruits to be consumed locally should be stored in a single layer of straw until they become yellow. For distant market, it should be packed in bamboo baskets lined with straw to avoid bruising or in CFB with fruits individually wrapped in newspaper.

Extraction of Papain

The latex or milky juice of the unripe green papaya fruit contains a large amount of digestive enzyme called 'papain' which is able to digest the protein in our feeds. Fully developed green large sized, hard papaya fruits which are about three months old are selected for tapping. The latex is obtained by making scratches or shallow incisions on the skin of the fruit. The incisions are about 0.3 cm deep. Usually not more than four incisions per fruit at equal distance are made every day. To cover the whole surface around the fruit not more than five tappings at intervals of four or five days would be necessary.

Non-metallic instruments should preferably be used in tapping and collecting, as the juice acts upon metals and gets discoloured. Papaya growers in Tamil Nadu use specially made poly-tarpaulins around the tapping tree to collect the latex. An ivory blade or a sharp edged piece of bamboo splinter may be used. The latex should be collected in porcelain glass or earthern containers. After about 2 to 4 hours, the latex is scraped out from the tray and dried in the sun. Tapping should be undertaken early in the morning so that drying in the sun can be done before mid-day. This makes the material sufficiently dry by the evening. When thoroughly dried, the latex becomes crisp and flaky. It may be then ground into a powder, preferably still warm. The dried papain is powdered and sieved in 10 mesh sieves. The cream coloured powder should be placed in air-tight bottles or polybags. Papain can be also dried artificially at temperature of 50 to 55°C which will attain better colour and quality. Potassium metabisulphite (KMS) at 0.5% may be added to it for better colour and keeping quality. The papain production is influenced by certain factors such as fruit size, fruit maturity, varietal factor etc. Potassium is known to influence the TSS content of the latex and the enzyme activity. Plant Protection: see table 16.18.

7. Guava (*Psidium guajava*) Family: Myrtaceae

It is the fourth most important fruit of India in respect of area and production. It is said to have been introduced from tropical America. It is grown in many parts of the world. It is also popularly called as apple of the tropics. It occupies an area of 1,50,900 ha in India with an annual production of 1710.6 MT. Half of this area is confined to U.P. and other important states are Bihar, Madhya Pradesh, Maharashtra, Andhra Pradesh, Kerala and Tamil Nadu. In Tamil Nadu it is largely grown in Coimbatore, Madurai, Ramnad and Kanyakumari districts.

It is popularly called the 'poor man's apple' or apple of the tropics' as it is a very rich and cheap source of vitamin C (100 to 260 mg per 100 gm of the pulp) and contains a fair amount of calcium. It makes an excellent jelly and does not lose the vitamin C in the preserved forms.

Besides *P. guajava*, the related species are

1. *P. guineese*, called Brazilian or Guinea guava which bears small fruits of poor quality
2. *P. cattleianum*, known as strawberry or cattley guava which produces very small fruits of 2.5 cm in diameter with attractive purplish red colour.
3. *P. friedrichsthalianum*-(Costa Rican guava or china guava) produces small fruit of globose in shape.

Soil and Climate

It is a hardy fruit which can be grown in poor alkaline or poorly drained soils without any manuring or irrigation. It can grow in soils with pH ranging from

Table 16.18 Plant protection in papaya.

Pests/diseases	Symptoms/damage	Control measures
Nematodes-*Meloidogyne incognita, Pratylenchulus reniformis*	Mainly in the nursery	Apply carbofuran 3 G @ 10 g/sq.m in the nursery. To control in the nursery, apply carbofuran 3 G @ 1 g/polythene bag after germination.
Collar rot (*Pythium aphanidermatum*)	Water soaked patches are found on the stem which gradually enlarge, girdle the base of the stem.	Ensure good drainage, drench 0.1% Bordeaux mixture before the monsoon season.
Papaya mosaic and leaf curl (Virus), Papaya Ring Spot Virus (PRSV)	Affected plants show stunted growth, leaves show severe curling, crinkling, prevalent in all parts of India.	Affected plants are removed and burnt.
Mealybug (*Paracoccus marginatus*)	Affected leaves show severe crinkling, curling, yellowing and drying of leaves. Affected plants show stunted growth. Fruit size and quality is reduced.	Spray methyl demetan 25 EC or dimethoate 30 EC @ ml/lit or chloropyriphos 20 EC 5 ml/lit or profenophos 50 EC 2 ml/lit using high volume sprayers.

4.5 to 7.5 but the best soils are deep, friable and well drained. It is a subtropical and tropical fruit which requires a distinct winter for developing good quality. It can thrive in semi arid tracts of India and enjoys cooler climate upto an elevation of 1000 m but it cannot withstand frost.

Varieties

Guava varieties are generally named according to the shape, colour or smoothness of skin or from their place of origin. Varieties can be classified as seeded varieties (highly seeded to less seeded types-diploids and seedless varieties (triploids). Important cultivars popularly grown in India are described in Table 16.19.

Propagation

In India, guava is commonly propagated from seed which germinates in about three weeks. Boiling the seeds for five minutes, soaking them in water for weeks prior to sowing or treating them in strong sulphuric acid for five minutes facilitates their germination. Propagation through this method is not desirable as the seedlings will take more time to come to bearing and seedling trees differ greatly from the mother plants. Vegetative propagation through layering is, therefore, recommended. Both air-layering and simple layering have been found to be successful. In about 45 days, layers can be separated from the mother plants. These separated layers should be planted in full size pots and they are hardened by gradually exposing them to direct sunlight. Such hardened layers are ready for planting in about six months. Though it is hard to root, semi hard wood cuttings, treating with IBA or NAA at 2000 to 5000 ppm root well under mist conditions. In some places budding techniques using forkert, shield, patch, chip etc have been tried with different success.

Recently, CISH, Lucknow recommends wedge grafting for rapid multiplication. This technique envisages growing of seedlings in polythene bags, grafting, capping and hardening of grafts. Seedlings of 6–8 months old, with 0.5–1.0 cm stem diameter are used. Shoots with apical portion of 3–4 months old, 15–18 cm long pencil thick with 3 to 4 buds are selected as scions. These shoots are defoliated on the mother plant, about 5 to 7 days prior to detachment and at the same time, the apical growing of the shoots is also beheaded. This helps in forcing the dormant buds to swell. The rootstock is beheaded at 15 cm high and the wedge grafting is performed. The graft is covered with 2 cm wide and 15–30 cm long poythene strips (150 guage). The cap is removed after 25 days and the grafts are transferred to net houses for hardening. Success of wedge grafting varies from 70–90% under green house and 33–77% under open condition.

Table 16.19 Important varieties in guava.

Name of the cultivar	Brief description
1. Allahabad Safeda	Tree vigorous, branching heavy with dense foliage, fruits round in shape, weighing 90-150 g, smooth skinned, white fleshed, most popular cultivar in Uttar Pradesh. TSS 9.5° Brix, Vit. C 140 mg/100 g of pulp.
2. Chitidar	Fruits almost round, white fleshed, smooth skinned with red spots on the skin, fruits weigh 80-140 g, rich in Vit. C (240 mg)
3. Lucknow - 46	Fruits have typical 'Pyriform shape' flesh white, good flavour, fruits are small (40-50 g) .
4. Lucknow - 49 (Sardar)	Fruits weigh 90-140 g, roundish to oval in shape, flesh meaty, TSS 9.5° Brix units, Vit. C 130 mg/100 g of pulp.
5. Seedless	Fruits oblong to globose in shape, flesh thick, creamy white in colour
6. Safed Jam (Hybrid between Allahabad Safeda × Kohir)	Fruits are bigger in size (120 g), soft seed, excellent fruit quality
7. Kohir Safeda (Hybrid between Kohir × Allahabad Safeda	Fruits are bigger (110 g) in size, with lesser seed content than parents
8. Nagpur Seedless	Fruits are medium (70-80 g) sized, with warty fruit surface, white pulp, good taste and flavour
9. Red Fleshed	Fruits roundish oval in shape, saffron yellow in colour with pink flesh colour. Keeping quality medium

(Contd.)

Table 16.19 Important varieties in guava.

Name of the cultivar	Brief description
10. Arka Amulya (Allahabad Safeda × Triploid)	Fruits are medium sized with white flesh, high TSS (12.5 Brix) and good keeping quality.
11. Arka Mridula	A seedling selection from Allahabad Safeda, fruits are medium sized with white pulp and few soft seeds and excellent fruit quality.
12. Lalit	Released from CISH, Lucknow, selection from Allahabad Safeda, fruits have attractive saffron yellow colour and red colour pulp with sugar acid blend.
13. CISH-G-1	A seedling selection from Allahabad Safeda, fruits are deep red coloured having attractive shape, few soft seeds and longer shelf life.

Planting

Pits of 0.5 m × 0.5 m × 0.5 m size are dug at a spacing of 5 m × 5 m. The layers with the ball of earth are planted in the centre of the pit. Recently, high density planting at 3.0 m × 6.0 m (555 plants/ha) and meadow orchard system with 1.0 × 2.0 m (5000 plants /ha) are recommended by CISH, Lucknow.

Manures and Fertilizers

Guava responds to the applications of inorganic fertilizers along with organic manures. Therefore for the bearing trees, 5.0 kg of FYM and one kg in each of N, P and K are applied per tree in two equal split doses, once during March and again during October. The manures and fertilizers are spread in the entire basin of the tree, 15 cm away from the trunk upto the leaf drip and incorporated by shallow digging. It also responds to foliar spray of nutrients and spraying of urea 1% + Zinc 0.5% twice a year during March and October increase the yield. Guava sometimes suffers from deficiency of micronutrients. Hence, a mixed spray containing $ZnSO_4$, $MgSO_4$, $MnSO_4$ @ 0.5% and $CuSO_4$ and $FeSO_4$ 0.25% plus a wetting agent @ 1 ml per 5 litre of solution at various stages *viz.* new flush, month after first spray at flowering and at fruitset are recommended.

Irrigation

Guava though can withstand drought, it responds to irrigation at interval of 10 to 15 days interval. Drip irrigation given at 60 per cent OPE replenishment gives maximum yield.

Training and Pruning

Open centre systems or delayed open centre is generally recommended. Ideal frame work consists of training to a single stem upto 80–90 cm and thereafer allowing four branches representing each quarter. Pruning consists of removal of suckers arising from the base of the trunk. Dried twigs and old branches are to be removed and the cut ends may applied with Bordeaux paste. Flowers are borne on the axils of current season shoots. Light annual pruning after harvesting promotes vegetative growth and flowering. In Tamil Nadu, it is recommended that the tips of 10–12 cm lengths of past seasons shoots are pruned during September and February every year to encourage more laterals. Pruned trees give large fruits and early ripening.

Recently, CISH, Lucknow has standardized training and pruning techniques to have multiple cropping under high density planting. To induce multiple cropping in a year, 'concurrent pruning technique' is advocated during May,

September and December. Matured shoots are headed back to 50 per cent in May for initiating emergence of new shoots below the cut end. After the appearance of flowers on the new shoots, which developed from the previous cut, semi hard wood portion of these shoot is pruned out in September for initiation of new shoots below the cut end. These shoots are repruned when the fruits attained 2–3 cm diameter in December. The new shoots, which resulted from the third cut, are found to be efficient in fruit production in the following season. Overlapping of fruit growth developed as a result of pruning, makes it possible to harvest 3 crops in a year.

When the trees become old, the branches are pollarded leaving 30 cm in length at their origin. The cut branches produce plenty of shoots and flowers and ultimately high yields. In the trees having upright and tall growth habits, the straight growing branches are bent and tied on the pegs driven on the ground. In the bent branches, dormant buds are activated and induced to produce flowers and fruits heavily. In certain parts of Maharashtra, root pruning is practiced to produce heavy yield. In this method roots are exposed and minute roots are cut away and irrigation withheld so as to allow the leaves to shed. Then, the basins are covered with the manures and soil and irrigated copiously.

Cropping

The fruit buds are borne on past season growth terminally or laterally. The flowers are borne on the current season growth in the axils of leaves. The flowers are solitary or in cymes of 2 to 3. The current season growth takes one or two months to bear flowers. The floral buds require 38–42 days for full development.

Layers generally take 2 to 3 years for fruiting. Guava flowers twice a year, first in April–May for rainy season crop and then in August–September winter season crop. In South India, there is a third crop with flowers in October. As the rainy season fruits are insipid and watery, do not keep well in certain parts of India, some practices are followed to avoid flowering and fruiting during rainy season so as to get large sized fruits of better quality during winter season. They are of (a) Bahar treatment-consisting of root exposure and or root pruning before onset of monsoon (b) Deblossoming of rainy season crop-spraying NAA 200–400 ppm and (c) withholding of water and removing the soil from around the upper roots during rainy season and covering it again with soil and manure mixture. Guava fruits should be picked immediately when it is mature and they should not be allowed to ripen in the trees lest the damage by birds and squirrels.

Recently, CISH, Lucknow recommends elimination of rainy season crop by spraying fertilizer grade urea 10–15% twice at the interval of 10 days during April–May to get good yield and quality fruits during winter. Individual hand picking is preferable to shaking the tree. Mature or half ripe fruits are mostly prefered for consumption than ripe over ripe fruits. Yield varies due to many factors. On an average 800 number of fruits weighing 20–25 kg may be obtained from guava tree. Production begins from third year with about 8 ton/ha and increases upto 25 tons/ha by seventh year. Though the trees can yield upto 40 years, economic life of the orchard is upto 20 years. Plant Protection: See table 16.20.

Humid Zone Fruits

These fruits are generally grown in places receiving more rain fall (> 1200 annually) and having higher relative humidity. This zone is distinct from subtropical zone in the sense that temperature during winter will not go down very low, similarly it is distinct from arid or semiarid zone as the atmosphere humidity will be always higher. This kind of peculiar climate is prevailing in South India in western and eastern ghat foot hills ranging 300 to 1200 m M.S.L and also in the valleys and foothills in North eastern hill regions. Important fruits under this category are:

Pineapple: (*Ananas comosus*), *Syn. A. sativus* Family: Bromeliaceae

The pineapple is said to have been introduced into India as early as 1548. The important pineapple growing countries of the world are Malaya, South Africa, Hawaiian Islands, Queensland, Singapore, Ceylon. Among them, the pineapple industry has been well developed in Hawaii.

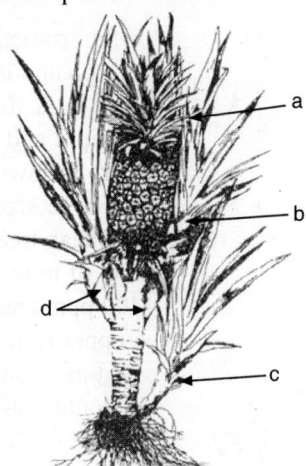

Fig 16.4 A Pineapple plant showing different types of planting materials.
a = crown, b = slips, c = sucker, d = peduncle.

Table 16.20 Plant protection in guava.

1. Tea mosquito bug (*Helopeltis antonii*)	Adults attack the flowers and the infested fruits show corky spots on the surface, fruits fetch low price.	Spray endosulfan 35 EC @ 2 ml/litre or Fenthion 100 EC or Malathion 50 EC @ 1 ml/litre or endosulfan 35 EC @ of 2 ml/lit or fenthion 100 EC @ 1 ml/lit or malathion 50 EC @ of 1 ml/lit or monocrotophos 36 WSC @ 2 ml/lit or neem oil 3%. Spraying should be done in early mornings or late evenings, at least four times at 21 days interval during fruiting season.
2. Fruit fly (*Dacus dorsalis*)	Dark greenish punctures on the surface of the fruit.	Remove the fallen fruits and bury them deeply with lindane 1.3% dusted over them. Spray endosulfan 35 EC or malathion 50 EC @ 1 ml/lit four times at 15 days interval. Stir the soil around the tree during pest incidence and dust Lindane 1.3%. Use polythene bags fish meal trap with 5 gm of wet fish meal + 1 ml. dichlorvos in cotton. 50 traps are required/ha, fish meal + dichlorvos soaked cotton are to be renewed once in 20 and 7 days respectively.
3. Guava wilt (*Fusarium solani and Macrophomina phaseoli*)	In the affected trees, branches wither and die one after another, serious in poorly drained soils having alkaline status.	Affected trees are uprooted, healthy trees are drenched with wet cerason.
4. Sooty mould (*Capnodium sp*)	Upper surface of the leaves and branches, exhibit dark, smoked appearance.	Spray starch solution-Maida 5% (1.0 kg maida or starch boiled with 5 litres of water and diluted to 20 litres of water).

The area under this crop in India has been estimated to be about 84,000 ha. Assam, Tamil Nadu, Bihar, Andhra Pradesh, Uttar Pradesh. Kerala, Karnataka, West Bengal and Tripura are the important pineapple growing states in India. The pineapple is a good source of vitamin A & B and is also rich in vitamin C. Besides being used as fresh fruit, it offers considerable scope for canning. The fruit can also be utilized for juice, jam, candy, alcohol and vinegar.

It is an herbaceous perennial. Throughout the length of the stem, it is surrounded by thickly with leaves. From the axils of the basal leaves, branches with closely arranged leaves arise, which are capable of reproduction of new plants. These are called 'suckers'. Just below the inflorescence, from the peduncle (fruit stalk) small branches with close arrangements of leaves arise and these are called 'slips'. A short piece stem with densely arranged leaves found above the inflorescence is 'crown' (Fig 16.4).

The inflorescence is a compact spike with 100–200 flowers. Flowers are hermaphrodite with functional pollen and ovule but there is self incompatibility, hence the fruitset takes place parthenocarpically. Fruit is a sorosis. The edible part of the fruit develops from outer zone of the axis tissue (peduncle), the ovaries, fused bases of the sepals and the bases of the bracts.

Climate and Soil

Pineapple adapts well to sub-tropical regions to tropical regions having humid climate (75–78%). It performs well in places receiving rainfall ranging from 100–150 cm. It requires an optimum temperature range of $21.0°C$ to $23.0°C$. It can be grown upto an elevation of 1100 m from sea level. The amount of sunshine plays an important role in plant growth and quality of the fruit. Large number of cloudy days retards growth and results in small sized fruits of poor quality. On the other hand, too much of sunshine causes sun-burning of mature fruits. However, shading could be provided to overcome this defect.

The pineapple plant comes up well in all types of soil except heavy clayey soil. It does not tolerate water stagnation and soil should be well drained. Throughout the tropical regions, pineapple is cultivated in the reddish brown laterite soils. The soil should be ateast 45–60 cm deep without hard rocks or stones. These soils contain considerable iron-oxides which help to develop the red colour. The pH of the soil should be 5.5 to 6.0.

Varieties

Important varieties which are grown in India are Giant Kew, Queen and Mauritius. Their distinguishing features are given in table 16.21.

In India, hybridization work taken up at Pineapple Research Centre of the Kerala Agricultural University (KAU) at Mannuthi has rsulted in the development of a high yielding hybrid, 'Amritha'. Its yield potential is about 85 tonnes per hectare. Each fruit weighs more than 2 kg. The fruit has single, small crown, and attractive golden yellow colour with desirable cylindrical shape. It has the added advantages of flesh colour, pleasant aroma, high TSS and total sugars and low acidity. The fruit yield and shape of Amritha is comparable with the highest yielding, cylindrical fruit of the female parent Kew and the fruit quality attributes such as flesh colour, flavour and sweetness matched those of the fruit of the male parent Ripley Queen.

Propagation

Pineapples can be propagated by almost all the parts except the fruits. Suckers, slips and crowns can be used, but the slips and suckers are found superior as planting materials in comparision to crowns. Neither too big nor too small slips or suckers are advisable. The best planting material is in ranges from 300– 450 g in weight while suckers of 300–500 g are the best. Suckers produce the first crop 15–18 months after planting. Slips take 20–22 months while the crowns take more than 24 months for the first crop.

Preparation of Field and Planting

The field should be ploughed to a depth of 30–40 cm to have a fine tilth. The field is laid out into trenches alternating with mounds for planting of pineapple. The depth of the trench should be about 22–30 cm. In hilly the trenches are to be made across the slope. Different systems of planting are being followed in various parts.

1. **Single-row system:** This method is recommended where labour is scarce and weeds are abundant. This system is very much useful for ratoons. In this system, rows are set at 60 cm apart and plants are set in at 30 cm spacing. The main disadvantage is the lack of support to the plants which may topple over especially due to the weight of the fruit. Direct sun rays may also cause 'sunscald' or sunburn on fruit.

2. **Double-row system:** Trenches are dug at 90 cm interval and in this two rows are set 60 cm interval and the plants are set at 25–30 cm in the two rows exactly opposite in zig zag manner. Double row planting permits high density planting. Trials conducted in many countries including India reveal that dense planting results in

   (a) High yields without affecting the size or quality of the fruit

   (b) Overlapping of the basal leaves provides shade and reduces evaporation loss as well as weed growth

Table 16.21   Distinguishing features of important varieties in pineapple.

Characters	Giant Kew	Queen	Mauritius
1. Leaves	Smooth, spiny at tip	Serrated	Short spiny
2. Fruits	Large	Small	Medium
3. Weight (kg)	1.6-3.0	0.5-1.0	1.0-2.0
4. Shape of fruit	Oblong,tapering slightly towards crown	Cylindrical	Oblong
5. Colour of ripe fruit	Yellow	Dark yellow	Yellow and red
6. Flesh colour	Light yellow	Golden yellow	Reddish yellow
7. Other characters	Almost fibreless very juicy with pleasant flavours	Sweetest of all varieties, most suitable for dessert	

(c) The leaves tend to twist and grow upright which provide the fruits natural covering to prevent sunscald and hence results in uniform, coloured and lustrous fruits.

If it is an irrigated crop, planting can be done throughout the year and in the hills (under rainfed), planting time should coincide with monsoon season *i.e.* from June to August.

Manures and Fertilizers

Pineapple is an exhaustive crop and it removes 123 kg of nitrogen, 32 kg of phosphorous and 308 kg of potash from one hectare of land yielding crop of 40 tonnes. Hence, it requires abundant supply of nitrogen and potash. Various recommendations available are given in table 16.22

Calcium nitrate (CAN) appears to be a good source of Nitrogen. Some times, zinc and iron deficiency appear which can be corrected by spraying 0.5%–1.0% zinc and Ferrous sulphate solutions at 15 days intervals. Boran deficiency, characterized by reduction in size of leaves, fruit cracking and hardening may be corrected by spraying 0.3% Borax.

Earthing up

It is a very important operation in pineapple cultivation.This gives better anchorage to the plants. Earthing up should be done for each application of fertilizers and also after weeding, hoeing and harvesting crops. This becomes more important for ratoon crop.

Flowering and Crop Regulation

Pineapple plants generally produce flowers after 12 months when they have about 35–40 leaves and the fruits take 4½ to 5½ months to ripen. Under normal conditions, only 40–50% of the population comes to bearing. Therefore, to induce uniform flowering, the following are recommended:

1. At 40 leaf stage, 50 ml solution containing NAA at 10 ppm +2% urea is poured into the crown.
2. 50 ml solution containing 2% urea + 0.04% Sodium Carbonate + 20 ppm Ethrel is poured into the crown.

The crop becomes ready for harvest after about 15–22 months depending upon the variety and type of planting material. To have spreading or staggering of fruit availability almost throughout the year, the following practices are recommended:

(a) Using different planting materials
(b) Planting suckers and slips at regular intervals from June–December.
(c) By using flower inducing chemicals such as Ethrel at 0.25 ml/litre or calcium carbide at 25 g/litre.

Fruits should be harvested when fully matured for canning purpose and also for distant markets. But for table purpose, the fruits are picked after they develop a golden yellow colour. Harvesting should be done with a sharp knife severing the fruit stalk with a clean cut in such a way the fruit is not damaged and the crown is retained. Such fruits can be stored without any fear of damages for 3–4 weeks after harvest in a well ventilated cool place.

Ratoon

Desuckering should be done immediately after harvest leaving only one sucker on the mother plant. All the slips should be removed. The plants mould be sufficiently fertilized and earthing up is done so as to provide good anchorage for the ratoon crop. In high density planting, ratoon can be taken successfully for 2–3 years and in traditional system for 4–5 years.

Yield

The yield in traditional system is about 8–10 tonnes per hectare. In high density planting, with a population of 28,000 to 70,000 plants per hectare, yield of 60 to 100 tonnes can be obtained.

Table 16.22 Fertilizer recommendations for pineapple.

Location	FYM (t/ha)	N	P	K	Remarks
			(g) plant		
Tamil Nadu					
(a) Irrigated	40–50	12	4	12	FYM is incorporated
(b) Rainfed	40–50	16	4	12	as basal dose at the time of field preparation while N, P, K are applied in two equal split doses at 6th and 12th months after planting.
North Eastern States	20–25	80-100	(kg)/ha 40-50	80-100	Applied in 3-4 split doses i.e. 3rd, 7th, 10th and 12th or 13th months after planting.

Table 16.23 Plant protection in pineapple.

Name of the pest and disease	Symptoms and damage	Control measure
Mealy bug (*Dysmicoccus brevipes*)	Wilting of leaves, commencing at the tip, with reddish yellow colour developing in the wilting area.	Phorate granules @ 1.7 kg a.i/ha Spray methyl demeton 2 ml/lit or monocrotophos 36 WSC 2 ml/lit.
Heart Rot and stump Rot (*Phytophthora parasitica*)	Green leaves turn yellowish green and tips turn brown. Central whorl of leaves will come out with a gentle pull, emits foul smell.	1. Provide good drainage 2. Pretreatment of the planting materials with any copper fungicide at 0.5%.

2. Jack *(Artocarpus integrifolia) Family: (Moraceae)*

The jack fruit is believed to be indigenous to India. Its tender fruit is used as a vegetable and the ripe fruit is delicious in taste. Apart from its fruit value, the timber is valued for furniture and construction and leaves used as fodder especially for goats. It is rich in carbohydrates (19%) besides vitamins and minerals. The jack is grown in Bangladesh, Ceylon, Burma and Malaya and also in Brazil. Assam has the largest area of 10,000 hectares followed by Bihar, Kerala and Tamil Nadu.

Jack is a large monoecious tree, producing cauliflorous spike with innumerable flowers. The trees develop a special fruiting branchlets called 'food stalks' which mostly on the trunk or on the scaffold limbs and on shoots of generally more than two years old. It makes about 7 to 10 cm of growth extension in about 15 days after which it produces leaves and bears spike at its terminal end. Pollination is by wind. The staminate spikes drop about 15 days after their emergence while the female ones set, develops into multiple fruit, called sorosis.

The other related species are (1) *A. altilis* (Bread fruit) and (2) *A. hirsuta.* The first one, commonly known as breadfruit, is cultivated and the second one is semi wild in western ghats and bears edible fruits known as 'Aini Pala'.

Climate and Soil

Jack comes up well under humid and warm climate of hill slopes/ plains. It is grown upto 1500 m MSL, but the trees grown above 1200 m MSL produce fruits of inferior quality. Jack is sensitive to frost and drought. The jack loves a good shade, but also thrives in open. Though it can be grown in a variety of soil, a deep rich alluvial or open textured loamy soil with a pH range of 6.0–6.5 and good drainage is desirable.

Varieties

The cultivated types are broadly classified into two groups-those producing fruits with 'firm flesh' and those bearing fruits with 'soft flesh'. The fruits of soft flesh type, when fully ripe, yield to the thrust of a finger easily. Its pulp is very juicy and soft. The taste varies from very sweet, sweet-acid to insipid. On the other hand, the fruits of firm flesh type will not yield to the thrust easily and when sounded produces a dull thud. The pulp is firm and crisp and the taste is variable in degree of sweetness.

Important varieties are:

(a) Singapore or Ceylon Jack: Introduced to Tamil Nadu in 1947, precocious (2½ to 3 years), fruits are medium size (7 to 10 kg), flesh sweet, crisp, carpels compact, yellow and firm with strong aroma.

(b) Hybrid Jack (Singapore Jack X Velipala) – fruits resemble Singapore Jack with bigger carpels and taste.

(c) Palur – 1 Jack – developed at TNAU centre, Palur, bears twice in a year (March – June and October – November), produces about 60–80 fruits, each weighs 12 kg, flesh yellow, crisp and sweet. The special feature is that the fully ripe fruits have flat stigmatic surface instead of a spiny surface.

(d) Palur-2 has a yield potential of 905 kg with 102 fruits /tree.

(e) PPI-1 Jack – developed at TNAU centre, Pechiparai, produces 105 fruits, each fruit weighs 17 kg, also bears in two seasons *viz.*, April–June and November–December. The flesh is sweet, crisp, tasty with pleasant aroma.

Propagation

It is commonly propagated by seed in many parts of India. Healthy and bigger seeds, weighing 4–6 g are selected from healthy fruits and seeds are to be sown immediately. Germination will take place in three to eight weeks. This method is now discouraged excepting for raising as rootstocks. In grafting, inarching and epicotyl grafting are found to be successful especially during rainy season. Seedlings of *A. hirsuta,* Rudrakshi and Jack fruit are used as rootstocks. Several methods of budding such as chip, patch are used with more success now on one year old seedlings.

Planting

Pits measuring 1m × 1m × 1m are dug at 9–12 m apart, one or two months before planting and allowed to wither. Grafts are removed from the containers and planted carefully so as not to disturb the ball of earth surrounding the roots. Planting is done during June–September in South West monsoon areas and extending upto December in other areas.

Manures and Fertilizers

Nutrient requirement for Jack has not been assessed scientifically. However. for better growth and development, the following doses are recommended per plant. These doses are applied in two split doses during June–July and September–October.

Manures and Fertilizers (kg)	*1ˢᵗ year*	*Annual increase*	*6ᵗʰ year and above*
FYM	10	10	50
N	0.150	0.150	0.750
P	0.080	0.080	0.400
K	0.100	0.100	0.500

Cropping

The jack tree requires considerable time for bearing. Seedling trees come to bearing 6–8 years after planting while grafts take 3–4 years. Flowers generally appear in December and continue upto March and the fruits ripen during summer. At higher altitude fruit growth may continue upto September.

The jack fruit is anemophilous and under normal condition about 75% fruitset is obtained. The fruits take 125–140 days from spike emergence. Tender jack fruits are harvested for use as vegetables. Maturity of fruit for ripe use is indicated by the presence of prominent spines on the fruit and also change in colour. Each tree can bear 20–30 fruits per tree. From one hectare, 30–40 tonnes of fruits may be obtained.

Table 16.24   Plant Protection in Jack.

Pest/Disease	Symptoms/damage	Control
1. Fruit borer	Caterpillars bore into shoots, buds, tender fruits or ripe fruit causing rotting at the site of boring	Apply carbaryl 0.4% during flowering season
2. Rhizopus rot	Attack the male spikes and young fruits resulting in premature shedding of tender fruits	Spray 0.5% Bordeaux mixture or copper oxy chloride (0.2%)

3. Mangosteen: (*Garcinia mangostana*) Family: Guttiferae

Fruits of the mangosteen, commonly known as the "Queen of the fruits" are sweet with a pleasant blend of acid. This is the only fruit in which glucose is in readily available form for giving energy. The fruit possesses a number of uses in medicine. The powder of the fruit shell is used as a remedy for chronic diarrhoea and dysentery.

It is originated from Indonesia or South East Asia. It is found growing wild in Malaya, Burma, Thailand, Moluccus, Cambodia and Vietnam. It has been planted on a large scale in Ceylon, the Philippine Islands and South India. Thailand is the world leading country in Mangosteen production.

The mangosteen is a broad–leaved evergreen tree and its fruit is a berry, having a mean weight of 80–150 g. The fruit contains 4 to 8 carpels or 'Quarters' of creamy or snowwhite with a very fine network of brownish or crimson streaks. The development of the fruit is parthenogenesis.

Climate and Soil

It requires a tropical climate with high humidity, high temperature and abundant rainfall. It grows well at an altitude of 400m to 900 m above MSL in Nilgiris, Courtallam hills of Tamil Nadu. It is not exacting in its soil requirement and is found growing in almost all types of soils. However, it prefers a deep soil with a high organic matter content and good drainage. The water table below 2 m from the ground is ideal.

Varieties

There are no named varieties available.

Propagation

The best and most common method of propagation is through seeds which are not formed out of gametic union but through parthenogenesis, arising from the inner wall of the ovary. This is the reason for the uniformity of the seedling trees of tbe mangosteen. Seeds immediately after extraction from the fruit should be sown, as they loose viability when the seeds are dried. The seed bed should be rich in organic matter with perfect drainage. The seedlings exhibit normally slow development of growth after germination. Seedlings at two leaves stage may be transplanted to earthern pots of 25–30 cm deep and 10 cm in diameter. Care must be exercised to avoid any injury to the taproot. Vegetative propagation through cuttings, layering, budding and grafting have been reported to be failure or with little success.

Planting

Seedlings of 2 year old are planted in the pits of 1.2 m × 1.2 m × 1.2 m spaced 7.0 m apart either way during May to November. After planting, shade is to be provided during the first year of growth.

Manures and fertilizers

Little information is available on the nutrient requirement of mangosteen. In Tamil Nadu, 75 kg of cattle manure, 1.0 kg N, 0.8 kg P_2O_5, 3.0 kg K_2O are applied per bearing tree during June–July.

After cultivation

Irrigation of trees during the bearing season will help to get higher and big sized fruits.

In South Indian hilly slopes, it flowers twice a year. The blossoms of April–May yield a crop of fruits from July to October, while the flowers of October–November yield a summer crop during March to May. The trees normally start bearing 7 years after planting in the main field or 9 years after sowing. An adult tree may yield 500 to 1,500 fruits per year. The optimum stage of harvest of fruits is when the colour of the fruit changes from a green-brown to a dark-brown or reddish purple. Under normal conditions of storage in dry, warm and closed place, the fruits can be stored for 20 to 25 days. The long storage normally hardens the rind and renders the opening of the fruit difficult.

Gambog is a common physiological disorder characterized by the exudation of an yellow gum from the fruits. It is particularly severe in the regions receiving a heavy and continuous rainfall with bright weather for about 2 to 8 weeks prior to ripening. The fruits harvested during monsoon rains in July–August are severely gamboged, while those harvested in other months are relatively free. Splitting of fruits has also been found to be common, especially in west coast of India after the commencement of monsoon. Such cracked fruits are often swollen. The exact cause for its occurrence has not been determined.

Avocado: (*Persea americana*) Family: Lauraceae

Avocado, commonly known as 'butter fruit', is indigenous to 'tropical America'. It was introduced into South India about 80 years ago. It is commercially cultivated in California, Florida and Hawaii states of USA as well as in several south east Asian Indian countries, South Africa and Australia. In India, it is grown in small pockets as scattered trees around Bangalore and in the hill slopes of Tamil Nadu, Kerala and Karnataka. The composition of the avocado is remarkably different from that of other fruits. It is rich in fat (25%) containing mostly unsaturated fatty acids only.

Avocado fruits is an excellent source of Vit-B and Vit-A. Many hybrids have been developed between the varieties of different races (Table 16.25). The important one among them is Paradenia Purple Hybrid. Its fruit is roundish borne in clusters of 2 to 8 with a mean weight of 335–450 g, deep purple skin when ripe, creamy yellow pulp with delicious flavours. At Tamil Nadu Agricultural University (Horticultural Research Station, Thadiyankudsai), a new clonal selection *viz.*, 'TKD-1' has been bred. Fruits are round, medium sized (640 g) with a yield potential of 394 fruits weighing 264 kg. Fruits are early maturing (6–7 months from flowering).

Climate and Soil

It comes up well in tropical and humid tropical area. If proper race and cultivars are chosen, avocado can thrive and produce well in condition ranging from tropical to warmer parts of the temperate zone. Avocado can be grown successfully in the regions with an annual rainfall of 125–180 cm. Rich loamy soil of good drainage but with ample soil moisture supply is necessary. It is tolerant to stagnant water for few days only. It comes up in soil having pH 5.5 to 7.5.

Propagation

Avocado is commonly propagated by seed. Treating the seeds with GA at 100 ppm in deionised water for 24 hours results in earlier germination. Air-layering is also successful. Mexican race root best and West Indian root poorly while the inter-racial crosses vary in their rooting behaviour. T-budding, side grafting, veneer grafting and cleft grafting are also reported to be successful. The rootstocks of Mexican types are frost resistant. For grafting purposes, the rootstock should be 2–3 months old, 30 cm high and growing vigorously.

Planting

Before planting, pits of 60 × 60 × 60 cm at a spacing of 5 × 5 m have to be dug and filled with leaf mould, top soil and well rotten farm yard manure.

After Cultivation

No systematic pruning methods are generally followed. Limbs resting on the ground, devitalized branches, (due to heavy crop) broken or diseased limbs should be removed. The best time to prune is during the dormant period of winter after fruits have been picked. Irrigation during dry months alone is advisable. Avocado requires heavy fertilization. Deficiency of major nutrients especially nitrogen restricts the growth and reproduction. Under South Indian condition, each bearing tree is manured with 40–45 kg of Farm yard manure and 5 kg of castor cake or 1 kg each of Urea, Superphosphate and Muriate of potash.

Cropping

Seedling avocadoes start bearing at 5–6 years whereas vegetatively propagated plants usually bear earlier. The flowering behaviour is unique and is termed as 'protogynous diurnally synchronous dichogamy. Though the female and male parts mature at different times, in each flower female part matures earlier (dichogamy-protogynous); however it synchronizes in such a way that all the opened flowers on a tree are female at one time and male at a different time. The synchronization is diurnal for each tree. The flower is functionally male at one part of the day and functionally female during another part of the same day. Hence, cross pollination is the rule in avocado. Based on this unique flower behavior, the cultivars can be grouped into two group's *viz.*, A and B. In group A, first opening takes place in the morning, second opening during the afternoon of the following day. In group B; first they open in the afternoon and then again in the next morning. Therefore, every morning 'A' pistils can be fertilized by B pollens while during afternoon 'B' pistils are ready to receive

'A' pollen. Under South Indian conditions the trees normally flower from November to February and the fruits are available from August to September. Each tree can yield 200–300 fruits.

C. Arid Zone Fruits

Arid zone refers to places wherein deficiency of moisture restricts the plant growth. This kind of zone in India occupies nearly 12 per cent of the country's land surface comprising 3,17,000 km^2, particularly in the states Rajasthan, Gujarat, Andhra Pradesh, Punjab, Haryana, Karnataka, Maharashtra and certain pockets in Tamilnadu. The main features of this zone are:

1. Rainfall is low to very low,
2. Prevalence of higher solar radiation and high wind velocity resulting in more evaporation/transpiration,
3. Atmospheric humidity is low,
4. The soils are of poor quality with low fertility level and poor water holding capacity and
5. The soils are also of saline and alkali nature.

Fruit trees for this zone should be able to sustain these difficulties. Besides, the orchard management practices should aim at utilizing the available water resources efficiently. This is achieved by adopting the following practices in these zones:

(a) Providing micro-catchment around the trees

Around each tree, a micro-catchment is formed which helps to conserve the rain water as arid soils have high infiltration rate as well as high run-off losses due to concentrated rains. Normally, the size of the micro catchment area will be 2 to 3 times the canopy cover area of a tree. Wider areas can be had for each catchments area but this may affect the population per unit area.

(b) Crescent bunding and opening of catch pits

Crescent shaped or semi-circular bunds with a diameter of 6–13 meters are prepared on the lower side while catch pits are dug on the upper side of the slope. The trees are planted in the center of the crescent. These bunds help to collect rain water while catch pits conserve the same.

(c) Planting on terraces in hilly areas or planting in trenches (0.5 to 1.0 m) across the contour.

(d) Provision of regular mulching around the tree basins.

(e) Use of anti-transpirants-these are chemicals which when sprayed on plants form a film which increases the diffusion resistance of water from stomata and thus reduces the transpiration loss of water e.g. kaolinite (3 to 5%).

(f) Adopting drip irrigation method.

Table 16.25 Features of different races in Avocado.

Characters	Maxican Race	Guatemalan Race	West Indian Race
1. Fruit size	Small (<250 g)	Large (upto 600 g)	Medium sized
2. Fruit development period (months)	6-8	9 to 12	9
3. Fruit skin	Thin, smooth	Thick, brittle and warty	Smooth, leathery and glossy
4. Seed	Large seed, fitting lossely in the seed cavity	Small seed held lightly in the seed cavity	Large, fitting loosely in the cavity
5. Oil content	High, upto 30%	Medium, 8 to 15%	Low (3-10%)
6. Resistance to cold temperature	More	Medium	Less
7. Important varieties	Duke	Taylor	Pollack, Fuschia

1. Ber (*Zyzyphus mauritiana*) Famlly: Rhamnaceae

Ber is reported to have originated from Indo-China region. Its fruits are eaten fresh as well as dried and processed into delicious candy. But fruits are very nutritious and rich in vitamin C (250 mg/100 g pulp), A and B complex and also sugars. It is estimated that there is about 12,000 ha area under ber in India. The major growing states are Madhya Pradesh, Bihar, Uttar Pradesh, Punjab, Haryana, Rajasthan, Gujarat, Maharashtra and Andhra Pradesh.

The ber genus Zyzyphus contains about 50 species of which 18 to 20 belong to India. Among these, only a few are of commercial significance. In India, Z. *mauritiana* and in China Z. *jujuba* are commercially cultivated respectively. Indian ber is an evergreen tree with spreading branching habit bearing round to elliptical shaped fruit. The Chinese ber is a small upright tree, deciduous and has oblong or ovoid fruits. It is highly resistant to frost.

Climate and Soil

Ber is found growing under tropical and subtropical climate all over India and thrives upto the elevation of l000 m above sea level. It is sensitive to freezing temperatures but can withstand extremely hot condinons by shedding leaves and entering into quiescence. Hence, it is called as 'King of Arid fruits'. It prefers atmospheric dryness for high quality fruits.

Ber has deep and extensive root system and hence it is capable of growing under wide variety of soils ranging from shallow to deep and from gravel and sandy to clayey. It can tolerate salinity, alkalinity and even waterlogged conditions to certain extent.

Varieties

There are more than 125 varieties available in India. Important cultivars are alone briefly described here.

Gola: Fruits round, weighing 21.5 g attaining light yellow-brown colour ripening; pulp 95.3 per cent, TSS 20 per cent, vitamin C 70 mg per 100 g pulp and yield 115 kg per tree.

Seb: Fruits ovate shaped, weighing about 19.6 g becoming yellow on ripening pulp 92.3 per cent sweet taste TSS 19.2 per cent vitamin C.62.3 mg/l00 g pulp.

Kaithali: Fruit ellipsoidal in shape, weighing approximately 20.45 g, becoming brownish yellow on ripening, skin little hard and thin, pulp 97 per cent; vitamin C 89.23 mg/100 g pulp and the productivity approximately 130 kg per tree.

Umran: Known for heavy production, fruits big, oblong in shape, weighing about 32.5 g becoming greenish yellow on ripening; pulp 96.4 per cent, sour TSS 19.5 cent, vitamin C 80 mg/ 100 g pulp, average productivity 150–200 kg/tree. Other popular cultivars are Mehrun, Darakhi, Banarasi, Dandan, Elaichi etc.

Propagation

Ber is generally propagated by budding. Though different types of budding methods *viz.*, shield, ring, patch and forkert have been used, the most commonly used method is 'shield budding' as it is easier to adopt and convenient. The wild species of *Zyzyphus* namely *Z. nummularia, Z. rotundifolia, Z. rugosa, Z. xylocarpa* are tried as rootstocks. However, *Z. nummularia* is slow growing and on budding incompatibility is manifested by formation of inverted bottle neck at the union. Therefore, *Z. rotundifolia* seedlings are alone used as rootstocks. The other two rootstocks are also known to be highly compatible.

Seeds for rootstocks should be collected only from fully ripe fruits and they are immersed in 17 to 18 per cent salt solution. The seeds with aborted embryos will float on the surface of water and they should be discarded and only sinkers should be used. The rootstocks are raised by

1. *In-situ* budding method-seeds are sown in well prepared pits at appropriate distance in March to April. Such plants if properly cared are ready for budding in July to September. This practice is highly suitable for arid conditions.
2. Nursery budding method-the seedlings are first raised in 25 cm long, 10 cm diameter 300 gauge polythene bags in April. Ninety days old seedlings are budded when they are still in bags and then transplanted in August in the field.

Budding time varies from place to place but June to September appears to be quite successful. Stooling and air layering have also been reported to be successful with the use of growth regulators like IBA and NAA. Excellent root formation in cuttings having 5–7 buds treated with 400 ppm of NAA has also been observed. Topworking by budding on neglected country ber with improved varieties are also found successful.

Planting

In subtropical region planting can either be done in spring (February to March) provided irrigation facilities are available or in monsoon (July to September). In tropics, however, the best time of planting is during July to August. The planting distance varies from place to place. Under rainfed and arid conditions, it could go upto 8 × 8 m and under irrigated conditions it should be planted at 10 × 10 m. For high density planting, the planting can be done at 5 × 5 m initially and thinning can be done after some years.

After the layout, pits of 60 × 60 × 60 cm are dug about a month prior to the period of planting and kept open in sun for about a fortnight. Irrigation must be done immediately after planting and repeated whenever necessary. In areas with salinity problem, application of gypsum @ 5 kg per pit is beneficial.

Training and Pruning

A ber plant normally tends to remain bushy and takes spreading form creating management problems. Therefore, proper training to develop balanced and strong framework is essential. This exercise should start from nursery itself. In the nursery, buddlings should be trained to have single stem. In the field, they are so trained to have a straight stem upto 30 to 45 cm depending on the climatic condition and then four to five strong, scaffold branches are allowed to develop. On these branches, three to four upright growing secondary branches are allowed. In second year, the secondaries should be trained to carry tertiary branches to complete the training job within 3 years of planting.

The fruits are borne in the axils of leaves on the young shoots of the current season. Therefore, pruning is required eyery year to induce maximum number of new healthy shoots which bear good quality fruits. Pruning consists of heading back of previous season's growth to 1/2 to 3/4 and thinning out of weak, diseased and interwoven branches. It should be done during the hot and dry season when the tree sheds its leaves and becomes dormant. More recently, pruning to remove past season's primary shoots to 6th secondary was found best.

Manures and Fertilizers

Under Gujarat conditions, a dose of 60 kg FYM, 400 g N, 350 g P_2O_5 and 450 g K_2O are recommended per tree for irrigated orchards while the quantity has to be reduced for rainfed trees to 30 kg FYM, 100 g N, 50 g P_2O_5 and 50 g K_2O per tree in July. Under South Indian conditions, 50 kg of FYM, 500 g each of N, P_2O_5 and K_2O are applied per tree above 2 years old. It will be desirable to apply total quantity of fertilizers and manures with monsoons in rainfed orcharding. However, the fertilizer application can be split into two in irrigated orchards, first half in July and second half in October. Foliar spraying with 2 per cent urea during fruiting is beneficial for good yield and better fruit size.

Cropping

Under North Indian conditions, flowering is from August or September to November with full bloom in October. Fruitset ranges from 2 to 18% and this can be increased by the application of GA 20 ppm and 2, 4, 5-T 20 ppm at full bloom. After this set a large number of fruits drop due to various reasons. It varies from 50 to 95 per cent of initial set. Preharvest spray with 2, 4-D at 10 to 20 ppm is the most effective control against this drop. It takes about 150 days for fruits to be ready for harvest after anthesis.

The fruit is harvested from November to March in different parts of the country which is earlier in South and late in North India. A grafted plant begins to bear in the first year itself but optimum production starts after 5 –6 years of planting. They remain productive for 15–20 years. The fruit should be picked when still crisp but has developed a light yellowish scarlet or brownish colour on the outer skin. The average yield of different varieties at their prime age of bearing vary from 80 to 200 kg per tree. The yield is lower i.e., 50 to 80 kg per tree in dry areas.

Table 16.26 Plant protection in ber.

Name of Pests and Diseases	Symptoms/damages	Control measure
1. Fruit fly (*Carpomiya vesuviana*)	Adult female lays eggs singly by inserting its ovipositor in the developing fruits, the larvae hatch out and start feeding on the pulp making galleries; the infested fruits become deformed and unfit for consumption.	1. Ploughing of orchard during summer months to expose the pupae to death through damage or birds. 2. Application of lindane 10 D @ 25 kg/ha during summer months to kill the pupae. 3. Dropped infested fruits should be collected and burried deep in the soil. Destroy infested fruits. Dig the soil under tree canopy to destroy pupae and incorporate Lindane 1.3% D @ 30 g /tree. Spray malathion 50 EC or endosulfan 35 EC or quinalphos 25 EC @ 2 ml/lit. Use polythene bags fish meal trap with 5 gm of wet fish meal + 1 ml dichlorvos in cotton @ 50 traps/ha, fish meal and dichlorvos soaked cotton are to be renewed once in 20 and 7 days, respectively.
2. Powdery mildew (*Oidium sp.*)	White powdery growth on flowers and fruits. Affected fruits either drop-off prematurely (or) become corky.	Spray 0.2% Karathane or Morestan.

2. Aonla (*Emblica officinalis*) Family: Euphorbiaceae

Aonla is in cultivation in India since time immemorial. It finds mention in many ancient literatures. It is indigenous to Central and Southern India. It is more popular as a backyard fruit throughout the country. However, commercial orchards can be seen in states like U.P. and Gujarat. Aonla fruit is very rich in vitamin C (600 mg/100 g of pulp) and pectin, therefore, regarded very important for medicinal value in Ayurvedics.

This fruit is useful in hemorrhages, diarrhea, dysentery, jaundice, anemia, dyspepsia and cough. Fruits are commonly used for preserve (murabba), pickle, candy, jelly, jam etc. It can be dried and powdered to be subsequently. It is also used in the preparation of inks, hair dyes, hair oils. It is a great health and vitality restorer.

It is a medium height tree, behaving as evergreen in tropics and deciduous in subtropics. Aonla has phyllanthoid branching habit with two types of shoots *viz.*, determinate (short) and indeterminate (long) shoots. Flowers are borne in the axils of the leaves on determinate shoots axillary cymules. Basal cymules are of male flowers followed by nodes, each with a cymule of one central female flower (rarely two) and several lateral males. Fruit is capsular (drupaceous) with a fleshy exocarp.

Climate and Soil

It thrives well in tropical region from sea level to 1800 m altitude and also in subtropical regions with distinct winter and summer. It is also grown now in arid and semi arid regions. It is not very exacting in its soil requirement and grows well in sandy loam to clay soils in India. It has tolerance to acidity and salinity including higher range of 6.0 to 8.0 successfully. For higher productivity, it requires deep and fertile soils.

Varieties

The existing varieties are recognized on the basis of size, colour or after names of places. The important cultivars are briefly described here:

1. 'Banarasi-Erect, spreading, medium to tall tree, shy and slightly alternate bearer. Fruit is roundish weighing about 38–50 g, whitish green, slightly fibrous, scanty to medium juicy.
2. 'Francis-Erect and tall tree, good and regular bearer. Large fruited (40–60g), oval roundish, light green, slightly fibrous, moderately juicy.
3. Chakaiya-Spreading, medium height, prolific and regular bearing tree. Medium size fruit (26–35 g) flattened at base and round apex, greenish colour.
4. Krishna-Moderate bearer, fruits medium to large (40–50 g) and fibrous flattened, conical, flesh fibreless.

5. Kanchan-Prolific bearer, spreading habit, fruits small to medium (30–45 g) flattened, oblong, skin smooth, yellowish in colour.
6. BSR-1-Developed at TNAU Agrl. Research Station, Bhavanisagar, trees are medium in height, spreading, fruits are medium in size (27 g).

Propagation

Aonla is propagated by seed as well as by vegetative methods. Seed propagation is essential to raise the rootstocks. Seeds are collected from matured fruits and sinkers are alone used for sowing. Fresh seeds generally give good germination. Seedling can be raised in seedbeds or polythene bags and they take about 4 months to attain buddable size. Both inarching and budding methods can be used for raising plants but budding is more successful. Forkert and patch budding generally done during June to September give good success. For rejuvenation of old orchards, T-budding is preferred. Inferior trees can be headed back to a height of 1.2 m from the ground and the stumps produced can be budded 3 months later. Even soft-wood grafting can also be used successfully for which seedlings are raised *in-situ* and headed back severely in the month of May for forcing new vigorous side shoots. On these new shoots, soft wood grafting method can be employed by way of wedge grafting.

Planting

Grafts or buddlings of aonla are planted in the beginning of monsoon and in areas with irrigation facilities, planting can also be done during February to March. Since the tree grows to a huge size a distance of 8 to 10 m both ways is recommended. A pit size of 1.0 × 1.0 × 1.0 m are made for planting well in summer and kept open for about a fortnight.

Training and Pruning

The branches of an untrained tree often break off, carrying heavy crop load due to brittle nature of wood. Hence, they are trained to single stem upto the height of about 1 m and then primary branches can be allowed at regular space all around the trunk. The pruning consists of removal of 50% of the erect growing shoots to induce new growth and also manage the size of the trees.

Manures and Fertilizers

No systematic research based recommendation is available on the manurial requirement of aonla. However for a mature tree, every year 0.9 kg N, 0.25 kg P_2O_5 and 0.75 kg K_2O are to be applied. These fertilisers are to be applied in two equal split doses once during September–October and again during April–May. Micro nutrient spray especially borax 0.6% thrice in the months of Sep–October at about 10 days interval is recommended to control fruit necrosis which develops due to the deficiency of boron. The typical symptom is that necrosis starts with the browning of inner most part of meso tissues at the time of endocarp hardening. Browning of mesocarp extends towards the epicarp resulting into brownish black areas on the fruit surface.

Irrigation

Aonla trees are hardy and stand very well against drought. Therefore, hardly any irrigation is practiced. However, the crop shall be benefited giving two/three irrigations at the time of full bloom and fruit set. During summer, when the tree is dormant, there may not be any benefit to irrigate the trees. Recently, many growers adopt drip irrigation and get 2–3 flushes of flowering under Tamil nadu conditions.

Cropping

Flowering takes place on determinate shoots appearing in spring and the blooming period lasts for 3 weeks. The fruitset may vary from 12 to 18%. This initial poor set is due to low sex ratio and lack of pollinators or polliniser or both. Normally the fruits take 8 months from flowering to fruit maturity.

Seedling trees may take 10 to 12 years to begin bearing while a vegetatively propagated tree starts fruiting commercial crop after 6 to 8 years of planting. Productive life of tree is estimated to be 50 to 60 years under good management.

Generally aonla fruits are ready for harvest in November to December. Their maturity can be judged either by the change of seed colour from creamy white to black or by the development of translucent exocarp. Maximum vitamin C content is observed in mature fruits while immature fruits are acrid and low in vitamin C content and minerals. On an average, a full grown tree yields 100 to 150 kg per annum. Plant Protection: See Table 16.27.

3. Custard Apple (*Annona squamosa*) Family: Annonaceae

Custard apple is one of the most delicious fruits mainly consumed fresh. However, in recent years, it is being used in ice–creams and puddings. The fruits are rich in sugars (13.5% per 100 g of pulp) and are also rich in calcium (17 mg/100 g pulp), phosphorous (45 mg/100 g pulp) and iron (1.5 mg/100 g pulp). It is very hardy, tolerant to drought, salinity and saline irrigation water to certain extent. They are generally found growing spontaneously in wild forests, scrub jungles, rocky slopes etc. Well maintained orchards are seen in Gujarat, Andhra Pradesh etc.

Table 16.27 Plant protection in Aonla.

Name	Symptoms/damages	Control measure
1. Gall caterpillar (*Bethusa stylophora*)	Young caterpillars bore into the apical portion of the shoot during rainy seaon and make tunnel. Due to this, apical regrowth is checked, side shoots develop below the gall and subsequent growth in following season is greatly hampered.	Cut away the infected apices and prophylactic spray of systematic insecticide like dimethoate 0.03 per cent prophylatically.
2. Bark eating caterpillar (*Indarbela tetraonis*)	Damages stem and branches of grown up trees by eating bark.	Affected portion should be cleared of frass and a few drops of kerosene should be applied in holes to keep this in control.
3. Rust (*Ravenellia emblicae*)	Raised rust appears as circular reddish pustules on leaves and also on fruits.	Spray 0.2 per cent mancozeb at an interval of 7 to 28 days from July to September.

The genus Annona contains more than seventy species of woody shrubs and among them five are commercially important like *A. squamosa* (Custard apple), *A. cherimola* (cherimoya), *A. reticulata* (Bullock's heart), *A. glabra* (Pond apple) and *A. muricata* (sour sop). Of all these, *A. squamosa* is probably the most common in cultivation in India. The plants of this species are shrubs or small trees with fairly stout and smooth trunk reaching to a height of 5 to 6 m.

Climate and Soil

It is a tropical plant, can be grown in mild sub-tropical also. It is sensitive to frost. It prefers dry climate during flowering but fruitset is aided by high humidity which begins with the onset of monsoons. Annual rainfall of 50 to 75 cm is optimum, though it can stand higher rainfall or drought. *A. muricata* thrives well under moist humid conditions while *A. cherimola* prefers a subtropical climate in the tropics at higher elevations. Well drained loamy soil is suitable. However, it can be grown in marginal land or in shallow and poor classes of soil which are available in arid zones.

Varieties

A number of local cultivars are available and some of them are Balnagar, Barbados, Mammoth, Israel hybrids, Red Sitapal, Washington etc., Fruits in these varieties vary from 125 to 360 g in its weight. Tamil Nadu Agricultural University centre, Aruppukottai has developed an improved cultivar *viz.*, APK-1, which has a yield pontential of 6.9 t/ha, drought tolerant and suitable for arid tracts. Fruits are medium (208 g) with TSS of 24.2° brix. Besides, IIHR, Bangalore has developed an inter-specific hybrid *viz.,* Arka Sahan by hybridizing Island Gem (atemoya) and Mammoth (custard apple). Fruits of Arka Sahan custard apple are remarkable for their sweetness (> 32° Brix, 22.8% total sugar), scarce small seeds (9/l00 g fruit weight) and slow ripening (5–6 days). The pulp is snow-white, mealy and juicy with a mild pleasant aroma. Nutritionally, a 100 g pulp of Arka Sahan contains 2.49 g crude protein, 42.29 mg phosphorus and 225 mg calcium compared with 1.33 g, 17.05 mg and 159 mg respectively, in common custard apple.

Propagation

Custard apple is generally grown from seeds, either directly sown in the field or raised in polybag nursery. Among the vegetative methods of propagation, grafting and budding have been reported to be successful. Trials conducted in Tamil Nadu showed that A. reticulata is a good rootstock for A. squamosa and A. cherimola.

Planting

Pits of 0.60 m^3 are made in summer and filled with soil and FYM (1:1). Planting distance is generally 5 × 5 m. Planting is done in the beginning of monsoons.

Training and Pruning

Newly planted grafts or seedlings are staked properly and trained to develop proper shape. A bearing plant is amenable to pruning and responds favourably in terms of better size fruits. Pruning consists of heading back and judicious thinning out of weak and diseased branches. This is done 10 days prior to expected sprouting period depending on the climatic conditions.

Irrigation

Custard apple does not require irrigation to produce fairly good crop provided effective water harvesting is effected in rhizosphere by developing micro-catchments in rainy seasons. However, in post-monsoon period, 2 to 3 irrigations help in better quality fruits and higher production.

Manures and Fertilizers

In Gujarat where well organised orchards exist, application of 10 to 15 kg FYM, 2 to 3 baskets of river silt, 500 g of castor cake, besides 50 g N, 50 g P and 50 g K per plant is applied. Under Maharashtra, application of 250 g N, 125 g P_2O_5 and 125 g K_2O per plant are recommended. In Tamil Nadu 15 kg of FYM, 375 g of N, P and 1000 g of K are applied per bearing tree.

Cropping

It takes about three years for plants to come into bearing and five years for developing full canopy. In subtropics, flowering period extends from spring (February to March) to rainy season while in tropics it is during rainy season. An average plant shall bear about 2,000 flowers and the fruitset is about 2 to 3 per cent. Dichogamy (protogyny) exists, therefore, it is benefited by hand pollination. IIHR, Bangalore has developed a hand pollination technique which is quite simple, fast and economical, as practiced in apple, cherimoya, atemoya etc. Pollen should be gathered from 6.00 to 7.00 AM by tapping and rubbing the flowers between the fingers in a clean paper cup. Best set can be obtained if hand pollination is carried out in the morning hours before 9.30 a.m.

Use of custard apple (sitaphal) pollen rather than that of Arka Sahan gives excellent results, hence it is recommended to raise about 10 plants /ha to serve as a pollen source. One can pollinate 150–200 flowers/ hour and in each tree about 150 flowers should be hand pollinated to achieve an expected yield of 25 tonnes/ha. Fruits developing from artificial pollination will command premium price in the market making this practice commercially viable. Fruitset is better when humidity is high. Set can be improved by spraying GA 50 ppm at bloom.

It is a climacteric fruit and should be harvested mature which can be judged on the basis of change in colour between tubercles and their falling apart. They should be harvested firm for distant marketing. A five year old healthy tree would yield about 50 fruits/year which can go upto 100 fruits/ year. Optimum productive life is about 25 years.

4. Pomegranate (*Punica granatum*) Family: Punicaceae

Pomegranate, a native of Iran, is found growing from Kanyakumari to Kashmir but commercial plantation exists in Maharashtra and Gujarat and to a limited extent in U.P., A.P., Karnataka and Tamil Nadu. Pomegranate is largely used as a dessert. The fruits are good source of sugars (14–16%), minerals (0.7–1.0%) and fair source of iron (0.30–0.7 mg/100 g). It is also used as a salad and in beverages. The roots, rind and seeds are important medicinaly especially against diarrhea.

This genus *Punica* has 3 to 4 more species but granatum is alone important economically. It is bushy plant bearing three types of flowers, *viz.*, male, hermophrodite and intermediate. Both self and cross pollination take place. Fruit is a large globose berry called 'balusta'. It has a leathery skin, brownish yellow to red, surmounted with persistent caly; seeds numerous surrounded by pink juicy pulp. This acid pulp, known as 'aril' is the edible portion.

Climate and Soil

Pomegranate has a versatile adaptability to wide range of climatic conditions. The fruit trees grow well in semi-arid climate where cool winter and hot summer prevail. Under subtropics, it is deciduous in nature while in tropics it is evergreen. During the period of fruit development and ripening, hot, dry climate should prevail for quality fruits. High humidity if prevails, it lowers the quality of the fruits.

It grows under wide variety of soil and can tolerate even alkalinity and salinity to certain extent. However, best results are obtained in deep, heavy loam and well drained soils. It is sensitive to soil moisture fluctuation causing fruit cracking which is a serious problem of this crop. As it is winter hardy and drought tolerant, it makes an excellent choice under arid and semi arid conditions of India.

Varieties

In India, most of the cultivars are known by the names of the places where they are cultivated. Recently, few distinct cultivars have been evolved and released for commercial cultivation. Fruits are found oblate or oblate in shape and vary in diameter from 8 to 12 cms. The rind may be thick or thin, colour may vary from pale yellow to crimson. The seed pulp in superior types is thick, fleshy and very juicy, while in inferior types it is thin, the seed coat varies in hardness. Some of the softer seeded types are known as seedless (Bedana). Lack of lignification of the testa is the main cause of so called seedlessness in pomegranate. Some of the important popular cultivars are described below:

1. Dholka-Fruits are large with whitish or pinkish white, sweet flesh, seeds are very soft.
2. Alandi-Fruits medium in size, fleshy testa, blood-red or deep pink with sweet, seeds are very hard.
3. Kabul-Fruits are large, rind deep red, mixed with pale yellow, fleshy testa, dark red with slightly bitter taste.
4. Jodhpur Local-Fruits medium to large, hard rind, fleshy, aril light pink, sweet, juicy, seed moderately hard.

5. Bedana Seedless-Fruits medium to large, rind brownish green, fleshy, aril-light pink or whitish, very sweet, very juicy, soft seeded.
6. Ganesh-A selection from 'Alandi' medium sized fruit, prolific yielder, with soft seeded, pinkish agreeable taste-a popular cultivar, which revolutionized pomegranate cultivation in Maharashtra and other states.
7. Jothi (GKVK-l)-A seedling selection released for cultivation from U.A.S., Bangalore. It has medium sized fruit, with attractive yellowish red fruit colour, red aril and soft seeded.
8. Co 1-A soft seeded selection, fruits, are medium in size with purple arils and soft seeds.
9. YCD 1-A selection made at TNAU Horticultural Research Station, Yercaud, fruits medium in size, weighing 350–400g, with soft seeds as deep purple arils.

In Maharashtra, four clonal selections were made from Cv. Ganesh and are called G-107, 0-132, G-133 and 0-137. Out of these, G-137 was found to be the best in respect of fruit weight (269 g), aril colour, aril size, T.S.S. (15.5%), juice (65.2%) and therefore, it was released in 1986 for commercial cultivation.

Cultivars of pomegranate can be broadly classified into two groups (a) evergreen (b) deciduous types. Evergreen cultivars have spreading branches with floriferous nature and medium to large sweet fruits with white to light pink aril colour while deciduous types have upright branching habit and small, and very acidic fruits with deep red aril colour. As red arils are preferred, exotic types having blood red coloured arils such as Shirin Allar, Gulsha Red, Gulsha Rose Pink were used to cross with Ganesh in Rahuri in 1976 and hybrids were obtained. Out of many F_1 hybrids, only seven had deep red aril colour but hard seed and poor in taste than Ganesh. Back crossing with Ganesh also did not improve the quality attributes. However, some desirable recombinants in F_2 progeny raised from open pollinated fruits of these hybrids have been identified. The individual plant selections (Sel 5, 303 and 130) were found to be most promising.

Propagation

Propagation through seeds brings about variability and hence vegetative propagation is resorted for multiplication of superior varieties. Common method of propagation is by rooting hard wood cuttings. The cuttings are taken from fully matured wood of about one year old 20–25 cm long. Rainy season is the best period to achieve maximum success.

Planting

Pits of 30 × 30 × 30 cm are made well in advance and filled back with a mixture of FYM and soil. In subtropical climate, where plants remain dormant in winter, planting should be done in the beginning of spring season (February/ March). However, in places where the plant behaves evergreen, the proper time of planting is beginning of monsoons.

Planting distance depends on soil type and climate. In Maharashtra, a close spacing of 0.5 × 2.0 m is recommended to obtain higher yields. In Tamil Nadu, a spacing of 2.5 to 3 m on either way is adopted.

Training and Pruning

Pomegranate plant can be trained on a single stem or in multistem system. In our country, multistem training is more prevalent and useful. If the plant looses one stem it continues to survive. But under Tamil Nadu conditions, single stem with six scaffold branches give more yield. Pomegranate fruits are borne terminally on short spurs arising from mature shoots. These bear fruits for 3 to 4 years. Hence, the gradual growth of new shoots should be encouraged by restricted cutting back of bearing shoots. However, suckers are to be removed regularly to maintain the shape of the frame work and also to maintain the balance between vegetative and reproductive phases.

Manures and Fertilizers

Well rotten FYM at the rate of 20 kg per pit is usually applied at the time of planting. After this, 20 kg FYM is given every year at the break of monsoon with a small quantity of ammonium sulphate. Based on the experimental findings, the following doses are recommended for high yielding in pomegranate.

Age of trees	Quantity of nutrients (g) Plant		
	N	P_2O_5	K_2O
Below 2 years	250	125	125
2 –3 years	500	125	125
3–4 years	500	125	125
5 years	625	250	250

The time of application will vary with the crop. For Amber bahar December-January, for Mrig bahar May-June and for Hasta bahar, October-November are the appropriate periods.

Cropping

Pomegranate starts flowering in about 3–4–years of planting. There are three distinct flowering seasons *viz.*, June–July (Mrig Bahar) coinciding with monsoon, Feb–March (Amber Bahar) and September–October (Hasta bahar). For 'Mrig Bahar' treatment, withholding of water from December to May results in sufficient suppression of growth. Trees usually shed leaves by March end and remain dormant upto May. Then, the field is ploughed, manures and fertilizers are applied and first light irrigation is given in the middle of May and thereafter one or two light irrigations are given until rain sets in. Trees put up new growth, flowers and fruits with crop available in October to

Table 16.28 Plant protection in Pomegranate.

Name of the pests/ disease	Symptoms/damage	Control measure
1. Pomegranate butterfly or Anar butterfly (*Virachola isocrates*)	Female lays egg singly on calyx of flower or young fruit. On hatching the caterpillar bores into fruit and feeds. Subsequently, infested fruits rot due to bacteria and fungus and drop off.	1) Bag the fruits with polythene covers during flowering period to prevent egg-laying when the fruits are upto 5 cm diameter. 2) Spray neem oil 3% or NSKE 5% at the time of butterfly activity. Repeat it if necessary twice at an interval of 15 days. 3) Adopt ETL (5 eggs/ plant with bearing capacity of 60 fruits). Application of methyl parathion 50 EC 1 ml/litre or Carbaryl 0.2 per cent or Phosphamidon 0.03 per cent at fortnightly intervals. 4) Release *Trichogramme chilonis* @ 1 lakh/acre.
2. Fruit spot (*Cercospora* sp., *Colletotrichum gloeosporoides*)	Discolouration of fruits starts from lower part, sometimes from the sides. After few days, discoloured area becomes brown to black and subsequently rot.	These can be controlled by spraying Dithane M 45 or Captan 500 g in 200 litres of water after fruit formation

November. This is the most advisable approach in arid and semi–arid regions with limited resources. 'Amber Bahar' is also taken in Deccan where enough water is available during hot weather. No irrigation is given to the trees after rainy season and they shed their leaves by October–November when shallow digging or ploughing is done. The manures and fertilizers are given in the months of December– January, the first irrigation is given in January and the flowers appear within a month after irrigation. In dry regions, quality fruits are obtained from Amber bahar compared with fruits of Mrig bahar. For hasta bahar, the trees have to be made dormant during August–September bur this is always uncertain because of the rains which occur during this period and therefore, not usually practiced.

Depending on the availability of heat units, these flowers take 5 to 6 months to be ready as mature fruits. They should be harvested mature, which can be judged by change-in skin colour, which turns slightly yellow and fruit looks fully developed and gives metallic sound when tapped. Plucking is done with hands. A pomegranate plant can give 20–25 fruits from 3rd to 4th year and slowly increase its yield upto 100 to 150 fruits after 10 years in a good orchard. These plants if kept properly, continue to give good crop upto 30 years of age.

5. Datepalm *(Phoenix dactylifera)* Family: Palmae

The datepalm is believed to have originated around Persian Gulf and became naturalized in this country in very early times. It is a staple food of the population extending from Western Iraq across Arabia and North Africa. In India, it is grown to a limited scale in Rajasthan, Punjab, Haryana and Gujarat.

The datepalm is monocotyledonous plant with strong straight unbranched stem growing to the height of 300m and above. It is a dioecious species, male and female palms look alike, axillary panicle is a spadix, covered by spathy bract. Fruit is a one seeded berry.

A closely related species *P. sylvestris* is widespread in India and is an important source of jaggery and a drink called 'tadi'. *P. humilis* is an another species found growing wild in Western and Eastern ghat hills in middle elevations. Dates serve as a rich source of sugars and mineral contents (rich in iron, calcium and phosphorous). These are eaten fresh, dried or used as date syrup.

Climate and Soil

The date palm is very exacting in its climatic requirement which according to an Arab saying that date palm should grow with its feet in running water and its head in the fire of the sky. Its successful cultivation requires (l) a long summer with high day as well as night temperature, (2) a mild winter (upto 30°C with low relative humidity and plenty of sunshine for short periods without frosts and (3) absence of rain at the time of flowering and fruit setting. The ideal mean temperature during flowering and ripening of fruits is between 25–29°C. For successful maturation nearly 3000 heat units are required. The heat units available in most of the North-Western districts of India are suitable for date palm cultivation. In Indian continent, in sub-tropics, fruiting period is confined only to February to July when ripening period coincides with rainfall and therefore, fruits do not reach ripening stage and are to be harvested at earlier stage of fruit development. Datepalm can tolerate soils containing more alkali or salts compared to other fruit plants but growth and productivity are greatly reduced. Sandy loam 2 to 3 m deep with good water holding capacity and drainage are the best.

Varieties

The varieties of date are numerous and are generally divided into three groups according to moisture content in the fruit i.e. soft, semi-dry and dry.

1. Soft varieties have a relatively high water content, little or no sugars when fully ripe; also called 'invert sugar dates' e.g. Barhee, Halawy.
2. Dry dates or semi-dry dates possess a high sugar and low water content and retain more or less sucrose when fully ripe; also referred as 'cane sugar dates'. (e.g. Semi dry e.g. Dayari).
3. Dry e.g. Thoory.

Propagation

Dates may be propagated by seeds or offshoots (suckers). When from seeds the population shall be half male and half female (1:1) and no two seedlings shall be alike and a few of these are likely to produce fruits of good quality. A plant raised through 'offshoot', which develops axillary buds on the trunk of 3 years or more and weighing 10 to 25 kg is preferred. The offshoot must have good root system. To promote rooting, the base of the offshoot should be in touch with soil which can be achieved through mounding the soil.

The separation of offshoots from mother palm is a cumbersome operation. The suckers, to be removed, should have its leaves trimmed off, unopened central leaves are left untouched, young inner leaves are half headed back and outer one to two-third. This will not only facilitate separation of offshoots but avoid transpirational losses after separation and increase the percentage of survival. Separation is done with the help of sharp chisel as close as parent palm. The wound if any to mother palm should be preferably coated or painted with coal tar and earthed up at once after the the removal of offshoot.

This detachment can be done in February to March and August to September. These detached offshoots can be directly planted in the field or put in nursery transplanted after one year depending on convenience.

Planting

For planting datepalm offshoots, the field should be thoroughly ploughed, levelled and pits of 1 m^3 are dug at a spacing of 8 × 8 m a fortnight before planting and well decomposed FYM and the soil is allowed to set by irrigating it. Best time for planting is during rainy season.

After Cultivation

Training or pruning is required. Thorns on the basal part of the leaves are pruned to facilitate harvesting. Young tree should receive 0.5 kg N per tree upto bearing age. At bearing age, it should be increased to 1.0 kg. However, application of Potassium and Phosphorus should be monitored according to soil conditions.

Usually in sandy soils, irrigation needs to be given at 7 to 10 days interval during summer and 15 to 30 days during winter. Young plants are more demanding water and require frequent irrigations for better survival, quick growth and precocious bearing.

Being dioecious, it requires artificial pollination mainly hand pollination for commercial crop production. Pollination is done by hanging strands of male flowers in an inverted position. Alternatively, pollens are allowed to dry, collected and applied on cotton balls and these balls are placed in the female tree. Usually one male flower is sufficient for 40 to 50 females. Mechanical pollination can also be done with clusters but hand pollination gives best result.

Average yield under rainfed conditions in India is from 40 to 50 kg per palm while under irrigated conditions, upto 200 kg of fruits per palm can obtained.

6. Fig *(Ficus carica)* Family: Moraceae

It is extensively cultivated in countries around the Mediterranean and in India it is grown around certain pockets like in Pune of Maharashtra, Shrirangapatnam in Karnataka and Lucknow in U.P. It is a large shrub or low growing deciduous tree. The pear shaped fruits are variable in size and colour and have high sugar content. This species is characterized by tree bearing fruits of two distinct sex form (a) fruits with long styled pistillate flowers only and (b) monoecious with both short-styled pistillate and staminate flowers in the same fruits known as 'capri-fig'.

The pistillate forms have been classified into three types. (1) the common (2) the San Pedro and (3) the Smyrna type. The important characters of these three forms are

(a) **Common type:** Trees of this type may or may not produce a first crop borne on wood of the previous season's growth. They do produce a more or less profuse second crop or even a third crop late in the season in axils of leaves of current growth. Both the crops are produced by parthenocarpically.

(b) **San Pedro type:** The first crop is produced by parthenocarpically but second crop requires pollination or the crop will drop.

(c) **Smyrna type:** Trees generally produce main or summer crop only. Its crop is non-parthenocarpic and needs pollination and fertilization.

Caprifig produces atleast three crops annually being inhabited by the fig insect *Blastophaga psenes*. The first crop is initiated in late fall, remains on the tree during the winter with the insect in larval stage and matures in spring. The second crop is the most prolific crop, appearing on the wood of previous season's growth. It matures in early summer with the staminate flowers producing pollen profusely in nature the pollen become attached to bodies of the insects as they emerge and is then carried to receptacle of either Caprifig where the insects oviposits for short styled flowers or to other figs having long styled flowers unsuitable for oviposition. The process of transfer of pollens is known as 'Caprification' which help in the formation of seeds or achenes which have been classified as 'druplets'.

The fruit or receptacle borne by figs is known botanically as 'syconium' which may be defined as a form of inflorescence in which the flowers borne on the inner walls of a hollow receptacle. The fruits may be eaten fresh or dried, preserved, candied or canned. It can also be made into jam. Great bulk of the fruits is dried. Fruits are valued for their laxative properties and have great medicinal properties and are rich in protein and digesting enzymes.

Soil and Climate

It is a deciduous plant but does not require chilling to break dormancy. It is grown in tropics and sub tropics. In North India, it remains dormant during winter and puts up new growth with the advent of spring season. However, in Western India, during monsoon (Aug–September) it becomes dormant and puts up new growth in October. It can be grown in wide variety of soil but prefers light soils with low level of organic matter. It can tolerate salinity and waterlogging to certain extent. Medium heavy soils are considered better.

Varieties

Important varieties cultivated in India are 'Pune Fig', Black Ischia, 'Brown Turkey', 'Bangalore' and Marseilles.

Propagation

Fig is commercially propagated by hardwood cuttings, taken from two or three years old shoots with 1.5 to 2 cm thickness and short internodes. Propagation by air-layering, ring budding and side grafting on *F. glomerata* rootstock is also possible.

Planting and Aftercare

Spacing varying from 4 × 4 m to 6 × 6 m is followed in various parts of India. In north, planting is done in early spring when the rooted cuttings are still dormant and in tropical India in the beginning of monsoon.

Young plant should be trained properly to give proper head and scaffold branches. It can be trained on single stem *vs* multistem but three stems per plant give high production in initial years.

Pruning of fig is important. It generally gives two crops in a year, the first on previous years growth and the second crop on current season and therefore, pruning intensity and its type will depend on bearing and type of crop desired. Practice of pruning is greatly variable. In Karnataka, the trees are headed back to about 2 buds of previous year's growth every year to keep them dwarf. This reduces the productivity but quality of the fruit is better. In Tamil Nadu and Maharashtra, light pruning is practiced, which gives higher production but quality of fruits is poor. In India, pruning is done in December to a level of 3 to 4 buds on previous season's growth.

Annually for a bearing tree, 50–75 kg of FYM, 1.0 kg N, 0.25 kg each of PK should be applied. Time of manuring shall vary with place and accordingly in North India it is done in spring while in South and West India, September is the preferred period.

It is fairly drought tolerant but during summer when the fruits are developing and ripening, irrigation twice a month is common throughout India. This gives a large and sweet fruits. Excess of irrigation makes the fruits insipid.

In North India, the spring crop ripening in May is taken as the main crop. In central and South India, the fig crops during July to September and February to May. The latter crop is sweeter and highly valuable. Fruits are harvested when mature, which can be judged on the basis of colour development. Sometimes, fruits are allowed to fall from the tree on their own and allowed to dry and then collected. Fruits intended for distant market are picked when slightly immature. A good harvest may give 300 to 500 fruits per tree which depends on the size of tbe tree and method of training.

D. Temperate Fruits

The temperate fruit industry of India is presently concentrated in the following regions:

1. North Western Himalayan Region (NWHR)	Comprising of the states of Himachal Pradesh, Jammu & Kashmir and Hilly areas of Uttar Pradesh
2. North Eastern Himalayan Region (NEHR)	Comprising of the states of Arunachal Pradesh, Sikkim and hilly areas of Assam, Nagaland, Mizoram, Meghalaya and West Bengal.
3. Hilly areas of other states	Mount Abu (Rajasthan), Nilgiris Hills, Kodaikanal (Tamil Nadu) and hilly areas of Madhya Pradesh.

The temperate fruit industry in India is, however, more developed in the North Estern Himalayan region which contributes the major part of temperate fruit production in the country. These temperate fruits alone occupy 11.50% of the total area under fruits. The important fruits under this category are

1. Apple (*Malus pumila*) Family: Rosaceae

Among the temperate fruits, apple is the premier fruit of the world. More than 80% of world's supply is from Europe. It is believed to have originated in South West Asia and introduced in India by British settlers after 1870. Apple fruit is a symbol of health and 'An apple a day keeps the doctor away' is the old saying indicating the importance of this fruit in maintaining the health. It is rich in sugars (10–11%), potassium (120 mg) besides vitamins like thiamin, B_6, C and biotin etc. It is used for table purpose and in canning industry for making jams, jelly etc.

The genus *Malus* has about 25 species and several sub species of so called 'Crab apples', many of which are cultivated as ornamental trees for their attractive flowers and fruits. *Malus pumila* refers to the cultivated species and *M. sylvestris* and *M. baccata* is the wild crab apple found wild in the Himalayan region. Fruit is a pome and stone cells are absent in the flesh.

Climate and Soil

Apple can normally be grown in areas which experience 800–1600 chilling hours (the number of hours during which temperature remains at or below 7°C during the winter season). These conditions are mostly availahle at an elevation ranging from 1,600–2,300 m above M.S.L. However, in South India, the chilling hourse are below 500 and hence this climate is described as 'warm winter climate'. Extreme cold temperature on the other hand causes freezing injury. For optimum growth and fruitfulness, apple trees need 100–125 cm of rainfall, equally distributed over the growing season. Excessive rains and fog during fruit maturity phase result in poor quality fruits. Abundant sunshine is important in growing apple since it is largely responsible for good colour development.

Apple can be grown in a variety of soil, however, a well drained, slightly acidic (pH 6.0), loamy soil or red laterite soil with a good depth upto 60 cm is considered ideal.

Varieties

The available cultivars in apple may be classified into different groups.Based on (a) the time of maturity as early,mid and late season varieties (b) based on the ploidy level as diploid, triploid and tetraploid cultivars (c) based on the purpose as dessert, processing and culinary apples and (d) based on the chilling requirement as high chilling cultivars and low chiilling cuhivars.

The important cultivars recommended for North Indian hills are:

Himachal Pradesh: Starcking Delicious. Starkcrimson Delicious. Rich-a Red, Red Delicious, Red Gold, Golden Delicious, Granny Smith, Amber, Gold Spur, Tropical Beauty etc.

Jammu and Kashmir: Irish Peach, Cox'S Orange Pippin, Ambri, Golden Delicious, Red Delicious etc.

Uttar Pradesh Hills: Early Shanburry, Fanny Golden Deicious, Mcntosh, Red Delicious etc.

Tamil Nadu Hills

Early Varieties (April–May)	Irish Peach, Coonoor–3
Mid Season Varieties (June–July)	Carrington, Winterstein
Late Season varieties (Aug–Sept)	Rome Beauty, Parlin's Beauty and Kodaikanal Beauty

Kodaikanal Beauty is a clonal selection identified at Horticultural Research Station, Kodaikanal (Tamil Nadu Agricultural University). It is highly adaptable to warm winter climate of South Indian hills, fruit is medium (150 g) in size, with crisp flesh.

Propagation

The choice varieties are propagated through grafting/budding. The selection of the right type of rootstock is an important factor in apple propagation. The rootstocks for apple can be of two broad categories *viz.,* (1) Seedling rootstock-usually raised from the seeds of crab apple or commercial cultivars after stratification (2) clonal rootstock – multiplied through clonally by mound or stool layering. As the seedling rootstocks do not impart uniformity and are susceptible to woolly aphid infestation, clonal rootstocks are more popular. The East Malling Research Station (EMRS), England selected and developed a series of apple rootstock which were named as Malling stock (M) in 1912. These rootstocks have a wide choice for tree vigour i.e as dwarfing rootstocks (M 9), semi-dwarfing rootstocks (M 2) and vigorous rootstocks (M 12), but they are susceptible to woolly aphids. Hence, John Innes Horticultural Institute, Merton in collaboration with EMRS started releasing woolly aphid resistant stocks *viz.,* Malling Merton (M.M) stocks. Important among them are M.M.I04, 106, 109, 109, 778, 779 etc. Whip and tongue method of grafting is followed in Tamil Nadu while in North Indian hills, budding and tongue grafting are followed. One year old grafts are suitable for planting in the main field.

Preparation of the Field

Pits of 6 0 cm^3 are dug at 4m spacing in Tamil Nadu hills while in North pits of same size at a spacing of 6.0-7.5 m are recommended for semi-vigorous and vigorous cultivars. For high density planting, a spacing of 1.5 to 3.0 m on either way is recommended. The grafts are planted during June–July under South Indian hills while in the North it is planted during Dec–March. However, if irrigation facilities are available, it is better to plant after winter in South Indian hills also.

Irrigation

If the soil moisture is inadequate, young grafts planted should be provided with regular irrigation till the plants establish. Drip irrigation is increasingly used in most of the apple orchards in Israel where higher yields are obtained due to this type of irrigation.

Training and Pruning

In Tamil Nadu hills, the plants are allowed to grow to a height of 60–70 cm by removing the side shoots and are then trained to open center system. In the North, they are predominantly trained to modified leader system or spindle bush system. Pruning is done during the winter months to remove the dried and diseased shoots, criss cross branches, water shoots etc.

Manures and Fertilizers

Under Tamil Nadu conditions, application of 250 g N, 1 kg P_2O_5 and 1 kg K_2O per tree is recommended for bearing trees. In Himachal Pradesh, application of 700 g N, 350 g P_2O_5 and 700 g of K_2O is recommended. Fertilisers may be applied 30 cm away from the trunk in old trees. Time of application is October–November in south Indian condition while in the north, Phosphorus and Potash are applied during Dec–January and half of the dose of N during 2–3 weeks before flowering and the second dose one month after the first application.

The trees often exhibit deficiency of nitrogen, zinc, boron, manganese and calcium and they may be corrected as indicated below:

Element	Chemical	Concentration (%)	Frequency	Time
Zinc	Zinc sulphate	0.5	1–2 spray at 5 days interval	May–June
Boran	Boric Acid	0.1	1–2 spray at 5 days interval	June
Manganese	Manganese sulphate	0.4	1–2 spray at 5 days interval	June
Calcium	Calcium Chloride	0.5	2 sprays at 15 days interval	June–July

Many cultivars of apple will bear heavily resulting in smaller, low and unmarketable fruits and limb breakage. Thinning of fruits tends to maintain tree vigour and regular production of quality fruits. In biennial bearing cultivars, the thinning program during 'on year'.

Cropping

After the winter is over, the buds start breaking off causing flowering in apple. In the hills of Tamil Nadu where prevailing low chilling weather, sometime, may not be adequate to force the buds to break. This can be overcome by spraying sandolin A (Dinitro–Ortho cresol) at 0.5% in 2 per cent mineral oil emulsion. This has resulted in the breaking the dormancy of the buds and substantial increase in the yield of the fruits.

Most of the commercial varieties suffer from self-unfruitfulness and high level of female sterility. The flower structure in these cultivars does not facilitate pollination by bees and the duration of the stigma receptivity is also short. Considering all these odds, excellent pollination conditions are necessary for good fruitset. One of the suggested measures is to plant pollinizers. A pollinizer should produce viable pollens which should be compatible with the main cultivars. Besides, their bloom period should also coincide with the main cultivars, they should by themself be also good and quality cultivar. To provide adequate pollination, atleast 33 percent of the total population should be of pollinisers. Good pollinizers are Tydeman's Early, Red Gold and Golden Delicious. Besides, top working of a branch of commercial cultivar with pollinizer, providing 5 bee hives per hectare of orchard and placing flower bouquets of the pollinizers of Delicious cultivar trees also facilitate better pollination.

The problem of pre-harvest drop is more severe in early cultivars where 40–60 per cent of fruits drop. In the mid-season cultivars *viz.*, Red Delicious, Royal Delicious and Golden Delicious, the pre-harvest drop is about 15 to 20 per cent. To contrct pre-harvest drop, spraying of NAA at 10 ppm is most effective. The spray should be done about a week before the expected drop.

There is a problem of fruit setting in marginal apple growing areas. Miraculan at 0.75 ml/L or Paras at 0.6 ml/L or Biozyme or Protozyme 2 ml/L can be sprayed at bud swell and petal fall stages to improve the fruitset. Apple surface colour development is greatly hampered in low lying growing areas due to warmer conditions. Such fruits fetch poor price. At very high altitudes, the maturity is delayed and fruits fetch poor price due to glut at late arrival in the market and deep colour.

Many cultivars of apple will bear heavily resulting in smaller, low quality and unmarketable fruits and limb breakage. Thinning of fruits tends to maintain tree vigour and regular production of quality fruits. In the biennial bearing cultivars, the thinning can be done in order to have enough bloom in the following years. Application of NAA 10 ppm 7 to 15 days after petal fall or when average fruit length is about 15 mm is most effective for fruit thinning.

Harvesting

Apple grafts take 3–5 years for first bearing and commercial crops are obtained from 10th year onwards. The correct stage for harvest is judged by the disappearance of the green colour and the brightening of yellow or whitish area. Under cold storage, apples are stored at –1.0 to 2.0°C at 85–100 per cent relative humidity. On an average, a well grown tree under Tamil Nadu hill conditions give 15–25 kg of fruits per tree as against 100 kg of fruits per tree in North Indian hills.

Pear *(Pyrus communis)* Family: Rosaceae

Next to apple, pear is an important temperate fruit. Its origin is believed to be mountanous regions of Western China. They are rich in sugars and vitamins. The fruits are mostly used as table purpose only but in western countries, a major portion is used for juice, wine making, canning, drying etc.,

The pear belongs to the genus *Pyrus* which includes many species among which *P. communis* includes all cultivated types, *P. pashia* is the commonly used rootstock in the North and *P. pyrifolia* is the commonly used stock pear in South Indian hills.

Climate and Soil

It is a temperate crop grown at an elevation of 1,400 m to 2,200 m M.S.L. in South Indian hills while in the North; it can be grown from foot hills to hills (i.e. 600–2,700 m M.S.L) experiencing 500–1,500 chilling hours. Pear grows best in deep, medium textured and well drained soils having depth atleast 2 m.

Varieties

Existing varieties may be grouped as European types, Asian types and their hybrids. European types are mostly the descendants of *P. communis*. The important varieties grown in India are:

North Indian hills	Varieties
(a) High Hills	Early China, Laxton's Superb, Bartlett, Delicious, Conference, Doyenne and Ducomice
(b) Mid-Low Hills	Sand Pear, Kieffer and China pear
(c) South Indian Hills	Kieffer, Kotagiri, New Pear, William and Jargonelle

'Kieffer' is a well adapted, widely grown variety. The fruit is medium to large, brownish, gritty and hard. Bartlett" also known as 'Williams' or William Bartlett' is the most popular variety all over the world, fruit is large, ovate pyriform in shape, flesh is firm, melting, juicy.

Propagation

Propagation is generally through vegetative means especially through grafting. In the North, the rootstocks commonly used are the one year old seedlings of *P. pashia* and *P. pyrifolia.* Clonal rootstock of Quince *(Cydonia oblonga)* is also used and is having a dwarfing effect on its scion cultivars. In south India, rooted cutting of country pear *(P. pyrifolia)* is exclusively used as rootstock for all cultivars. Whip and tongue method of grafting is normally followed in pear during dormant months. The existing country pears are top worked successfully with improved varieties during December employing cleft grafting technique.

Planting

One to one and half year old grafts are ready for planting. Pits of 60 cm^3 are usually dug at 5 m interval and the grafts are planted June-July to October–November seasons.

Training and Pruning

A number of training systems have been evolved currently to increase orchard efficiency and secure early and high return. Some of the common systems are 'free standing', pyramid', 'spindle' and 'palmette'. In India, open centre system is generally followed.

Pruning consists of removal of dead, diseased, broken and criss-cross branches during November–December months. All the shoots may be headed back to half during dormant season so as to induce spreading of branches.

Manures and Fertilizers

In Tamil Nadu, bearing pear trees are manured @ 40 kg of FYM and 2.0 kg of Fertilizer mixture (9 : 9 : 9 NPK). Pears are sensitive to boran deficiency. Cracking of young fruits and pitting of older fruits are the common symptoms of boron deficiency. Therefore, boric acid spray at 0.1% can be given.

Cropping

Pear trees commenece flowering at the age of 5–8 years mostly on spurs and terminal shoots. By and large, cross pollination is considered for the production of good crop which is achieved by planting pollinisers at the time of planting itself. The cultivars are either self-fruitful or partially self fruitful or self-unfruitful. Even then the extent of self-fruitfulness is determined by the location, season and tree vigour. Even under optimum plant vigour and weather conditions, cross pollination results in a higher fruit set (70%).

Fully mature pear fruits but still firm and green are harvested for canning and distant market and they are picked at a slightly late stage for local consumption. Early varieties come to harvest in May–June and late varieties in July–October under South Indian hills. On an average, a tree will yield 100–120 kg of fruits while Cvs. William and Jargonelle yield 30–40 kg of fruits per tree.

3. Plums *(Prunus sp)* Family: Rosaceae

Plum is also one of the important temperate zone fruit crops. They are rich in sugars and carotenes. They are used either as fresh dessert fruit or cooked. Plums are grown on a commercial scale in Jammu and Kashmir, Himachal Pradesh and the hills of Uttar Pradesh and to some extent in the Nilgiris and Pulney hills.

Plums belong to the genus *Prunus* and based on the centre of origin, the following important types are recognized:

S. No.	Species	Common Name	Centre of Orgin
1.	*P. domestica*	European plums	Europe
2.	*P. insititia*	Damson plums	Western Asia
3.	*P. cerasifera*	Cherry plum	Western and Central Asia
4.	*P. salicina*	Japanese plum	China
5.	*P. americana*	American plum	North America

Bulk of the cultivated plums belongs to *P. domestica.* The fruit exhibits both yellow and green ground colours and also both red and blue skin colours. The cultivated varieties of European plums have been classified into:

(a) **Prunes:** Blue or purple in colour, the fruit is oval with bulging vertical side and compressed bilaterally. The distinguishing feature of this group of plums is the 'high sugar content' which makes them suitable for drying without removal of the pit. All prunes are plums but all plums need not be prunes.

(b) **Green gage plums:** Fruit is greenish yellow, round and of high quality. Fruit is characterized by yellow skin and flesh.

(c) **Lombard plums:** Fruit is purplish red in colour.

The fruits of Japanese plums are usually large and heart shaped, often with a pronounced apex which distinguishes them from other type of plums.

Climate and Soil

Plums exhibit variations in their climatic requirement and hence it is found growing from the subtropical plains to the temperate high hills of India. European plums demand comparatively more chilling requirement than Japanese plums. In south Indian hills, the plums thrive in hill station from 1,400 to 2,200 m above MSL.

Red laterite soils with good drainage and rich in organic matter is more suited.

Varieties

Majority of the cultivars belong to European group followed by Japanese plums and Damson plums in order.

Important cultivars are:

S. No.	European plums	Japanese plum
1.	Green gage	Kelsey
2.	Golden Transparent	Santa Rosa
3.	Victoria	Starking Delicious
4.	Stanley	Alubokharo, Rubio, Hale, Gaviota, Beauty, Shiro, Abundance.

Based on the season of maturity, varieties can be grouped as early, mid season and late varieties for South Indian Hills as indicated below:

Early (April–May)	Mid Season (June–July)	Late (July–August)
Alubokharo, Rubio	Hale, Gaviota, Abundance	Shiro, Kelsey

Propagation

Plums do not come true to type from seed and hence its cultivars are propagated by budding on the seedlings or vegetatively propagated rootstocks obtained by means of layering or cutting. Under Indian condition, plums are shield budded on common peach seedlings of 1–2 years old during November–January.

Myrobalan plum *(P. cerasiformae)* is commonly used rootstock in England and U.S.A for European and Japanese plums.

Preparation of Field and Planting

Pits of 0.60 m^3 are dug two months before planting. Budded plants of one year old are generally used for planting. A spacing of 4.0 × 4.0 m is usually followed in South Indian conditions while 5 to 6.0 m spacing on either side is followed under North Indian conditions.

Training and Pruning

The growing point is tipped at a height of 50–60 cm to allow side shoots to develop. The trees are usually trained to modified central leader systems or the open centre system. Japanese plum cultivars are mostly adaptable to the open centre system.

Pruning in the prebearing period is usually kept light and corrective. In bearing plums, pruning is directed towards increase the size of fruits, preventing the breakage of branches as a result of heavy crops and promoting continuous new growth. Japanese plums require heavier annual pruning as compared to the European plums as the former one tends to overbear and may set such heavy crops that the fruit size is considerably reduced.

Manures and Fertilizers

Under Himachal Pradesh the following doses of fertilizers are recommended per tree.

Age of the tree	CAN (g)	Super Phosphate	Muriate of potash
Newly planted plum tree	450	190	150
(The same dose is increased every year)			
Sixth year	2700	950	750

In South India, for bearing tree, annually during October–November, a dose of 30 kg of FYM and 1.5 kg of the fertilizer mixture (9:9:9) are applied per plant.

Cropping

The trees start flowering and fruiting 4–5 years after planting and a regular bearing is obtained after 10 years. The existing cultivars may be classified as self-sterile, partially self-fertile or self-fertile. Besides, flowers of plums exhibit heterogamy ie differences in the length of stamens and styles exist. Hence, plum cultivars grown commercially require pollination to set satisfactory crops. A cultivar is said to be self fruitful, if it exhibits an average of 30 per cent of fruitset which is considered adequate for a commercial crop. The cultivars that are self-unfruitful set only 1 to 2 per cent. Hence, these cultivars should be interplanted with pollinisers. Most japanese plum cultivars are self-unfruitful and require cross-pollination and some cultivars like 'Santa Rosa' and 'Beauty' are partially to fully self-fruitful and may bear heavier crops when planted among other cultivars. It is advisable to keep 5–10 bee hives per hectare on the orchard periphery to increase the fruitset.

Some plum cultivars tend to overbear and removal of part of the crop (thinning) enables the remaining crop to improve the size, quality and grade of plum fruits. Thinning is achieved by hand thinning and also by spraying growth regulators like NAA and 2,4,5-T at 25 and 50 ppm.

Harvesting

Plums should be picked when mature enough to continue it's ripening process of the tree and plums grown for distant markets should be picked when firm ripe. Plums are picked by hand and packed in shallow crates. Each well grown tree can bear 25–30 kg of fruits.

4. Peaches *(Prunus persica)* Family: Rosaceae

It requires relatively low chilling climate than plums, pears or apples. Its cultivation is confined to the mid-hill zones (1500–2000 m MSL) of the Himalayas and also in sub-mountainous regions and western parts of the country in Punjab, Haryana, Delhi and Western Uttar Pradesh and to a limited extent in the South Indian hills *viz.,* Nilgiris and Upper Pulney hills. Fruits are rich in sugars (6.0–9.5%) and are also a good source of vitamins and minerals. They are grown both for fresh market and processing.

Peaches are believed to have originated from China. It is a small to medium sized tree. Fruits are variable in size, shape, colour of the skin and flesh usually with a fuzzy skin. There is an another group of peaches called 'nectarines' in which they do not have fuzzy skin. Nectarines are usually smaller in size and have greater aroma. These might have developed from peaches as budsports.

Climate and Soil

Peaches like other temperate fruit crops require certain amount of chilling period below or at 7.2°C to break its dormancy. Low chilling cultivars require 200–300 hours of chilling and chilling hours of above 600 is much favoured by most of the other peach cultivars which produce quality fruits. Warm to hot summer is favourable for the development of quality peaches. In South Indian hills, it is grown from 1,300 to 2,200 m above MSL. Rainfall is desirable during summer season for good peach crop but high humidity is conducive to the attack of pests and diseases. Red laterite soil with good drainage and high organic matter is suitable for peach cultivation.

Propagation

In South India selected varieties of peaches are shield budded on common peach seedling rootstocks. In the North, peach is propagated by shield or ring budding and occasionally by splice or cleft grafting. The commonly used rootstocks are the seedlings of peach, plum and apricots.

To raise peach seedlings for rootstock purpose, peach seeds are collected from mature fruits and they do not germinate unless they are after-ripened under moist conditions at 0°C temperature for a long period (70–80 days) i.e.

they are to be stratified. Under South Indian conditions the stones are collected during May–June and are allowed to after-ripen in a shady place for a period of 3–4 months and during September–October, they are sown in a trench of 60 cm deep in alternate layers with river sand. They are periodically kept moist. After a period of 3 months; the stones start sprouting and these sprouted seeds are set in the nursery at 15 × 15 cm spacing. They will reach a buddable stage within 10–12 months time. In the North the stones are sown in shallow beds first during October–November and the sprouting embryos are transplanted in the nursery during spring.

Varieties

Peach cultivars can be also classified as early (March), mid season (July) and late varieties depending upon their fruiting season as below:

S. No.	Varieties	Early	Mid-Season	Late
1.	North Indian Mid hills	Alton Early White Giant	July Elberta Kanto–5	J.H. Hale
2.	North Indian low hills and valley	Sharbati, Shan-i-Punjab		
3.	South Indian hills	Floridasun	Shah Pasand	Shan-i-Punjab Sharbati

Planting

Pits of 60 cm cube are dug at a spacing of 4.0 × 4.0 m in square system two months before planting. One year old budded plants are used for planting during June–July or October–November months.

Training and Pruning

The plant is tipped at a height of 60 cm and thereby 3–4 shoots are allowed to grow, thus trained to open centre system. Pruning in peaches is an important operation. It bears fruit bud in the past season shoot *ie* one year old shoot and not on the current season shoot. A branch which fruited once will not fruit in the subsequent year, thus leaving barren limbs on the tree. While pruning, therefore, old growth is removed, retaining one or two buds which may in succession produce fruiting area for the coming years. Young trees should receive more of heading back and older trees should receive more of 'thinning of shoots' to regulate the crop of high quality.

Manures and Fertilizers

Under South Indian conditions, bearing trees are manured 25 kg of cattle manure and I kg each of standard mixture (9 : 9 : 9) per plant during October–November. Peach orchards in North Indian hills and plains are applied with 50 kg FYM + 500 g N, 250 g P_2O_5 and 600 g K_2O per plant only once during December–January.

Cropping

Young peach tree should make more than 30–45 cm of new terminal growth annually and 30 cm of growth is sufficient to maintain a mature bearing tree in good vigour and full production. Peaches generally come to bearing within 3–5 years of planting depending upon the cultivars. Among the temperate fruit plants, peaches are the first to come to flowering under South Indian conditions and they do flower even in November–December depending upon the altitude. All commercial cultivars of peaches (excepting J.H. Hale) are self fruitful and hence will set good crop, even if planted in compact blocks. NAA 10 ppm spray reduces fruit drop. The fruits should be harvested when they start developing blush on pale background and decrease in firmness. The fruits are perishable. Trees of 10 years and old may yield around 15–20 kg of fruits.

Table 16.29 Plant protection in pome fruits.

S. No.	Pests/disease	Damage/symptoms	Control measure
1.	Woolly aphids (*Eriosoma lanigerum*)	The insects infest the roots and stems of the trees causing unsightly galls	Spray Methyl demeton @ 0.03% to 0.06% to control the aphids. Use resistant rootstalks M 778, 799, MM 104, MM 110, MM 112, MM 113, MM 114 and MM115. The parasitoid *Aphelinus mali* should be conserved in the field.
2.	Borer (*Zeuzera sp.*)	The caterpillar bores into the stem or shoot causing the plants weak or dying in case of severe infestation	Clear the hole with flexible wire and then insert cotton wick soaked with methyl parathion 0.2% or dichlorvos 0.15%
3.	Fruit fly (*Dacus zonatus*)	Flies lay eggs on the fruits and the maggots feed on pulp, soon rotting starts and the fruit is rendered unfit.	Spray malathion 50 EC @ 2 ml/lit or fenthion 100 EC @ 1 ml/lit. Set up methyl eugenol 1% trap with malathion 50 EC 1 ml/lit between 6 a.m. and 8 a.m. Use polythene bags fish meal trap with 5 g of wet fish meal + 1 ml. dichlorvos in cotton. 50 traps are required/ha, fish meal and dichlorvos soaked cotton are to be renewed once in 20 and 7 days respectively.

(Contd.)

Table 16.29 Plant protection in pome fruits.

S. No.	Pests/disease	Damage/symptoms	Control measure
4.	Lichens (Algal and fungal symbionts)	Various coloured radiating growth of lichen appear on the tree.	Spray quick lime 1 kg/10 litre of water after pruning.
5.	Apple scab (*Venturia inequalis*)	Typical scale symptoms appear on the above ground parts, particularly the leaves and fruits, severe infection leads to premature drop of leaves and formation of misshapen knotty fruits.	The following spray schedule is recommended: 1. Silver tip to green tip stage – Captan 0.2% or Mancozeb 0.4% 2. Pink bud or after 15 days – Captan 0.2% or Mancozeb 0.3% 3. Petal fall-Carbendazim 0.05% 4. 10 days after Petal fall – Captan 0.2% or Mancozeb 0.3% 5. 14 days after fruitset – Captafol 0.15%
6.	Peach leaf curl (*Taphrina deformans*)	Curling, puckering and distortion of young leaves occur. Infected portions develop pink or red bronze colour.	Spray copper oxychloride 0.3% or Carbendazim 0.05% during winter and spring months.

Table 16.30 E. Minor Fruits.

Common, botanical name and family	Varieties	Method of propagation	Spacing	Special practices if any	Season of bearing	Yield	Remarks
1. Strawberry (*Fragaria vesca*) Rosaceae	Phenomenal, Majestic, Sujatha, Labella	Runner, splits	45 × 45 cm	Remove the flowers and runners for 3-4 months, apply mulches	Can be allowed to bear 3 months after planting	About 1 kg/m²	A temperate fruit requires cooler climate, now can be grown from 1000 m to 2200 m MSL
2. Litchi (*Litchi chinensis*) Fa: Sapindaceae	Soharapur Dehradun	Seed or through buddings grafting	10 × 10 m	—	Starts bearing from 6th year onwards, ripe during summer months	80-100 kg/tree	A subtropical fruit
3. Durian (*Durio zebethinus*) Fa: Bombacaceae	Chanex, Deception	Seed or grafting or budding	10 × 10 m	—	Harvested during August-September	40-50 fruits/tree	A humid zone fruit, fruits when ripe emits an offensive odour
4. Rambutan (*Nephelium lappaceum*) Fa: Sapindaceae	Seematjan, BR-1	Seed or graft, budding	7 × 7 m	—	Harvested during August-September	25-50 kg /tree	A humid zone fruit, often called hairy litchi
5. Persimmon (*Diospyros kaki*)	Fugel, Jiro Hachiya	Inarching on its own root stock	5 × 5 m	—	Flower in February and	20-25 kg/tree	Fruit will not ripen by itself

(Contd.)

Table 16.30 E. Minor Fruits.

Common, botanical name and family	Varieties	Method of propagation	Spacing	Special practices if any	Season of bearing	Yield	Remarks
Fa: Ebenaceae		or on *D. lotus*			fruit available during August		and hence kept in air-tight boxes
6. Carambola (*Averrhoa carambola*) Fa: Oxalidaceae	Dah pon, Golden Star	Layering or seed	7 × 7 m	–	July to September	100-150 kg of fruits per trees	A humid tropical tree
7. Passion fruit (*Passiflora edulis f-flavicarpa*) Fa: Passifloraceae	Golden passion fruit and purple passion fruit, Kaveri	Seeds and cleft grafting	2.5 to 4.0 m on either way	Trained over trellises	Flowers throughout the year	80 fruits per vine	Kaveri is a F_1 Hybrid between purple and yellow passion fruit, developed at IIHR, Bangalore
8. West Indian Cherry (*Malphigia punicifolia*) Fa: Malphigiaceae	No named cultivar	Layering	2.0 × 2.0 m	Trained to get a bush stature	Throughout the year with peak bearing in December – January	10-15 kg of fruits from a 5 year old bush	Fruits are the richest source of vitamin (1400 mg/100 g of edible portion) and hence called vitamin 'C' tablet fruit.
9. Jamun (*Eugenia jambolana*) Fa: Myrtaceae	Ram Jaman	Seed, Inarching	10 × 10 m	–	Start bearing from 6th year, fruit ripe during June-July	50-80	–

Part 4

ORNAMENTAL HORTICULTURE

17

Ornamental Horticulture

GARDENING — AN INTRODUCTION

Importance of gardening

The importance of gardening is realized by every individual now in the globe. The gardens serve the purpose of public recreation and education. They serve to refresh the body and the mind. They are virtually a retreat for the public from the harsh strains and stresses of life. Gardens have become essential to modern civilization. Home gardening is an integral part of family life and it lends itself for the family to relax and enjoy the nature and beauty of garden. It also helps to develop some useful hobbies and skils in flower and dry arrangements. In the recent years, landscape architecture has come to play a vital role in the developments of home, public institutions, public places like bus stand, railway stations, towns and cities.

Landscaping the roadsides on cities and towns is also given much importance in tropical countries like India as they provide shade and beauty to that area. In foreign countries, landscaping of homes and public buildings is an important art and there are professional and qualified landscape architects who do the job. This field of gardening is slowly gaining momentum in our country, especially in big cities. The government has also established many parks and a few Botanic gardens in our country. In most of the big cities and towns, Agri-Horticultural societies have been established with the sole objective of promoting horticulture, more particularly gardening. These societies organize annual flower shows and garden competitions to promote healthy attitude among the public to develop gardening.

History and development of gardening in India

It is believed that the gardening in India is as old as its culture. Five to six thousand years before Christ, gardening was in vogue. It is understood that Indians were the first to choose gardens as the ideal atmosphere for meditation. Gardening has been mentioned in classical literature of India. 'Asokavana' of Ramayana was believed to be established with trees, shrubs, ponds, etc.

The use of 'Brindavan' (landscaped city) and 'Rajavanam' (large park where kings go for hunting) did find very frequent mention in ancient literature. Many trees were considered, 'sacred' to Hinduism e.g. *Ficus religiosa, Saraca indica, Aegle marmelos* and *Mimusops elengi*. During Bhuddha's period (about 563 B.C.), gardening received further encouragement. Gardening is considered as a source of peace and solitude. Imitation of nature and improvement over nature were considered as a primary goal in these types of gardening. King Ashoka could be considered as the 'Father of road side avenue planting' as he was the first king to order planting of road side avenues in India. From 6th to 10th Century, gardening received a great deal of encouragement from Indian kings.

During the time of king Somadeva (11th Century) and king Hamira (13th Century), gardening developed into a fine art. The 'Vanas' and 'Ashrams' gave way to gardens established by kings and nobles and gardening became virtually the monopoly of the feudals. The period between the beginning of the 14th Century to the end of 16th Century showed sudden change in the style of gardening due to the influence of Moghul emperors who ruled India during these centuries. King Feroze Shah and King Babar brought with them the finest art of gardening from their country and the Moghul style gardens were established in Delhi, Hyderabad, Agra, Kashmir, Lahore and Pinjore. Some of these gardens are still famous to this day, e.g.

1. Garden at Fatehpur Sikri (U.P.) by king Akbar
2. Moghul garden at Pinjore (Punjab) by king Fadai Khan
3. Shalimar Bagh at Lahore–by Shah Jahan
4. Tajmahal Garden in Agra

These Moghul gardens had a distinct style. They were laid out near rivers or rivulets, slope of a hill or river bank. The garden was square or rectangular with paths running parallel to water courses. Running water was the life and soul of the garden and the water channels were paved with tiles of brilliant blue colour to reflect the sky and give the impression of depth. Either side of the central channel of water was planted with flowering trees, shrubs and creepers like rose and jasmine. All the Moghul gardens had walls surrounding them.

During the 16th Century when British came to India, they introduced the styles of gardening of England and continental Europe. The first style developed by the British was the formal or symmetrical style of gardening. By 18th Century, this style became monotonous and then the informal style or natural gardens began to develop. The modern gardens developed during 19th and 20th Centuries, involved the combinations of the formal and informal styles. The important features in English gardens are lawn, rockery, mixed borders of herbaceous perennials, annuals and shrubs. Lal Bagh at Bangalore, Government Botanic Garden at Ooty, Sims Park at Coonoor and Bryant Park at Kodaikanal are examples of gardens established in British style.

Designing a Garden

A garden may be defined as an area embellished with plants, a valuable and pleasurable adjunct to a house. A mere collection of plants will not make a garden. It is the skillful arrangement and disposition of plants over area making a design or pattern or picture as it were that forms a garden. Therefore, gardening warrants apart from knowledge of the science of plant growing, an artistic taste on part of the gardener. Often the two terms *viz.*, 'botanic garden' and 'park' are confused. A botanic garden is a place where plants are assembled for study and display purpose. They serve as a centre of plant introduction. In the garden, the plants are so assembled to bring out a landscaping effect. A park on the other hand is a pleasure ground suitably landscaped with plants and play equipments for recreation. Normally, botanic garden is situated on a vast land while the park is raised in a smaller area, in the heart of the cities or towns, due to the pressure on land. The 'country park' is larger in extent and much of the natural vegetation as is possible is retained in landscaping it. It is also used to stock deer, rabbits and other animals in a natural setting (e.g. The Guindy park in Chennai), The term 'National Park' indicates a larger forest area set apart for the protection and conservation of fauna and flora facing extinction and hence it comes under forest activities. Hence, whether it is a botanic garden or a park, the following are the important garden components, parts or features.

1. **Lawn:** Lawn is an important component of a garden. No garden is complete without the presence of a lawn.
2. **Shrubs and shrubberies:** Groups of shrubs planted at corners will be useful in natural designs. Shrub borders of informal (allowed to-grow without trimming) or formal (trimmed to a height) can border the main walks and paths. A shrubbery is a border planted with different kinds of shrubs and a shrub border is where only one kind of shrub is used.

3. **Climbers and creepers:** Climbers and creepers are used to grow against or over walls, trellises, arches, pergolas, arbours, pillars or large trees. These climbers may be light or heavy depending upon the amount of wood they produce.

4. **Trees:** They form the main framework of the garden. They are generally planted along the boundaries. Masses of trees in a corner will help to give depth and perspective. A spreading tree is an ideal feature for 'picnic' ground in a large public garden. Trees with beautiful or fragrant flowers or handsome foliage or form and trees which provide adequate shade are grown in gardens.

5. **Flower beds and borders:** Several flowering annuals and herbaceous perennials can be grown in beds and borders. Flower beds of simple designs can be laid out on the outskirts of lawn along the foundation of buildings, in the path leading to the entrance of the house and on sides of foot steps. Borders are continuous beds of more length and width containing plants of heterogeneous characters as distinguished from flower beds which are composed of plants of one kind only. Borders are named as 'shrub border', 'herbaceous border' or 'mixed border' according to the plant materials used to fill them. These borders can be had on the sides of path, walks and drives or in front of shrubberies and trellises with climbers.

6. **Ornamental hedges:** A good live hedge is essential to enclose a garden (e.g. *Thevetia nereifolia* and Divi etc.). Ornamental internal hedges can also be planted inside the garden with attractive foliage or flowering shrubs. These are pruned to maintain a height of 50 to 60 cm. They help to divide the garden into a number of parts; each will have its own distinct features.

7. **Edges or edging:** These are the materials of any description which are used in gardens for dividing beds, borders etc. from roads, walks or paths, demarcating spaces allotted for particular purposes as flower beds. These can be either dwarf growing plants (Eupatorium, Alternanthera) which would stand frequent trimming or they may be made of bricks, stones or concrete slabs.

8. **Drives, roads, walks and paths:** All these should occupy minimum space and not be too many in number. They should serve to link one part with the other part. Paths may be made up of earth, brick and concrete or be paved. Paved paths are particularly effective in formal gardens. Paving can be done by flat stones or concrete slabs or bricks. Sometimes paving with irregularly sized stones to create an odd pattern will result in a 'crazy path'. The inter spaces can be planted with ground spreads.

9. **Rockery:** This is intended to bring together in a short space an idea of a mountain or alpine garden with plants growing in the crevices of rocks. This is an elevated structure resembling a miniature mountain range or the slope of a hill with a few dominant peaks or valleys.

10. **Carpet beds:** In large public gardens, close growing plants like Verbena or Altenanthera are used to form certain designs or letters of alphabets. Foliage plants are better suited than flowering plants as they stand severe clipping much better. Carpet beds are troublesome to maintain in good health. They require constant attention. The plants should be trimmed now and then, not allowing them to overgrow.

11. **Topiary:** Certain plants are often trimmed to shapes of animals, birds, seats, etc. The shrubs which are amenable for bending and withstanding frequent trimming are suitable for developing 'topiary'. Cupressus, Casuariana and Bougainvillea are suitable for topiary work.

12. **Trophy:** It refers to the arrangement of potted colourful foliage or flowering shrubs and flowering annuals or herbaceous perennials around a tree or any central object such as a statue. These potted plants are often arranged in tiers.

13. **Conservatory or green house or fern house or fernery:** There are certain ornamental plants with beautiful foliage or flowers or both which cannot thrive in the open, exposed to direct sun or wind. Such plants can be grown in conservatory or green house or fernery wherein required shade, humidity and cool environment are provided. By having a small pool inside, the conservatory is rendered cool and humid. Shade is provided by growing a creeper over the roof which will not shut out light completely. Ferns, Anthurium, Dieffenbachia, and Peperomia are some of the examples of plants which are commonly grown inside a green house.

14. **Sunken garden:** This is formed taking advantage of a natural depression. The garden goes down through a series of terraces to a small pool or a fountain at the bottom. In the terraces, flower beds and strips of lawn are laid out. It breaks the monotony of flat ground in a garden.

15. **Garden adornments:** There are several garden adornments and accessories such as fountains, statues, garden seats, ornamental pots and pillars, arches and pergolas, trellises, hanging baskets, tubs, vases and urns with plants which make the garden more enjoyable. Playing of a fountain is an interesting feature in garden and the water in the cistern should be kept clean. Garden seats made up of stones, concrete or metal are placed under the tree. Handsome tubs, vessels and urns are utilized to display plants in conspicuous places. Arbours, arches,

pergolas and trellises serve as support to several beautiful plants and to dispel monotony in garden. Arbours are usually open on all sides, Very often a long wall or the end of a pergola leads to an arbour. Arches are generally erected over walks, usually at the entrance and are usually two metres in height. Pergolas are series of arches connected over a walk.

Landscape Gardening

Landscape gardening involves integration of space, plants and other accessories in an aesthetic manner to meet the needs of man. This branch of horticulture is an important field in which architecture, art and horticulture (knowledge of plants and their cultural requirement) are combined to achieve beauty and utility. The space here refers to the land as well as vertical space. Landscaping mainly concentrates on spatial design which depends on the land. Land is the surface on which the landscape designs are executed. The land forms may be classified as plains, plateau, hills and mountains. Generally, plain land is featureless and one with natural slopes is a blessing to the eyes. Accessories include mountains, rivers, natural ponds, statues and other garden adornments. Plants include all type of plant materials viz., trees, shrubs, creepers and climbers, annuals, biennials, herbaceous perennials, bulbous plants, ornamental palms, lawn grasses, etc. The primary requisite before landscaping a place is the 'plan' to a scale indicating the location of permanent features such as road, rocks, large trees, buildings or the central object viz., the house, etc.

'Aesthetic' refers to 'sense perception of beauty' and is chiefly concerned with visual perceptions only in the garden. The components of beauty of landscape are colour, shape, texture, pattern, line and point.

1. **Colour:** It is an important component in landscape design. A light spectrum gives seven colours viz., violet, indigo, blue, green, yellow, orange and red (VIBGYOR). In this, blue, yellow and red are called 'primary colours' and orange and violet are called 'secondary colours' since the latter colours can be formed by combining two primary colours as indicated below: (Fig. 17.1)

Primary colours	Secondary colours
Blue + Yellow	Green
Yellow + Red	Orange
Red + Blue	Violet

When primary and secondary colours are mixed, intermediate colours are formed. Colours are sometimes known as 'hot' or 'bright', (e.g. red, yellow and orange) and 'cool' or 'light' ones (e.g. green, blue, white). The predominant colour in nature is green and it is due to the presence of green pigment called 'Chlorophyll'. Various shades and intensity of green colour commonly occur in plant kingdom depending upon plant species.

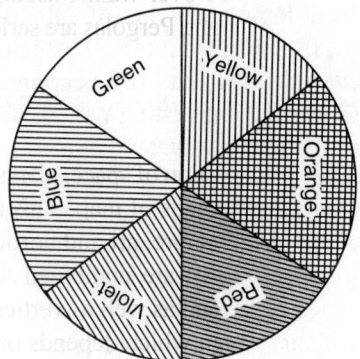

Fig 17.1 Colour wheel.

The major contribution of other colours in landscape design is by flowers. Flowers are available in almost all colours except black. The available colour can be combined in a judicious manner in landscape design. When we use a single colour, it is called 'monochromatic' but it often results in monotony. When we combine two colours it is called 'dichromatic'. Two opposing colours of the colour wheel are employed to contrast each other, or by combining a dark and a light colour or a hot and a cool colour. For instance, in a lawn its monotony can be well broken by growing red salvia either at the centre or at the fringes. In 'triads' or trichromatic colour schemes, primary colours are combined together or secondary colours are alone combined together. When various colours are combined, it is called a 'polychromatic scheme'. In a mixed flower bed, polychromatic effect is created by growing annuals which are available in different shades.

2. **Shape:** Shape in a landscape garden refers to the symmetrical or asymmetrical growth feature of plants - trees, shrubs, creepers etc. Symmetrical shapes are often created by pruning and training.

3. **Texture:** Texture refers to the surface structure of plants with reference to their leaves, bark, flowers and other organs. Often conifers which have needle shaped leaves provide a fine texture whereas the broad leaved plants give an effect of coarse texture. Similarly, varying textural forms of bark such as smooth, rough, split or peeling are

interesting and provide visual contrasts. In a lawn based on the type of grass used, we experience smooth or coarse surface.

4. **Pattern:** In the natural arrangements of plant parts, a definite scheme or model is seen and is often referred as 'pattern'. The feathery leaves of the palms, the drooping branches of Salix or *Ficus benjamina*, the fan arrangements of leaves as in travellers palm, the tiered branching of the *Terminalia catappa* and the bearing of fruits in the trunk (*Crescentia cujete*) offer interesting pattern in garden.

5. **Point and line:** Theoretically, a point is a speck on the surface and when an innumerable number of points are arranged continuously, a line is formed. In a landscape, the lines formed by the trunk and branches or the outlines formed by borders, beds, edges or roads and walks are important. This may be straight or curved. Straight lines are generally featureless while smooth and gentle curves are having pleasing and aesthetic effects. Straight lines are often common in formal gardens.

Beauty principles

The beauty components by themselves do not convey beauty unless or otherwise they are combined together in a proper manner. The following principles are to be followed to combine the above components to derive a maximum aesthetic value.

1. **Balance:** The balance in landscape design is a visual equilibrium of different garden components. In the formal gardens, balance is achieved by positioning plants and other landscape objects at equal distance from a real or imaginary plane or axis. In the informal or naturalistic designs, balancing is achieved mainly by the neutralizing effect. For example, in a garden, if a big tree is there in one side, it can be balanced by positioning a small tree opposite to it and here balancing is achieved by contrasting effect.

2. **Proportion:** It refers to the relative composition of the different components to the whole of the landscaping unit. In other words, all the garden components should be in the right required proportion and one unit should not play a dominant or major role. For instance, if a garden has more area under lawn with a minimum area for other garden components such as tree, shrubs, annuals etc., then it is not proportionate.

3. **Scale:** Scale is a relative term. In a big lawn, if a very small water pool garden is created, the lawn will be dominating and rendering the pool ineffective, owing to the differences in their dimensions. This can be overcome by adopting appropriate scale.

4. **Rhythm:** Rhythm generally refers to the measured cyclic repetition. In landscaping, rhythmic effect is created by repeating colours, shape etc. at measured intervals. This helps to break the monotony of its flat surfaced hedge.

5. **Hormony**: It refers to the pleasing effect obtained due to the proper arrangement and combination of the various garden features. No part of the garden should detract the value of the others and one part should try to merge with others so as to produce a hormony as whole.

Lay out of garden

In laying out a garden, several factors like taste and judgement of the owner, position of the house, size of the house, the extent of ground, availability of water, topography of the land and labour availability are to be taken into consideration. In the history of garden making, there have been two styles. The 'natural style' which is also known as 'landscape style' aims at imitation of nature inside the garden and strives to produce a rural effect with large open lawns. These lawns may be provided with clumps of trees and shrubs on its fringes. The other style, *viz.*, 'formal' or 'geometrical' or 'symmetrical style' aims at providing harmony and contrast in colour and a balance as a whole. In this type of gardening, one half of the design is a counterpart of the other. Sometimes, these two types of styles may be combined to enjoy the benefit of both the styles and this combined style is known as '*artistic*' or 'free' or 'picturesque' styles. This style dispenses the 'formality' and flatness of the geometric style and offers much great freedom in the layout.

The following are the important principles to be taken into consideration in laying out a garden:

1. The garden should produce an attractive picture in the landscape with the house.
2. Mass planting has to be preferred to individual planting to have greater range and variety of forms, shades and texture.
3. Each garden part should be distinct by itself and set off from every other area. But these different units should contribute finally one strong and homogeneous effect.
4. The building and the garden should merge each other. This is achieved by maintaining a continuity by 'base planting' around the buildings by keeping plants in the verandah and in rooms.
5. If a lawn is established, it should be centrally situated in front of the building. The lawn must be proportionally spacious according to the size of the building.
6. The size of the various features of a garden should be in proportion to the size of the ground. The size of the building, the number of

paths, the length and width of shrubs, borders, flower beds etc., should be all in proportion to the extent of the garden. Similarly the trees selected should be suited to the size of the buildings. A small low house should not be dominated by large, tall trees. A tall story building should not have small, short tree.

7. The landscape plan for a home gardening must take into consideration at least four general uses of land.

    (a) The public areas, the frontage are viewed from the street or by one approaching the front door. This area is put into limited use by the family. It is desirable to keep this land minimum in area. Lawn, trees to frame the house and some shrubbery may be planted in this area.

    (b) Foundation area includes the small area surrounding the house. It is very important and this should receive much and early attention. This area may be assigned to plant attractive shrubs or herbaceous perennials.

    (c) The utility or service area includes driveways, cloth yards, a place for garbage, tennis/shuttle court, etc.

    (d) Outdoor living or private area is planned for privacy. It is usually away from the approach area. It may be located to either side of the house with screening accomplished by fence, shrubs or trees. It usually occupies a larger area.

8. In our South Indian conditions, home landscaping should primarily aim at providing shade. Home ground may be assigned judiciously for outdoor living, parking the vehicles; children play area, corners for growing sacred flowers and flowers for personal decorations and vegetable gardening.

9. Locating or arranging the plant materials is the final step in developing the horizontal plan or blue print. The functions or purposes of plant materials include shade, screening, privacy, background, accent, framing, wind break and specimen plants. Shade is obtained primarily from broad-leaved evergreen trees in our country. The shape and ultimate size of trees for shade are important criteria of selection and placement. Screening is frequently attained by effective use of shrubs, although vines on trellises and fences are also effective screening materials. In some places, clipped hedges are also used as screening materials. Privacy in the home grounds can be obtained from buildings, fences and plants. Background plantings generally consist of trees or shrubs along the back or distant end of the grounds. They give a feeling of depth and soften the harsh lines produced by man made structures. They add coolness to the surroundings during summer.

Accent points of the landscape are those which have some distinct features and should catch the 'eye' immediately. Framing plants are those which encircle the main point of interest, the house in a home plan. The trees and shrubs on either side both in the back yard and fore ground form the framing view of the house. Wind breaks are usually of tall growing trees with strong branches and are placed on the boundaries to reduce wind velocity.

18

Lawn

A lawn has a charm all its own. It serves to enhance the beauty of a garden, be it large or small. It enhances the beauty of surrounding objects, whether they are trees, shrubs or flower beds.

Grass being one of the most hardiest perennial herbs, it is not difficult to keep a lawn in good condition provided one takes good care of it. The view of the lawn from the verandah and the windows of the house should be free and uninterrupted. A spacious lawn may be provided with beds of flowering plants, a group of shrubs or trees of attractive form or specimen shrubs as *Thuja orientails* or traveller's palm or large succulents and there in pockets made in the lawn to form colonies of themselves and bear their beautiful flowers. Roof garden lawns on terraces have become a common sight and feature in almost all cities today. Before laying a roof lawn, the drainage aspect of the roof and the load which the roof can withstand are to be considered. Good water proofing treatment has to be given to the roof along with slope to drain water into seepage collecting troughs on the terraces itself. To reduce the weight of the soil on the roof, lighter soil media like peat, vermiculite, coconut husk powder, etc. are to be mixed with farm yard manure and leaf mould. The depth of medium may be around 12-15 cm. Terrace lawns, apart from their aesthetic value are reported to lower the room temperature considerably.

In making a lawn it is essential that the ground be prepared properly, the surface being thoroughly dug and uniformly leveled, coarse stones, roots and other obstacles being removed and surface made smooth. Provision for drainage of excess rain water should be made if the ground is not sloppy. Let the ground settle down during the first showers. Any weed that may come up may be removed.

Methods of lawn making

After the ground has been prepared, the lawn may be laid out by adopting any of the following methods:

(a) From seed

In our country, lawn is seldom made by sowing seeds. A lawn from seed is thought of only when grass roots are not available. About 30 kg of seeds may be necessary for a hectare. The soil should be reduced to fine tilth and given a light rolling. The seeds should be sown on a windless day, evenly and thinly and covered with fine light soils. The ground should be rolled again and watered liberally with a water-can or with hose pipe fitted with nozzle. The seeds take about 5 weeks for germination. For the first few times, the grasses are cut with a scythe. Lawn mower may be used when the roots have established and are spreading.

(b) By turfing

Turfs are pieces of earth with compact grass on them. These turfs should be cut uniformly thick in squares from a place where the grass is short, compact and free from weeds. They should be spread upon the prepared ground side by side and beaten down flat with a turf-beater. Any cavities in between should be filled with fine soil. Then the entire turfed area should be rolled and watered liberally. This is the most expeditious way of making a lawn.

(c) By turf plastering

The roots should be cut into bits, 2 to 4 cm long. In a pit, a mixture consisting of two parts of these roots and a part each of well decomposed horse or cow manure, fresh cow dung and red earth or loam should be made and rendered into a paste by stirring it with required quantity of water. The paste is evenly spread over the prepared ground which has previously been watered. The ground is covered with a litter or a layer of coarse manure to minimize evaporation and protect the roots from the heat of the sun. If there is no rain for the next two days, it is watered liberally. Grass will shoot up in a fortnight. It is cut with the scythe to start with and after three months, the mower is used.

(d) By dibbling roots

This is the cheapest but slowest method. Small roots should be dibbled about 15 cm apart into the prepared ground when it is wet after rain. The roots spread and grow underground in the course of six months making a fairly compact lawn by frequent mowing, rolling and watering.

After Care

After making a lawn; rolling, moving, watering and restoration of patchy places should receive regular attention.

1. Weeds should be pulled out as soon as they appear otherwise they soon spread, seeds multiply and overpower the grass. The gaps occupied by the weeds are filled with grass roots and fine soil.
2. In the absence of rain, it is watered regularly at 10 days interval.
3. Mowing should be done at brief intervals and never allowed to produce seed stalks.
4. Heavy roller should be used frequently but not when the ground is either too wet or dry. A mower should not be employed until a firm green sword has been formed. The grass is first cut with a sickle and the surface is then rolled.
5. To have a perfectly green lawn, it is fed once a month with a liquid manure prepared by dissolving 45 g of Ammonium sulphate in 5 litres of water.
6. Once in a year, it is raked before the rain, top dressed with a rich mixture of well decomposed manure and sand. This will stimulate the grass with a new vigorous growth.

A spacious lawn though beautiful will often be monotonous. So, to break the monotony, some beautiful tree or shrub is recommended as 'single specimen' in the lawn.

Plants Suitable for Planting in Lawns

Trees:

1. *Amherstia nobilis*
2. *Callistemon lanceolatus*
3. *Magnolia grandiflora*
4. *Cupressus macrocarpa*
5. *Pinus longifolia*
6. *Thuja orientalis*
7. *Araucaria excelsa*

Shrubs and creepers

1. *Agave americana*
2. *Furcraea gigantea*
3. *Musa superba*
4. *Bougainvillea spectabilis*
5. *Cestrum nocturnum*
6. *Dombeya spectabilis* and.
7. *Hibiscus sp.*

Sometimes mounds are created in the spacious lawn to break the feeling of monotony. Monotony is also overcome by planting flowering or foliage shrubs on the fringes of the lawn or by having flowered beds or borders on the selected area inside the lawn.

Commonly employed grass species to make lawn are given below:

Scientific Name	Common Name	Texture	Situation
1. *Cynodon intermedius*	Hariyali	Medium fine	Open sunny location, drought resistant
2. *Stenotaphrum secundatum*	Buffalo grass	Coarse texture	Shady situation, requires frequent watering
3. *Sporobolus tremulus*	Chain grass	Fine	Saline, open sunny situation
4. *Poa annua*	Blue grass	Medium fine	Prefers acid soil, higher elevations
5. *Pennisetum clandestinum*	Kikuyu grass	Rough	Acid soil conditions, higher elevations
6. *Zoisia japonica*	Japan grass	coarse grass, slow in growth	Poor sandy soil, open sunny situation
7. *Cynodon sp.*	Bermuda grass or Hyderabad grass	Fine texture	Open sunny situation, needs mowing

Astroturf

It refers to a synthetic lawn popularly used in developed countries in roof garden and stadium. It dispenses the normal maintenance usually required for normal lawns. However, it requires constant water sprinkling to bind the synthetic fibre to provide a surface akin to a lawn carpet.

19

Arboriculture

ARBORICULTURE

Arboriculture refers to the cultivation of trees and by convention; it means cultivation of trees for the purpose of science, education, recreation and landscaping. This is different from 'Silviculture' which means growing trees for forestry purpose. In a botanic garden, the place where different groups of trees are planted and maintained for scientific study is called an 'arboretum'. The cultivation of trees is of ancient origin and mention has been made of it in a number of ancient writings. Historical records show that as early as 200 B.C., planting of trees in avenues was done by the Great Emperor, Asoka.

Specific Benefits of Cultivation of Trees

1. A judicious planting of trees contributes much to the beauty, variety and the enjoyable features of gardens.
2. It contributes to ecological balance. A good group of trees support many a number of other flora and fauna in nature and leads to ecological balance.
3. They act as weather conditioners by purifying and humidifying the air, screening the dust carried by wind, checking the radiation of heat and obstructing glare.
4. They limit the environmental pollutions, prevent noise and dusts etc. Each 30 m width of trees can absorb about 6–8 decibels of sound under city. Neem and tamarind trees are highly prized for their ability to filter the noise, dust and light.

5. Trees afford shelter and shade and make summer times more pleasant.
6. Some trees fill the air with the delightful fragrance of their flowers and some others provide fruits.

In a garden, the trees form are essential component. Along with the shrubs, they form the framework of garden and are permanent, easy to grow and require very little attention. Trees are either deciduous or evergreen. Deciduous trees generally produce their blooms when they have shed their leaves or just after or are being clothed with fresh foliage. In India, the period between February to June is remarkable for the flush of blooms of many of the trees. Some trees flower during specific months. Others do flower throughout the year. A due proportion the flowering trees selected in such a way that one or the other of them is in flower throughout the year is very much desired. With regard to the nature of usage in a garden, trees can be grouped into the categories:

(a) Those trees which are grown for their beautiful or fragrant flowers
(b) Trees mainly grown for the richness and attractiveness of their foliage or form or both
(c) Trees which are grown for their shade.

Selected examples under each group are given in the Tables, 19.1 and 19.2 respectively.

Trees in Landscaping

Woody plants like trees are the backbone of a landscape. Trees provide skyline to the landscape, background for highlighting architecture, screening of less pleasing sights, as roadside avenues as screens for privacy from roads and flats opposite. One of the places wherein trees can be extensively planted is on the roadside which is known as avenue planting. This has two aspects namely beauty or utility and both. These aspects can be combined by careful planning and judicious selection of the right tree species. Roadside plantations help in noise abatement. To reduce the noise generated by high speed traffic on national highways to tolerable limits, about 20 to 30 m wide belts of trees and shrubs may be necessary. To reduce the noise generated by moderate speed in the cities, 7 to 15 m wide belts of trees and shrubs may be required. Evergreen trees are better for noise abatement than deciduous which do not afford a barrier to sound when leafless. Recently, planting in strips along railway lines is also becoming popular. The main objectives are stabilisation of railway track and protection of railway track against erosion and checking of the shifting sand in desert areas getting on to the railway track. From the point of view of the safety of railway track, the first row of trees should not be planted very close to the railway track. The first row of trees is accordingly recommended to be planted at a distance of about 7.5 m from the centre of the track. There are many designs which can be adopted for planting roadside trees.

1. **Formal planting:** Here, there are definite planting patterns, almost always in lines parallel to the road, either in single or double or multiple rows depending upon the space available. Planting consists of a single variety (homegenous pattern Fig.19.1 and 19.2) or more than one variety consisting of contrasts of colour, form and foliage (Fig. 19.3). For every few hundred meters pattern be altered to break the monotony. Similarly, contrasts in form like spreading and tall growing trees can be altered; the taller to provide skyline and the former to provide shade.

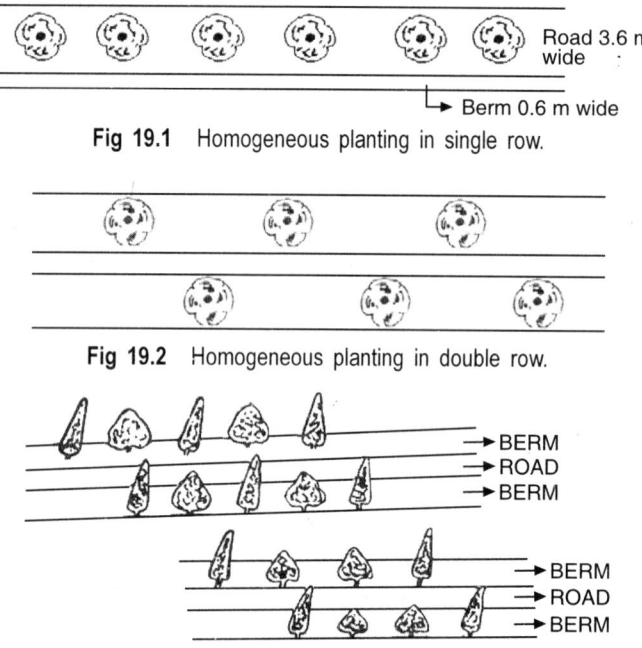

Fig 19.1 Homogeneous planting in single row.

Fig 19.2 Homogeneous planting in double row.

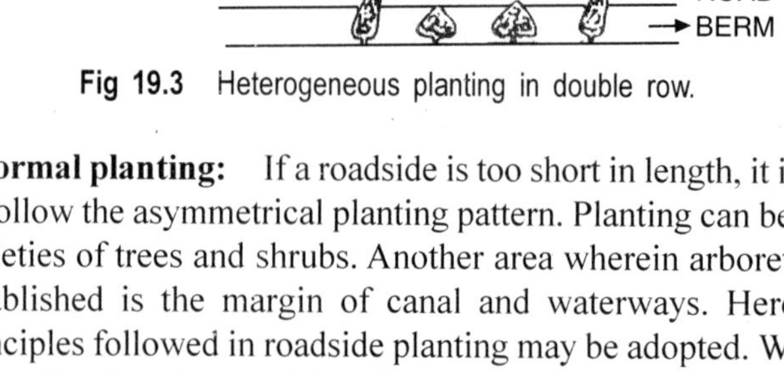

Fig 19.3 Heterogeneous planting in double row.

2. **Informal planting:** If a roadside is too short in length, it is advisable to follow the asymmetrical planting pattern. Planting can be done with varieties of trees and shrubs. Another area wherein arboretum can be established is the margin of canal and waterways. Here also, the principles followed in roadside planting may be adopted. Water loving plants like *Bassia longifolia, Terminalia sp., Syzigium cumini* may be selected for the above purposes.

Trees may be also established in large waste lands known as 'woodland planting'. This sort of tree planting serves to conserve rare trees and associated insects and birds, besides providing recreation and education. The principle behind establishing such woodland is mainly to provide pleasure and utility to the community. It is established that for every 1000 population, one acre of woodland must be provided. In woodland, single trees can be grown or grouping of trees can be had. Grouping may be based on bark, foliage, flower characters or height of the plant or based on their soil and climatic requirements.

Propagation of Trees

Most ornamental trees are propagated by seeds, some by suckers and a few by cuttings and layers. Germination of seeds may be a problem in some species. Tree species like *Lagerstroemia, Mimusops elengi* exhibit hard seed coat imposed dormancy and this can be broken by soaking in water for 24 hours or mechanically breaking open the hard seed coat. In other group of tree species like *Ficus,* the treatment of the seeds includes grinding with sand or acid scarification. In tree species like *Thespesia populnea* and *Erythrina indica,* vegetative propagation is obligatory. The following hints may be borne in mind in the maintenance of a good arboretum:

1. Proper initial planning before actually taking up the planting as the established trees should not pose a hindrance to the underground pipe system or electric or telephone system.
2. The growth habit of the trees; certain trees like *Peltophorum ferrugineum, Melia azedarach, Delonix regia* and *Azadirachta indica* are fast growing and establishing quickly. The tree species like *Pterocarpus santalinus* and *Tamarindus indica* are slow in their growth rate and take considerable time to establish.
3. Dead and old branches are sources of accidents and wherever possible such branches must be removed. Sometimes, big limbs which interfere with the telephone or electric lines need to be removed. In all cases, a clean cut should be given.
4. In certain old trees, a big cavity is normally seen in the trunk. Such cavity is to be treated by scraping the internal surface to remove remnants of dry rot fungi. These surfaces are then painted with a fungicide coating. The cavity may be filled with inert materials like jelly, sand, cement or petroleum tar and paddy husk.

Table 19.1 Selected flowering trees.

S. No.	Botanical name and family	Common Name	Flowering time	Colour of flowers	Characters
1.	Acacia dealbata (Leguminaceae)	Silver wattle	Feb-March and July–August	Yellow	A small tree, finely cut leaves underside silver white, propagated by seeds, suckers.
2.	Amherstia nobilis (Leguminaceae)	Queen of flowering trees	April-May	Vermillion colour	6 to 12 m high, leaves coppery green, clustered and hanging.
3.	Bauhinia purpurea (Leguminaceae)	Mandharai	Throughout the year	Purplish rose colour	Good sized trees, fragrant flowers
4.	B. tomentosa	Thiruvatti	Throughout the year	Sulphur yellow	Small tree, 2 to 3 m height
5.	Bignonia magapotamica (Bigoniaceae)	Trumpet flowers	March-April	Light pink	Handsome deciduous tree 7 to 9 m height
6.	Calophyllum inophyllum (Guttiferae)	Punnai	May–June	Fragrant white	A beautiful evergreen tree with large racemes, round fruits, raised from seed.
7.	Cassia fistula (Leguminaceae)	Sarakonnai or Goldenshower	February–May	Bright yellow	Medium size beautiful tree with long pendulous racemes, propagated by seed and suckers
8.	Cassia grandis	Pink shower	March April	Rose pink colour	Spreading quick growing tree, pinnate leaves, pods are long.
9.	Delonix regia (Leguminaceae)	Gul Mohar May flower	April–May	Orange scarlet	Tree beautiful in bloom. Raised from seed

(Contd.)

Table 19.1 Selected flowering trees.

S. No.	Botanical name and family	Common Name	Flowering time	Colour of flowers	Characters
10.	*Erythrina indica* (Leguminaceae)	Kalyana murungai or Indian coral tree	March–May	Scarlet red	Propagated by seed and cuttings
11.	*Jacaranda mimosaefolia* (Bignoniaceae)		March–May	Blue colour	Deciduous often grows to 10 m height, pretty foliage, feathery
12.	*Melia azedarach (Meliaceae)*	Persian Lilac	March–May	Lilac	Deciduous 15 to 20 feet height, bipinnate leaves small heliotropic scented lilac flowers in large panicle.
13.	*Michelia champaca* (Leguminaceae)	Shenbagam	April–May and Sept–October	Light yellow whitish and red	Seedling takes three years to bloom.
14.	*Parkia biglandulosa* (Leguminaceae)	Badminton ball tree	April–May	White	Pretty foliage, pinnate leaves, small white flowers with long peduncle, propagation from seed
15.	*Peltophorum ferrugineum* (Leguminacea)	Copper shield	April–May	Pale yellow colour	Quick growing tree, fine graceful feather foliage, pinnate leaves, raised from seed.
16.	*Spathodea campanulata* (Bignoniaceae)	Indian Tulip tree	June–July	Orange, scarlet	Leaves are glossy and bright green raised from seed.
17.	*Callistemon lanceolatus* (Myrtaceae)	Bottle brush tree	April–May and Aug–Sep	Brilliant crimson scarlet	A small tree with narrow lanceolate leaves, propagated from seeds or layers

(Contd.)

Table 19.1 Selected flowering tree.

S. No.	Botanical name and family	Common Name	Flowering time	Colour of flowers	Characters
18.	Eucalyptus ficifolia (Myrtaceae)	Flowering gum	December-March	White to pink, scarlet or deep red	A popular medium tree seen in South Indian hills
19.	Lagerstroemia flosreginae (Syn. L. speciosa) (Lythraceae)	Pride of India	April-May	Mauve	A large deciduous tree, easily propagated from seeds.
20.	Magnolia grandiflora (Magnoliaceae)	Champa tree	May–June, occasionally during the other months also	Creamy white coloured scented	A medium to big but slow growing tree, propagated by layering
21.	Millingtonia hortensis (Bignoniaceae)	Indian cork Tree or Tree Jasmine	June-November	Fragrant pure with jasmine like	A tall stately rapid growing tree, propagated from seed or by sucker,
22.	Saraca indica (Leguminaceae)	Asoka tree	February-May	Yelowish orange to orange scarlet	An indigenous evergreen, slow growing, small tree, propagated from seeds.

Table 19.2 Selected foliage trees

Botanical name and family	Particulars
1. Adenanthera pavonina (Leguminaceae)	Popularly called 'Bead tree' or Kundumani, a quick growing shade tree, producing red beads like seeds.
2. Albizzia lebbek (Leguminaceae)	Tamil: Vahai, quick growing shade trees with thin feathery foliage
3. Araucaria excelsa (Coniferae)	A very tall tree, conical in shape, short, slender, horizontal branches, starting from the trunk with a certain regularity.
4. Azadirachta indica (Meliaceae)	Tamil: Vembu, medium sized ever green tree with foliage light green, serrated leaves.
5. Casuarina equisetifolia (Casuarinaceae)	Tamil: Chavukku, a lofty tree, rapid growth, slender branches, ornamental, fuel, value, propagated by seeds.

(Contd.)

Botanical name and family	Particulars
6. *Cupressus macrocarpa* (Pinaceae)	Cypress, evergreen tree suited to higher elevations propagated by cuttings and seeds.
7. *Ficus elastica* (Moraceae)	Indian Rubber Tree, quick growing often reaching 15 to 60 m height, smooth shining leaves.
8. *Filicium decipiens* (Burseraceae)	Popularly called 'Fern leaved tree' grows 5-10m high, producing bright green compound leaves resembling fern fronds, propagated from seeds.
9. *Grevillea robusta* (Proteaceae)	Silver oak, reaches 8 to 11 m height, handsome foliage, propagated from seed.
10. *Kigelia pinnata* (Bignoniaceae)	Spreading, moderate sized shade tree, popularly called 'sausage tree'. It produces fruits measuring 30-60 cm length.
11. *Pinus longifolia* (Coniferae)	A big tree thriving at 500 to 2000 m elevations above sea level, leaves are pendulous, needle-like.
12. *Polyalthia longifolia* (Annonaceae)	An elegant evergreen tree, with lance-shaped polished leaves, propagated through seeds.
13. *Ravennala madagascariensis* (Scitaminae)	Popularly called 'Traveller's palm bearing distichous leaves resembling those of banana with an appearance of a gigantic fan, propagation through seeds or suckers.
14. *Samanea saman* (Leguminacese)	Rain tree, large wide spreading tree propagated by seed.
15. *Swietenia mahagoni* (Meliaceae)	Evergreen, good-looking, tall large, very slow growing, good shade tree, popularly called 'Mahogany Tree'.
16. *Tectona grandis* (Verbenaceae)	TeaK: it is truly a grand and majestic tree with large leaves.
17. *Terminalia catappa* (Combretacae)	The Indian Almond tree-quick growing tree reaches about 10m height, handsome stately growth with branches 10 m spreading horizontally and coming from the main stem or axis in whorls, propagation from seed.
18. *Swietenia mahagoni* (Meliaceae)	Mahogani tree: evergreen good looking tall tree with good shape and attractive foliage.

20

Shrubs and Climbers

SHRUBS

Shrubs are plants generally with woody stems, smaller than trees and bigger than most herbaceous plants. Shrubs are either deciduous or evergreen. They can be classified under three heads for horticultural purpose:

1. Those grown for their attractive flowers (Table 20.1)
2. Those grown for their handsome foliage (Table 20.2) and
3. Those grown for their attractive ornamental berries (Table 20.3).

Importance of shrubs in garden

(i) Being permanent, they form part of the frame ork of the garden.
(ii) They form the chief features of landscape gardening placed in front of tall trees and joining the spacious lawn etc.
(iii) Shrubs which are amenable for frequent training are chosen for topiary work.
(iv) Tall growing shrubs often serve as screen.
(v) They are useful as a single specimen in the lawn.
(vi) They can be trained to form standards i.e., trained to single stem and allowed to branch out and form a handsome head only above a particular height. e.g. *Bougainvillea, Ixora, Murrya exotica.*

Table 20.1 Shrabs for attractive flowers.

S. No.	Botanical Name	Family	Key features of the flowers
1.	Acalypha hispida	Euphorbiaceae	Long drooping spikes of crimson red flowers
2.	Allamanda grandiflora	Apocynaceae	Yellow funnel shaped flowers
3.	Artabotrys odoratissimus	Annonaceae	Scented flowers, green and turn to yellow on ripening
4.	Barleria cristata	Acanthaceae	Flowers are pink, violet, or yellow
5.	Bauhinia tomentosa	Caesalpinaceae	Sulphur yellow flowers
6.	B.acuminata	Caesalpinaceae	Sulphur yellow flowers
7.	B.galpinii	Caesalpinaceae	Bright scarlet flowers
8.	Bougainvillea sp.	Nyctaginaceae	Single or double in red, rose, pink, yellow or white colour, including variegated leaves.
9.	Camellia japonica	Theaceae	Double flowers with white, rose or pink colour.
10.	Cestrum nocturnum	Solanaceae	Scented flowers at night
11.	Clerodendron inerme	Verbenaceae	White flowers
12.	Dombeya spectabilis	Sterculiaceae	Cream/deep pink flowers
13.	Duranta plumieri	Verbenaceae	Blue flowers
14.	Hamelia patens	Rubiaceae	Orange red flowers
15.	Hibiscus rosa-sinensis	Malvaceae	Mostly red in colour, yellow pink and orange colours also available
16.	H. mutabilis	Malvaceae	Flowers fade from pink to crimson
17.	Holmskioldia sanguinea	Verbenaceae	Peculiar orange red flowers
18.	Mussaenda erthyrophylla	Rubiaceae	Red, pink, yellow and white
19.	Nerium oleander	Apocyanaceae	Single or double in rosy pink colour
20.	Pentas carnea	Rubiaceae	Red, pink or violet colour
21.	Poinsettia pulcherrima	Euphorbiaceae	Red or orange colour flowers
22.	Tabernaemontana coronaria	Apocyanaceae	White flowers
23.	Tecoma stans	Bignoniaceae	Yellow flowers
24.	Thevetia nereifolia	Apocyanaceae	Yellow colour flowers

Table 20.2 Shrubs grown for foliage.

S. No.	Botanical Name	Family	Remarks
1.	*Acalypha sp.*	Euphorbiaceae	Red and green coloured leaves
2.	*Aralia sp.*	Araliaceae	Large pinnated / pinnatifid, variegated trilobed leaves
3.	*Codiaeum sp.*	Euphorbiaceae	Variably coloured and shaped leaves
4.	*Eranthemum elegans*	Acanthaceae	Blotched with white, green grey and bronze coloured leaves
5.	*Graptophyllum hortense*	Acanthaceae	Variegated blotched creamy white
6.	*Panax fruticosum*	Araliaceae	Feathery, tripinnate leaves
7.	*Phyllanthus nivosus*	Euphorbiaceae	Small mottled pinkish white leaves
8.	*Pisonia alba*	Nyctaginaceae	Pale yellow foliage

Table 20.3 Shrubs grown for attractive berries.

S. No.	Botanical Name	Family	Remarks
1.	*Duranta plumieri*	Verbenaceae	Yellowish orange coloured berries
2.	*Ardisia crenata*	Myrisinaceae	Greyish red or crimson berries
3.	*Nandida domestica*	Berberideae	Red colour fruits

Hedges

For several reasons, a garden should be enclosed by a good hedge or fence. It provides protection from cattle, shelter from wind and privacy. The best plant material for forming such a hedge would be a quick growing hardy shrub with attractive foliage and, or handsome flowers, drought resistant and should stand trimming to shape and capable of being quickly and easily raised from seed or from cuttings to fill up the gaps promptly.

Decorative internal hedges are formed of small-growing shrubs or under-shrubs which have handsome foliage and bear in some cases handsome flowers as well. The usual height for an ornamental internal hedge varies from 30 to 65 cm. Its objective is to seemingly divide the garden into a number of parts, each part containing a distinct feature of its own as a rosary or flower bed or collections of bulbs etc. As the boundary or the screen or the ornamental hedges are meant to be permanent features of a garden, proper care should be exercised in planting them. Trenches of 30 to 40 cm wide and 40 to 45 cm deep should be dug and refilled with top soil, farmyard manure and red earth. Hedges are planted either with seeds or cuttings in the rainy season. Shoots should be tipped as they grow to induce them to branch out and side shoots should be cut back to the desired dimensions. A compact and thick hedge is possible if it is trimmed as often as necessary, cutting back the overgrowing shoots strictly to the desired dimensions.

A. Plants Suitable for Ornamental Internal Hedges

1. *Acalypha sp.*
2. *Barleria sp.*
3. *Bougainvillea sp.*
4. *Cupressus macrocarpa*
5. *Duranta plumieri*
6. *Eranthemum sp.*
7. *Hamelia patens*
8. *Hibiscus rosa-sinensis*
9. *Lantana camera*
10. *Malpighia coccigera*
12. *Meyenia erecta*
14. *Pedilanthus tithymaloides*

B. Plants Suitable for Boundary or Tall Hedges

1. *Acacia farnesiana*
2. *Acalypha*
3. *Agave americana*
4. *Aralia*
5. *Bougainvillea*
6. *Caesalpinia pulcherrima*
7. *Casuarina equisetifolia*
8. *Carissa carandas*
9. *Pithecolobium dulce*
10. *Punica granatum*
11. *Tecoma stans*
12. *Thevetia nereifolia*

Edges

Edges are plants which are employed in gardens for dividing borders etc. from roads, walks or path or demarcating spaces allotted for particular purposes, *say* flower beds. Mechanical edging made of bricks or rough or irregular stones embedded in the ground are also often used in gardens.

Edging plants are of dwarf growing in habit with handsome foliage and are amenable for regular trimming. The edging plants should be perennial, hardy, easily propagated and should have lasting foliage or flower or both. The height of an edge should be about 15 to 20 cm. Constant trimming is necessary to maintain proper shape and height.

A. Foliage plants suitable for edging

1. *Alternanthera*
2. *Coleus sp.*
3. *Eupatorium cannabinus*
4. *Iresine sp.*
5. *Santolina chamaecyparissus*
6. *Echeveria*
7. *Cineraria maritime*

B. Flowering plants suitable for edging

1. *Alyssum*
2. *Amaryllis*
3. *Gerbera*
4. *Lobelia*
5. *Gazania splendens*
6. *Plumbago capensis*
7. *Zephyranthes*

Climbers

Climbers and creepers are plants which attach themselves to supports by their rootlets, hooks, tendrils or by the stem and leaves. Shrubs which have long scandant branches and require support are also included in this group. Walls, trellises, arches, pergolas, arbours, pillars are best adorned by growing climbers and creepers against or over them. Climbers can be broadly grouped into:

1. Heavy climbers which require support such as arches, pergolas or pillar or a tree
2. Light climbers which are best suited for growing against wire netting or trellises.

Most climbers are perennials but there are a few annuals which can be grown in pots with balloons or in hanging baskets or in low trellises (e.g. *Cobaea scandens, Ipomoea, Nasturtium, Thunbergia and Clitoria*). Climbers are propagated by layers and cuttings. They are planted in well prepared pits. Liberal watering and manuring are necessary for good growth.

Proper training is essential for climbers to be at their best. Climbers which are to be trained over screens and trellises should be induced from the base of the plant to cover them completely. This is achieved by pinching off the terminal bud to induce lateral branching when they are about 30-40 cm high. In the case of climbers over arches, pergolas and trees, one or two leader shoots may be trained to the top to branch out.

Uses of climbers in garden

1. Certain climbers are grown in gardens for their attractive foliage. e.g. *Asparagus sprengeri, Ficus repens, Hedera helix, Scindapsus aureus*.

2. Some light climbers can be trained as 'screens' in gardens e.g. *Bignonia venusta, Jacquemontia violaceae, Passiflora edulis*.

3. Climbers like *Allamanda, Antigonon, Aristolochia elegans. Solanum saeforthianum* can be used on arches, bowers and pergolas.

4. Heavy climbers like *Bougainvillea, Scindapsus, Petrea volubilis* can be trained over strong pergolas or on trees which look very attractive. Selected list of common ornamental climbers and creepers suitable for garden is given below:

S. No.	Botanical Name	Family	Colour of the flowers
1.	*Allamanda grandiflora*	Apocynaceae	Bright yellow
2.	*Antigonon leptopus*	Polygonaceae	Bright pink and white
3.	*Aristolochia elegans*	Aristolochiaceae	Dark purple and creamy yellow
4.	*Asparagus sprengeri*	Liliaceae	Leaves are attractive
5.	*Bignonia venusta* Syn: *Pyrostegea venusta*	Bignoniaceae	Golden orange
6.	*Bougainvillea sp*	Nyctagineae	Various colours
7.	*Ipomoea palmate*	Convolvulaceae	Purple
8.	*I. tuberosa*	Convolvulaceae	Yellow
9.	*Jacquemontia violaceae*	Convolvulaceae	Bright blue
10.	*Jasminum grandiflorum*	Oleaceae	White, tinged pink scented
11.	*Monstera deliciosa*	Araceae	Leaves are attractive
12.	*Petrea volubilis*	Verbenaceae	Purple blue
13.	*Porana volubilis*	Convolvulaceae	White
14.	*Quisqualis indica*	Combretaceae	Pale pink and white
15.	*Scindapsus aureus*	Araceae	Variegated leaves

(Contd.)

S. No.	Botanical Name	Family	Colour of the flowers
16.	*Solanum seaforthianum*	Solanaceae	Purplish blue
17.	*S.wendlandii*	Solanaceae	Lilac blue
18.	*Tecoma jasminoides*	Bignoniaceae	White with rose purple streak in the throat
19.	*Thunbergia grandiflora*	Acanthaceae	Bluish, purplish or white
20.	*Vallaris solanaceae*	Apocynaceae	White scented flowers

21

Annuals, Biennials and Herbaceous Perennials

Annuals are a group of plants which attain their full growth from seed, flower and die in one year or one season. Mostly they complete their life history in 3 to 6 months. They comprise of several of the most beautiful and easily grown plants widely varying in form, habit of growth and colour.

Annuals find a variety of uses in landscaping

1. Flowerbeds of simple design can be laid out on the outskirts of lawn, along the base of buildings, in the path leading to entrance of houses and on sides of foot steps.
2. Certain annuals are useful as edging (e.g. dwarf marigold, alyssum and candytuft).
3. Certain annuals are useful in hanging baskets (e.g. petunia, verbena and alyssum).
4. Certain climbing annuals are useful to cover trellis work (e.g. tall nasturtium and *Cobaea scandens*)
5. Some annuals are useful for massing in beds, e.g. Asters, phlox, salvia, zinnia and verbena.
6. Few other annuals are useful for planting in shrubberies in vacant spaces; they are sunflower, hollyhock, tall growing species of amaranthus, tithonia etc.

7. They serve as perennial sources of supply of cutflowers for indoor decorations.

Classification of Annuals

Annuals may roughly be grouped as follows:

1. Rainy season annuals, which can stand more rain than others and therefore grown to flower during the rainy season. The time for sowing them would be from April-May in most of the places.
2. The cold-season or winter annuals thrive and bloom best during winter. They are sown in September to October.
3. Hot weather annuals which are required for blooming from March to May are sown in December–January.

Hints in Raising Flowering Annuals

1. The seeds are sown in seed pan or raised beds. Thin sowing is necessary to get good sized vigorous seedlings. In the seed pan, a pot mixture consisting of two parts of well sifted soil, two parts of leaf mould and one part of sand may be used.
2. Fine seeds may be mixed with 3 to 4 parts of sand before sowing.
3. Annuals which do not stand transplanting like Calendula and Antirrhinum are sown broadcast in beds.
4. Watering the nursery may be done with rosecan.
5. After the seeds germinate completely, over crowded seedlings are thinned out.
6. As transplanting often results in heavy casualties, the seedlings are pricked before transplanting. Pricking is the practice of transplanting young seedlings into small pots individually or in the nursery beds with richer soil giving wider space (10 to 13 cm). Pricking is normally done when the seedlings have produced 2 to 4 leaves. This helps to increase the fibrous root system and to develop vigorous plants.
7. One month after planting when the seedlings have produced six to eight leaves, they can be transplanted into main beds.
8. Generally a spacing of 30 × 30 cm may be given for most of the annuals.
9. Tall growing annuals like hollyhock may be provided with stakes.
10. The terminal buds of seedlings are pinched-off after they establish and when they are about 25 to 30 cm high. This encourages lateral growth and a more bushy shape.

List of important flowering annuals is given in Table 21.1.

Biennials

Biennials are plants which grow in one season, flower, fruit and die in the next season. Generally, the period of growth is 6 to 9 months. Biennials are grown in the same way as annual and are put to similar uses. Example of biennials are canterbury bulb, Scabiosa, Gladiolus etc.

Herbaceous Perennials

Herbaceous perennials are those perennial plants with soft succulent stems (as compared to shrubs which have woody stems). They are propagated by seeds, cuttings, offset and slips. They are useful as herbaceous or mixed borders or for pot culture. The following are the examples for herbaceous perennials:

1. **Chrysanthemum:** Flowers are single or double available in attractive colours. Perennial species include *C. frutescens* and *C. maximum* and its varieties, propagated easily by suckers.
2. **Michaelmas Daisy:** *Aster amellus*, called also as perennial asters. Flowers are single with white, rose, blue, lilac and purple colours. Easily propagated by division of the clumps and occasionally from seed. They thrive best in the cold and the rainy seasons.
3. **Solidago:** Popularly known as 'golden rods' producing erect feathery rod-like trusses crowded with pretty golden yellow flowers. They are suitable for mass planting in beds and borders in and adjoining lawn. They are raised by suckers.
4. **Gerbera:** Stemless perennial herbs with radical stalked leaves, flower heads are solitary, large and sterile with varying colours. Propagation by division of clumps or from seed.
5. **Gazania splendens:** Perennial plant about 20 cm high, with pointed leaves with silver and bearing beautiful daisy like flowers, in yellow orange shades; useful in beds, borders for edging and carpet bedding and on rockeries, propagated by seed or suckers.
6. **Penstemon (*Fa: Scrophulariaceae*):** Has a large erect spikes of tubular, open-mouthed, gloxinia–like flowers, which are available in several shades of colours, a good bedding plant, propagated by seeds, cuttings or division.
7. **Pelargoniums (*Fa: Geraniaceae*):** Commonly known as geraniums, popular herbaceous perennial pot plants grown for the beauty of their flowers which are borne in large trusses propagated by cuttings or from seed.

Table 21.1 List of important flowering annuals.

S. No.	Name and family	Height	Colour of flower	Time of flowering	Method of propagation	Blooming period	Remarks
1.	Althaea rosea (holly hock) Malvaceae	1.20 to 1.80 m	Various colours	Aug–Sep.	By seeds & transplanting	9 months after sowing	Large single or double flowers useful for screens, borders and for background suited to hills
2.	Antirhinum (Snapdragon) Scrophulariaceae	15 to 45 cm	Various colours	Dec - Feb	By seeds & transplanting	2 months	Bedding or pot or border plant Pink, rose, Apricot orange, crimson, white, yellow flower
3.	Balsam (Impatiens balsamina) (Balsaminaceae)	25 to 130 cm	Rose like and variegated	Aug–Sep.	By seeds & transplanting	2 months	Can be grown throughout the year suited for borders also
4.	Celosia spp. (Cock's comb) (Amaranthaceae)	25 to 60 cm	Fasciated flowers of varying colours	Throughout the year	By seeds	3½ - 4 months	Pretty annuals with terminal fascinating flowers of varying colours. Useful as borders and mixed borders
5.	Cosmos spp. (Cosmos bipinnatus) (Compositae)	5 to 15 cm	White, crimson rose and purple	Aug–Sep Dec–Feb	Seeds	2-2½ months	Popular rainy season annual with graceful foliage. It can be grown throughout the year
6.	Dianthus spp. (Fairy queen) (pinks and carnations)	–	–	–	–	2-2½ months	Popular rainy season annual with graceful foliage. It can be grown throughout the year

(Contd.)

Table 21.1 List of Important flowering annuals.

S. No.	Name and family	Height	Colour of flower	Time of flowering	Method of propagation	Blooming period	Remarks
(i)	Indian pink or Chinese pink Sweet William (D. barbatus)	25 to 30 cm	Various colours	Aug–Sep Dec.–Feb.	Seeds	–	Useful for pots and borders also
(ii)	Sweet William (D. barbatus)	25 to 30 cm	Various colours	Aug–Sep Dec.–Feb.	Seeds	–	Useful for pots and borders also
(iii)	Carnation pink (D. caryophyllus) (Carryophyllaceae)	30 to 45 cm	Pink white Crimson and others	Aug–Sep Dec.–Feb.	Seeds and cuttings	–	Suited for pots particularly Carnation-Marguite the most successful in plains
7.	Gerbera (Compositae)	15 to 30 cm	Various colours	Aug. Sep. Dec.–Feb Apr.–May	By divisions of suckers	Through out the year	Suited for beds, borders
8.	Gomphrena globosa (Globe Amaranth or Bachelor's button (Amaranthaceae)	3 to 6 cm	Pink, purple orange & white flowers	Through out the year	Seeds	1 month	Suited for beds, borders and as cut flowers. Thrives well in all garden soil
9.	Helianthus sp. (Sunflowers) (Compositae)	5 to 20 cm	Yellow with brown (dark) colour	Aug–Sep, Dec.–Feb Apr.–May	Seeds & cuttings	2–2½ months	Staking the plants is essential in the case of tall and unbranched varieties
10.	Marigold (Compositae)	60 to 90 cm	Bright yellow, lemon yellow, orange	Aug–Nov.	Seeds & cuttings	2–3 months	Suited for beds and broders
11.	Tagetes erecta Petunia sp. (Solanaceae)	45–60 cm	Various colours	Sep.–Oct. Dec.–Jan	Seeds	3–4 months	Suited to flower beds mixed borders, pot plants, window border and hanging baskets

(Contd.)

Table 21.1 List of Important flowering annuals.

S. No.	Name and family	Height	Colour of flower	Time of flowering	Method of propagation	Blooming period	Remarks
12.	Phlox (Polemoniaceae)	30 cm	Various colours	Sep.–Oct. Dec.–Jan	Seeds	3–4 months	Suited for beds and pots, period of bloom lasts for over a month
13.	Salvia splendens (Labiatae)	60–90 cm	Scarlet blue purple pink	Aug–Sep. Dec–Feb	Seeds	2–2½ months	Can be grown through out the year. Suited for beds and borders. Pinching back the shoots in early stages builds up better plants.
14.	Tithonia speciosa (Mexican sunflower) (Compositae)	1–25 to 1–80 m	Reddish orange flowers	May–Sep Oct and Dec. Jan	Seeds	3 months	Reddish orange flowers on long inflated stalks, can be grown throughout the year valuable pot planting in long borders and in shrubberies.
15.	Zinnia elegans (Youth and old age) (Compositae)	30 to 75 cm	Various colours	Dec.–Jan Apr.–May	Seeds	2 months	Hard flowers in profusion for a long period, single or double flowers borne on long stalks. Attractive in borders and beds. The first flower buds should be nipped off for allowing the plants to grow bushy and bear numerous flowers.

22

Special Group of Garden Plants

(A) SUCCULENTS AND CACTI

The term 'succulents' refers to a certain specialized forms of plants which store moisture in their foliage, or in stem or in rootstock enabling them to live through periods of drought. Their leaves are fleshy with plenty of water-holding tissues, often reduced in size, covered with a thick epidermis with only a few stomata and are often coated with a whitish or blue wax or wooly hairs.

The cacti are one type of succulents which are exclusively belonging to the family: Cactaceae. They are perennials, bearing spine cushions called 'areoles'. Interestingly, all cacti are succulents but all succulents need not be cacti. Cacti have usually globular columnar structure from where the spines, hairs and flowers grow out. Most of the cacti do not have leaves but the succulents have leaves. The cacti usually bloom annually and are beautiful and large.

Most of the succulents originate from the hot and arid or semi-arid regions of Asia and America (especially Mexico). Some of the important succulents used in garden are given in Table 22.1

Table 22.1 Important succulents used in gardening.

Sl. No.	Botanical Name and family	Brief Description
1.	*Adenium obesum* Fa: Apocynaceae	Handsome plant with succulent swollen stem at base, shrub growing upto 12 m, very large showy funnel shaped—white pink to crimson coloured flowers during dry months.
2.	*Agave americana variegata* Fa: Agavaceae	Handsome plant with leaves having dark green in the centre and margined with rich yellow, and with stout spines at apex.
3.	*Aloe abyssinica* Fa: Liliaceae	Ornamental soft succulent thick leaves, often prickly, often arranged in a rosette, bears small bright vermillion coloured flowers.
4.	*Furcraea watsoniana* Fa: Agavaceae	Ornamental foliage plant resembling Agave, assumes giant proportions in growth, leaves are variegated, yellowish white, white and green, measuring 1-1.5 m long.
5.	*Kalanchoe sp.* Fa: Crassulaceae	Flowering shrub with thick fleshy leaves from which they are propagated, flowers showy provided in terminal clusters.
6.	*Pedilanthus sp.* Fa: Euphorbiaceae.	Often used as a hedge border plant, produce variegated leaves.
7.	*Sansevieria zeylanica* Fa: Agavaceae	Produces erect strap-shaped leave 0.25 to 1.0 m long, which is green striped with grey brands. Other species of garden importance are *S. cylindrica* which produces green cylindrical leaves and *S. trifasciata* which develops like rosette, with spirally arranged leaves around the stem.

Some important Cacti are:

1. **Cephalocereus**—small columnar forms and the most distinguishing feature is the mass of hair from the areole which looks like a thick bunch of wool, depending on the species, either the growing point is fully covered with the wooly hairs or the body also, hence popularly called, 'Old man cactus'.

2. **Cereus**—curious looking, long stemmed vigorous growing, thorny very hardy plants.

3. **Echinocactus**—popularly known as 'Barrel cacti' having unusual spination on various shaped barrels.

4. **Echinocereus**—low growing plants forming groups of clusters the columnar or prostrate stems having numerous ribs with small comb-like spines.

5. **Mamillaria**—popularly called the 'Nipple cactus' or Elephant tooth cactus, dwarf plants with leafless cylindrical or globular stem, bearing evenly over their surface, small tubercles, some what resembling the teats of animals, each tubercle being crowned by a rosette of stars of hairy spines.

Like any other plant, cacti can be raised from seeds but the process is very slow. Hence the common practice is to grow them through vegetative buds. The container for cacti is filled with about 10 cm of gravel and above that a thin layer of sand is spread to provide drainage. The best soil mixture consists of sandy loam, little cowdung manure and some broken bricks. Watering by sprinkling should be done once in or twice in a week. Repotting is necessary with fresh soil once in two or of three years, usually in the same containers, since the plants grow very slowly.

(B) ORNAMENTAL PALMS

Palms are a special group of plants used extensively in landscape gardening, due to their beautiful trunk, leaf colour, shapes, plant structure and growth habits. Palms include about 200 genera and three thousand species. Their roots, trunks, (stems), fronds (leaves), flowers and fruits are different and distinct in appearance and structure from others of the plant kingdom. Based on the trunk and its different manifestation, palms can be divided into four groups *viz.,*

(a) Solitary palms—Single erect trunk (e.g. *Cocos. Phoenix, Elaeis).*

(b) Clumping palms—Multiple trunks arising from at or just below the ground level (e.g. *Areca lutescens, Rhapis).*

(c) Branching palms—Branching can occur above ground (e.g. *Hyphaene indica*) or below ground (e.g. *Nipa fruticans).*

(d) Trunkless palm—have very much reduced trunk called acaulescent (e.g. *Phoenix acaulis).*

The leaves of the palms called 'fronds' vary in form. They may be grouped into 'fan leaved type' and 'pinnate leaved type' palms.

Use of Palms in Landscape Garden

1. Palms are well suited as single specimens in lawn (e.g. *Areca triandra).*
2. Palms are excellent specimens for avenue planting in the gardens (e.g. *Roystonea regia*).
3. They are also suitable for decoration of conservatories, verandahs, stair-cases, for indoor decorations as potted plants.

Hints to Grow Palms

1. Palms can be propagated from seeds or division of clumps (e.g. *Rhapis*)
2. Palm seeds are to be harvested at fully ripe stage and are to be immediately sown in raised beds having more proportion of sand.
3. Depending upon the species, the seeds start germinating from 3 months even continue upto 2 years in certain cases.
4. The seedlings may be lifted at first pair of leaf stage and potted off simply in small pots.
5. Palms generally prefer pot bound condition and thrive even in undersized pots.
6. Repotting to the next large sized pots may be done at the stage when the roots increase and fill the pots fully, almost forcing the pots to open or crack.
7. Liquid manuring with oil cakes and Ammonium sulphate may be given to the palms once in fifteen days.
8. Regular watering especially on alternate days is essential for the palms grown in the pots.
9. Potted palms intended for keeping indoors should be acclimatized sufficiently before keeping them indoors, so that they retain their fresh appearance and lusture.

The following are the some of the attractive palms for the garden:

1. *Areca triandra* —an elegant single trunked palm
2. *Caryota urens* —Toddy palm, produces a long drooping flower spike.
3. *Hyphaene indica* —Indian doum palm - unique branching habit
4. *Livistonia decipiens* —Solitary, fan shaped leaves with stout thorns on their leaves.
5. *Pritchardia pacifica* —it has ornamental large broad flabellate plaited leaves, often 10 m, broad.
6. *Rhapis excelsa* —Dwarf-rattan like palms, slender stem and fan shaped leaves, produce large number of suckers.

(C) BULBOUS PLANTS

The term 'bulbous plants' refers to all seasonal plants with underground modified stems containing stored food for the development of the seasonal aerial shoots of stems, leaves and flowers. The modified stems include bulbs, tubers, corms, rhizomes etc., Bulbous plants are grown for their flowers or foliage or both. There are a number of them, varying in habit of growth, form, colour etc. Bulbous plants are characterized by three stages in their growth *viz.*, the growing, the blooming and the resting periods. Generally, the bulbous plants after bloom enter into rest period. At this time, the bulbs are taken out and placed them in moist sand. After 3–4 months, the dormant buds swell and push out the shoots. When sufficient growth is made, the bulbs may be planted. All bulbous plants generally thrive at higher elevations, there are some kinds which do not thrive and bloom at medium elevations and many do not perform well at low elevations. Important bulbous plants are given in Table 22.2.

Orchids

Orchids are the most fascinating and beautiful of all flowers and they exhibit wide range of diversity in form, size, colour and texture of flowers beyond the imagination of human mind. Certain flowers mimic the form of birds, spiders, scorpions, moths, butterflies and several other insects. The flowers are either solitary or are borne in clusters, spikes, racemes or panicles. Their flower last very long; as long as three months in some kinds. Orchids belong to the family 'orchidaceae' which includes hundreds of genera and thousands of species and varieties. They are widely distributed in several parts of the tropical zone.

Orchids may be generally Classified into Four Groups

(a) Terrestrial or ground orchids—which grow more or less in soil and have their roots embedded in it from which they absorb the nourishment necessary for the growth. Most of the terrestrial orchids belong to temperate zone.
(b) Epiphytic orchids—which grow upon the branches of trees having their roots exposed to air, from which they imbibe the entire nourishment requisite for the growth. Though they cling to trees and grow on them, they are not parasitic. Most of the epiphytic orchids belong to tropical zones.
(c) Semi terrestrial orchids—they send roots to the ground and derive partial nourishments besides they also develop aerial and adventitious roots, with which they cling to trees for support and get their chief nourishment by absorbing moisture and food from the air.

Table 22.2 Important bulbous plants.

Sl. No.	Common Name	Botanical name & Family	Brief Description
1.	Blue African lily	*Agapanthus companulatus* Fa: Liliaceae	Produces umbel from March to June, containing 10–30 blue flowers with funnel shaped, suitable for 1000–2000 m above MSL, propagated by offsets.
2.	Cannas	*Canna indica* Fa: Scitaminae	Produces large trusses of flowers of larger size, based on height, classified as dwarfs, medium and tall.
3.	Crinum lily	*Crinum sp.* Fa: Amaryliidaceae	Bears usually white or red tinted flowers, mostly in summer.
4.	Gladiolus	*Gladiolus sp.* Fa: Iridaceae	A popular cut flower, producing single or double spikes, propagated through corms or seeds.
5.	Day lily	*Hemerocallis sp.* Fa: Liliaceae	They bear single or double large and attractive flowers on tall scapes.
6.	The garden amaryllis or trumpet lily	*Hippeastrum sp.* Fa: Amaryllidaceae	Spectacular flowers, having various shades of bright colour.
7.	Dahlia	*Dahlia sp.* Fa: Compositae	Most gorgeously coloured, free blooming, available in all colours except blue.
8.	Red-hot poker	Kniphofia *sp.* Fa: Liliaceae	A handsome plant, bearing immense spikes, closely covered with brilliantly coloured tubular flowers in orange, rose, salmon scarlet shades.

(d) Saprophytic orchids—They live on dead matter and are very rare. Some orchids grow on rocks and are called lithophytes.

The structure of an orchid plant has an intimate bearing on natural conditions which they obtain in their native homes. Most of the epiphytes are exposed to long drought period in a year, during which they get no food or water and hence they develop pseudobulbs (thickened secondary stems with one or two internodes) and thick leathery leaves in which they store food and water as reserve supplies to sustain them during the period of drought. Further, they also send long aerials shoot which not only help to cling to trees but also help them to absorb moisture from the air and food from the dust.

Important genera of orchids which can be grown by garden enthusiasts are:

1. **Calanthe:** Mostly terrestrial, easy to grow in pots, bear erect many flowered capes.
2. **Cattleya:** Very beautiful flowering evergreen, epiphytic genus, flowers in single or in clusters usually at the apex of the pseudo bulb, largely cultivated for cut flower.
3. **Cymbidium:** Intermediate in character between true epiphytes and terrestrial orchids, growing in nature in trees and in places there is accumulation of dead leaves. Flowers are mostly white or yellow to dark purple-maroon in colour, borne on long spikes.
4. **Cypripedium:** An interesting large genus of hardy terrestrial plants, bloom in winter, flowers attractive in pink, green, brown or white with labellum forming an inflated pouch, resembling a lady's slipper, hence popularly called as 'Lady's slipper orchid'.
5. **Dendrobium:** One of the largest and most decorative genus, includes both evergreen and deciduous species, flowers are borne singly in two or three or in pendulous clusters, gracefully dropping.

Besides, *Epidendrum, Phalaenopsis, Spathoglottis and Vanda* are also important genera producing attractive flowers.

Hints to Grow Orchids

1. One should try to place the orchids in conditions similar to those obtained in their native homes.
2. Orchids are grown either on logs or blocks of wood or attached to trees or in wooden or wire baskets or in pots or in ground according to the habit of the kinds in question.
3. Terrestrial orchids are, as a rule, cultivated in pots, pans and hanging baskets filled with soils composed of knobs of charcoal bricks broken into small sizes, coarse decayed leaf mould and fibrous loam and peat. Epiphytic orchids are best grown on logs of woods by placing moss initially on the log and then the plant is placed on it with it roots spread on it and the plant is kept in position by passing round the log and over the moss, strand of copper wire or strong thread.
4. Potting is generally done after the flowering season and repotting is done only when necessary.
5. True epiphytes require no manure. Terrestrials are to be liquid manured; prepared from cowdung of organic cakes. Orchids may be also fed with nutrients in liquid forms, diluted sufficiently.

6. Both the plants and the growing media should be sprayed with the nutrient solution.

7. Care has to be exercised in watering orchids. Meager watering injures and destroys them and over watering especially during rest period is certainly harmful.

8. Propagation is effected by division of clumps of pseudobulbs or stems after flowering. Seed propagation is necessary to develop hybrid seedlings and is effected through culturing them under aseptic conditions in agar medium supplemented with nutrients required by the plant.

23

Rockery and Water Gardens

ROCKERY

A rockery is intended to bring together in a limited space an idea of a mountain or alpine garden with plants growing in the crevices of rocks. The term 'rockery' in our country is associated with usually in shady tree, a large mount of earth heaped up under it with a number of boulders embedded in and jutting out of mount. A few plants mostly hardy ferns peep through the plants between the rocks. But essentially, a rock garden should have wide variety of plants, dwarf herbaceous perennials, hardy shrub, ferns, colorful cacti and succulents.

A rockery should simulate natural rock formation with plants growing on them. The outlines should be rugged. The rockery should be an elevated structure with a few projecting work and winding path and if possible a shallow, irregular pool with water-loving plants, somewhere inside, all looking as natural as possible.

A rockery may be laid out in the open or under the shade of a well established tall growing tree. The contour of the intended rockery is marked out on the chosen site and good garden soil enriched with manure and leaf mould is heaped up to the required height and well firmed by moistening with water and beating with turf-beater. Rocks of irregular shapes are then fixed, sloping backward into the mound, so made commencing from the edge with larger stones leaving pockets or spaces between the stones for accommodating plants. The spacing between the stones should be varied in size to suit growing in them single plants or clumps of plants and plants with large or small roots. In a high rockery with steep slope, terracing is advisable to prevent the wash-off the soil during rains.

The after-care of a rockery consists of weeding, thinning out crowded clumps of plants, applying each season a top dressing of leaf mould or compost to pockets containing plants, regular watering in summer and rainless periods and frequent removal of dead leaves and shoots.

The following group of plants can be used in a rockery:

Annuals and Herbaceous biennials

Calendula, Dianthus, Gaillardia, Nasturtium, Petunia, Phlox, Portulaca, Verbena and Vinca.

Dwarf Shrubs

Barleria, Crossandra, Clerodendron, Eranthemum, Iresine, Lantana and *Dracaena.*

Creepers

Pilea, Philodendron and *Tradescantia*

Small Shade Loving Plants

Bilbergia, Alocasia, Anthurium, Coleus, Cyperus, Maranta, Sansevieria Rheodiscolour and Caladium.

Lilies

Crinum and *Zephyranthes*

Terrestrial Orchids

Spathoglottis, Epidendrum

Cacti and Succulents

Tall succlents like *Agave, Aloe, Cereus, Opuntia, Yucca* and *Furcreae.* Small succulents like *Bryophyllum, Crassula* and *Kalanchoe.* Very small succulents like *Echinocactus. Echinopsis, Cotyledon, Sedum* and *Sempervivum.*

WATER GARDEN

A water garden will provide conditions for growing any or all of the following plants:

1. True aquatic plants
2. Bog or marsh plants
3. Moisture loving plants

A shallow pond or small natural lake can be converted into an ideal natural water garden. The water in this pond should not dry up during summer or overflow during rainy season The plants are set out as in their natural environment and should merge with surroundings.

In the shallow water near the edge, *Nymphaea* can be grown. Towards the center where the depth may vary from 15 to 50 cm, Lotus (*Nelumbium speciosum*) and *Victoria regia* may be planted. The edges of the pond may be planted with *Veronica palustris* and *Calla palustris* which will send their roots into the soil and spread out over the surface of water. Beyond the margin of the pond in the soil, which is saturated with moisture, water-loving plants such as grasses, ferns, Colocasia, Cyperus can be grown. Moisture loving trees and plants such as bamboos, pandanus, palms etc. are planted beyond the water loving plants as a back ground in such a way that the taller ones provide shelter to the smaller one in front.

Lily Pond

In the absence of a natural water garden, an artificial pool may be constructed. A rectangular or square or round pond is a welcome feature on formal gardens. A pond more than 2 m in diameter will be impressive. The following are some of the important points to be kept in mind in laying out a lily pond:

1. After construction, water is filled and left in the lily pond at least for four weeks. This helps to remove the harmful chemicals in the cement.
2. Addition of crystals of potassium permanganate to water will reduce the toxicity due to harmful chemicals.
3. Bottom of the pond may be filled with a mixture make up of five parts of silt and one part of cattle manure.
4. Once the aquatic plants are planted, water is let in only to cover the crown of the plants. Gradually the water level may be raised as and when new growth is made by these plants.
5. Fish and water snail may be added in the pond, which help to keep the water free of mosquito larvae and waste materials.
6. After some days the water in the pond may become green and unsightly. This may be got rid off by the addition of 30 gm of potassium permanganate to every 1,25,000 litres of water.

A. Plants suitable for marginal planting and for the bank of the pond

1. *Alpinia*
2. *Areca lutescens*
3. *Cyclanthus*
4. *Hedychium*
5. *Costus speciosus*
6. *Colocasia*
7. *Xanthosoma*
8. *Raphis flabelliformis*

B. Moisture loving trees and plants for the back ground

1. *Bamboo*
2. *Tamarix*
3. *Salix*
4. *Acacia farnesiana*

C. Plants suitable for water gardens are given below:

S. No.	Common Name	Botanic Name	Remarks
1.	Water lilies	*Nymphaea coerulea* (Nymphaeaceae)	Narrow petalled, sky blue flowers, open in the day and close in the evening fragrant also.
		N. lotus	Night bloomer, pure white and scented.
		N. lutea	Showy yellow flowers.
		N. sulphurea	Pale to deep yellow flower
2.	Lotus or sacred lily	*Nelumbium speciosum* (Nymphaeaceae)	Very beautiful with large flowers of pink or white colours.
3.	Giant water lily	*Victoria regia* (Nymphaeaceae)	Leaves are about 1 m in diameter.

D. Plants suitable for planting margin of water garden are given below:

S. No.	Common Name	Botanic Name	Remarks
1.	Sweet flag	*Acorus calamus* (Aroideae)	Aromatic plants, propagated by seeds or division of rhizomes
2.	Arrow head	*Sagittaria sagittaefolia* (Alismaceae)	Large sagittate leaves. Flower are white, deepening to pale blue at the base of the perianth. A good oxygenerator and hence useful in a pond in which fish are bred.
3.	Umberella plant or plam	*Cyperus alternifolius* (Cyperaceae)	It is a rush-like perennial herb. 0.75 m high with a compact habit of growth with numerous, erect, dark green joint less angular stems.

24

Indoor Gardening

INDOOR GARDENING

Indoor gardening refers to beautifying the area inside the house with plants. Unlike outdoor gardening, the house plants are grown generally in containers made of earthen pots, ceramic pots or any container of cheap cost. In some places the plants are grown on 'totum pole' which is a support covered with sphagnum moss. Big sized bottles with narrow mouth are also sometimes used to grow the house plants and this technique is popularly known as 'terrarium'. Shade loving plants are generally preferred as house plants. The pots should be raised off the ground and placed on neatly arranged and concealed bricks or on wooden panels. The indoor garden can be established in the following zones of the houses:

1. **Open zone:** This is available in roof terraces. This zone is very warm especially during summer in inland plains. Plants like Agave and cacti, which can tolerate reflected heat, can be selected for the above purpose.

2. **Shade of a tree in front of a house:** Such places near the eastern side of the building may be considered for growing certain house plants which can easily come up under shade. Most of the foliage plants like Crotons, Graptophyllum, Eranthemum, Dracaena and Asparagus are preferred as potted plants in the area.

3. **Varandah of a house:** This area normally gets only diffused light and the air movement is also good. The plants best suited for growing in varandahs are palms such as *Livistonia, Thrinax, Caryota, Areca lutescens*, ferns and Begonias etc.

4. **Living room, drawing room etc.:** In these places, we can keep the plants either near the window or away from it. Near a window, plants with brighter foliage and occasionally herbaceous flowering plants are preferred, while plants with dropping foliage like *Zebrina, Sedum, Mesembranthemum* are preferred in the former cases. The house plants may be arranged in symmetrical or asymmetrical style, but care must be taken to avoid over crowding. The important plants which are preferred as indoor plants are given below:

(a) **Foliage plants:** Dieffenbachias, Philodendron, Aglaonema, Monstera, Anthurium, Colocasia, Caladium, Alocasia, Maranta, Aralia, Panax, Heliconium, Begonia, Asparagus and all kinds of ferns.

(b) **Flowering plants:** Ixora, Graptophyllum, Hibiscus, Eranthemum, Nerium, Zebrina etc.

Care of Indoor Plants

1. **Repotting:** The house plants soon fill the pots with its roots and often need a larger pot for satisfactorily continuing its growth. It is then shifted to a pot a little larger on size with its root and soil intact. This process is known as 'repotting'. Plants are in a dry state. In this condition, they should be watered some time before potting is commenced, so that the surplus water will have to drain away. Soil used for potting should be sufficiently moist to hold together when pressed in the hand, but not wet.

2. **Watering:** Pot plants require to be watered much more carefully than those growing in the ground. Applying too little or too much of water is undesirable. When the leaves droop it is a sign that the plant is either in need of water or suffering from water stagnation at the root. Always soft water must be used for watering the potted plants.

3. **Syringing:** It refers to the operation of spraying the plant surfaces through a fine stringe nose. It is essential to the health of plants growing under cover where the natural rain does not reach them. Syringing creates a moist atmosphere, cleans the leaves and thus assists in promoting their functions.

4. All plants growing in a varandah should be frequently turned round in their position so as to equalize the effect of the light otherwise their growth will be lop sided.

5. Too much light is just as detrimental as too little and the majority of indoor plants should not be placed in full sun. Sun scorch or leaf burn will soon make them very unattractive. Requirement for light vary with the individual plants. Most flowering plants require considerable

light to bloom, while most foliage plants need diffused light. The only exception to this rule is plants that have variegated leaves. When a variegated plant is placed in a dark corner, the few green cells present in the leaves cannot manufacture enough food to maintain a healthy growing condition. Flowering plants require more sunlight for profuse flowering and hence they can be better located near the windows.

6. The humidity of air in the house or room is very low. Many house plants require a higher humidity than is normally present. It is often difficult to provide the necessary humidity in the room. The humidity around the plants can be increased by the following ways:

    (a) Use trays of pebbles in which water is poured to just below the tops of the pebbles. Place the pots on the pebbles, being certain that the bottoms are not sitting in the water.

    (b) Another method is that the single plant may be top-dressed with sphagnum moss and the moss kept nicely damp.

Hanging basket: Hanging basket refers to the practice of growing suitable plants in certain kinds of plant baskets and other containers of pleasing design and suspending them, after placing them in wire frames in conservatories, corridors, rooms and under shady trees to have great ornamental value. Hanging baskets are usually made of galvanized wire or wood. As single plant in such baskets, those which are erect growing, intermediate in size with graceful foliage may be selected. Similarly, plants which are intermediate or dwarf in growth habit with drooping graceful foliage with or without blooms are also a fitting material for massing in these baskets. Selected plants should have the ability to grow and bloom in hanging basket.

It is necessary to prevent the soil in the basket from drying rapidly. For this purpose, the basket is lined with moss or gunny bag or coconut fibre and then filled with soil which can hold moisture for sometime. The soil for hanging baskets consists of two parts each of leaf mould and rich loam and one part of well rotten cow dung. The plants in these baskets may be started either with seeds or seedlings or cuttings depending upon the species. Regular watering and feeding the plants with liquid manure are the two important practices to sustain the growth of the plants for quite a long period.

The following is a selected list of plants suitable for growing in hanging baskets. *Adiantum sp., Asparagus sprengeri, Begonia recumbens, Impatients repens, Nasturtium sp. (Tall kinds), Nephrolepis (a kind of fern), Orchids such as Vanda, Phalaenopsis, Saccolabium, Pilea microphylla, Petunia hybrids. Sedum sp., Setcreasea, Tradescantia, Verbena, Vinca variegata, Zebrina. Wedelia trilobata, Peperomia etc.*

Bonsai

Bonsai refers to the Japanese art of growing miniature trees and shrubs in containers. It involves techniques of extreme dwarfing. The optimum size of bonsai may be only 30 to 60 cm in height, but miniature sizes of below 25 cm have also been preferred. Bonsai of minimum 10 years old are preferred, but of 100 and even 200 years of age are available and are highly valued as 'venerable' specimens. Bonsai requires special types of containers. They should be usually shallow with 5 to 7.5 cm deep (except for cascade type of bonsai for which deep pots can be used.) Round, hexagonal or square shaped containers with 25 to 30 cm diameters are preferred.

Containers with specific colors like mosaic, sky blue, terra cotta, grey or jungle green are well suited for bonsai culture. Plants adaptable to extreme dwarfing like juniper, pine, elm, maple, cypress are suitable materials for bonsai culture. But in tropical places like India, the tree species like *Manilkhara*, Sapota, *Bassia*, Tamarind and *Ficus* spp. and shrubs like West Indian cherry are well suited. Trunk and branches may be bent, forced and tied by coiling them with heavy wire. This wire is removed after several months when training to shape have been accomplished. The bonsai plants may be trained to different shapes like twisted trunk, upright, S-shaped, semi-cascade, cascade, slanting or any other formal shape (Fig. 24.1). Pruning of both tips and roots is usually done at planting time and periodically pinching the tip and removing the excess side shoots are essential to maintain the general outline of design. Annually, the plants are either repotted or lifted from the pot, root pruned and reset.

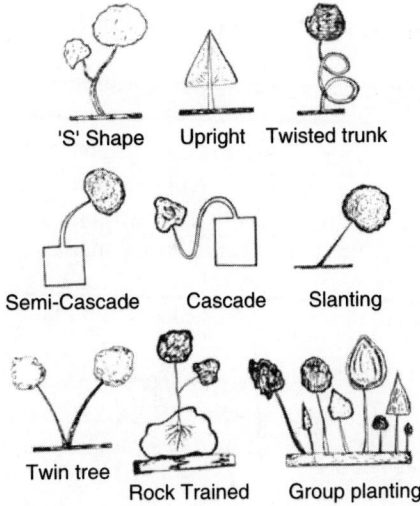

Fig 24.1   Various forms of training bonsai.

The bonsai plant is fed sparingly of weak fertilizer solution containing major and minor nutrients. Sometimes, extracts of oil cakes may be also added. Similarly, the plant is given only minimum requirements of moisture. It is advisable to water twice a day to plants kept in shallow containers and once a day to plants kept in deeper pots.

Selection of hardened woody plants that have been subjected to adverse conditions is a good starting point. Such kind of planting materials may be collected from rock crevices or from the walls of any buildings. Old seedlings kept in containers from any nursery or cutting, grafted plants or layers may be also utilized in bonsai culture.

25

Economic and Cut Flowers

(A) ECONOMIC FLOWERS

Economic flowers are those flower crops which are grown on a commercial scale as field crops in certain selected areas of the state for large scale supply of flowers to markets. These flowers unlike those grown in gardens are used by people for personal decoration, for religious offerings etc. It is estimated that about 24,000 ha is under economic flowers in India and in Tamil Nadu alone 11,237 ha area is available under economic flower. Jasmine and marigold are the most important loose flowers and dominating the traditional flower market in India. Tuberose, scented rose, chrysanthemum, crossandra, gompherena, nerium, barleria, hibiscus are the other important traditional flowers cultivated in considerable area. In addition to their use in social and religious functions, traditional flowers are being commercially used for perfume (jasmine, scented rose and tuberose) and natural dye (marigold) extraction.

The cultivation of important flower crops is dealt here.

1. Jasmines

Of the several species of jasmines occurring in India, the following are usually cultivated as field crops in South India.

    (a) *Jasminum auriculatum*–Mullai

(b) *J. sambac*–Malligai

(c) *J. grandiflorum*–Jathimalli or Pitchi

Mullai and Malligai are cultivated largely in Ramanathapuram, Salem and Coimbatore districts. Since the flowers are perishable within a short time and will have to be disposed off in the markets within a few hours of picking them, the areas of cultivation are concentrated round about the cities and towns. Recently jasmines are airlifted from Coimbatore to Bangalore and Chennai and even to Gulf countries and the US every day.

Varieties

For commercial cultivation the following high yielding varieties are recommended: Malligai–Ramanathapuram Gundumalli.

Mullai–Parimullai, Co-1, Co-2, Pitchi–Co-1, Co-2

Parimullai is a selection from a medium pointed type, resistant to gall mite and its yield potential is 8 t/ha. Co-1 Mullai has bold flower buds with long corolla tube with a yield potential of 10 t/ha. Co-2 Mullai has flower buds having long corolla tube and longer buds with a yield potential of 11.1 t/ha. Co-1. Pitchi has a yield potential of 11 t/ ha and its flower buds are pink in colour, suitable for fresh flower market and also for extraction of floral concrete (0.29% recovery). Co-2. Pitchi is a high yielding mutant with long corolla tube. Arka Surabhi (Pink pin) is a pitchi type, released by 11HR, Bangalore. The average flower yield is 10 t/ha.

Soil and climate: All the three species of Jasmines are suitable for growing in the warm plains of Tamil Nadu. Jasmines prefer a well drained red soils and sandy loams.

Season of planting: Jasmines can be planted between June to November, taking advantage of the monsoon period.

Propagation: Layers or rooted cuttings are used for planting. All the species of jasmines are normally propagated by layering. This involves more time and use of large shoots which restrict the number of plants to be propagated from the bush. It has been now established in all the species the terminal cuttings when rooted in mist chambers (where relative humidity is maintained at more than 95% always) percentage of rooting is also high and more number of rooted cuttings can be multiplied from a single bush within 45 days.

Preparation of field: The land has to be ploughed twice or thrice to a fine tilth and pegs are marked at required spacing in the long furrows of 0.3 m wide. Pits of 30 cm cube are to be dug out at the peg-marked points and the pits are filled with top soil mixed with compost in equal proportion. The spacing and manurial requirements of different jasmines are given in Table 25.1.

The fertilizers are applied in equal split doses once in November after pruning and again in June–July.

Table 25.1

	Mullai	*Malligai*	*Jathimalli*
Spacing	1.5 × 1.5 m	1.25 × 1.25 m	2.0 × 1.5 m
Number of planting materials required/ha	4400	6400	3350
Manuring:			
Farm Yard Manure (kg/plant)	30	15	30
Urea (g/plant)	150	75	150
Super Phosphate (g/plant)	750	375	750
Muriate of potash (g/plant)	200	100	200

Pruning: After flowering stops in October, the bushes will start to shed the leaves. Pruning at this stage is recommended to encourage the flower bud initiation, differentiation and ultimately the production of flowers in the next season. Pruning in all the three species involves removal of all the shoots above 45cm height from the ground level. Pruning may be done during the last week of November for Malligai and last week of December for Mullai and Jathimalli. Pinching of new shoots produced after pruning helps to stagger the flowering and thereby enhance the productivity. However, pinching delays flowering by about 17 days in *J. grandiflorum* and 14 days in *J. auriculatum*.

Irrigation: Irrigation is important for jasmines. Constant and adequate water supply (irrigating twice a week) during the flowering season from March to October is essential for good yield of flowers. After flowering is over, water supply can be gradually cut off.

Harvest and yield: Plants will start flowering in the second year after planting and the yield will be optimum from the third year onwards. Mullai and Malli produce flowers from March to October and Jathimalli produces throughout the year. Fully developed unopened flower buds should be picked in the mornings. It requires about 30–40 persons to pick flowers from an acre. The plants will yield economic yield upto 15 years of planting and afterwards replanting is recommended.

Yield: On an average about 10,000, 8750 and 11,000 kg of flower buds/ha can be obtained from Mullai, Malligai and Jathimalli respectively.

Care in handling plays a critical role in post harvest flower quality since even the slightest mechanical damage may lead to discolouration and loss in concrete yield or marketable flower quantity. Delicate handling during harvest, handling and transit is emphasized by the fact that non-adherence to such care could render the entire produce unmarketable. Shelf life of Jasmine flowers,

from harvest at tight bud stage to withering varies inherently between species
i.e.,

J. multiflorum	48–60 hours
J. sambac	28–30 hours
J. auriculatum	28–30 hours
J. grandiflorum	24 hours

Different post harvest management practices like pre-cooling, packaging, chemical treatment, cold storage etc. have been found to be beneficial in extending the shelf life of Jasmine flowers and increasing valuable time available for air lifting to foreign markets.

Treating *J. sambac* flowers with sucrose, boric acid, copper sulphate, aluminium sulphate and silver nitrate remain fresh upto 72 hours. Similarly chemical solutions $Al_2 (SO_4)_3$ (0.1%), silver nitrate (0.01%) and Sodium benzoate (0.1%) cause extension of shelf life upto 72 hours in *Jasminum* spp. Though no standard grades are available for jasmines, flower buds should be graded according to their shape, size and freshness before packing in the boxes or baskets for marketing. Corrugated cardboard boxes are the proper packaging material for distant markets while bamboo baskets may also be used for local markets. Packaging should be functional, economical and attractive besides being acceptable in markets. Harvested flowers should be given cold treatment before packing.

Jasmine Concrete

The flowers of jasmine are commercially valued for another purpose *viz.,* production of jasmine concrete. It is a wax-like substance containing the natural flower perfume together with some plant waxes, albuminous matter and pigments. Extraction of jasmine concrete from *J. grandiflorum* flowers has been practiced in France, Sicily, Morocco, Turkey and Tanzania. In India though cultivation of *J. grandiflorum* has been done for several centuries, the use of this flower concrete extraction has been started only in recent years.

Jasmine concrete is extracted from opened flowers picked in the early hours of the day (6 to 9 a.m.). The flowers are soaked in hexane (food grade hexane having boiling point of 70°C) at the rate of two liters per kilo of flowers for half an hour and filtered. During this time the perfume substance along with the wax and pigments get dissolved in hexane. Then the solvent is evaporated over a water bath at 75°C. When all the solvent is evaporated a wax like sediment known as concrete will be found at the bottom of the container. The recovery of concrete for the commercial jasmine species are given in table 25.2. However, concrete extraction on commercial scale is limited now only with *J. grandiflorum.*

Table 25.2 Concrete recovery in Jasmines.

S. No	Name of species	Percent recovery concrete
1.	J. auriculatum	0.33
2.	J. grandiflorum	0.27
3.	J. sambac	0.19

Plant Protection: Bud worm may be a serious pest in Jathimalli and Mullai and this can be controlled by spraying 2 ml. of monocrotophos per litre of water. Leaf spot disease may be prevented in Jathimalli during the monsoon period by spraying of mencozeb at monthly intervals at the rate of 2 g/lit. of water. In 'Malligai' the leaves often turn yellow and this yellowing is caused by three factors viz., iron deficiency, nematode infection and root rot disease. Spraying of 0.5% ferrous sulphate at monthly intervals is recommended against iron deficiency and drenching the soil around the plant with Copper Oxy Chloride (2.50 g/l of water) is recommended against root rot. Application of 10 g of aldicarb granules near root zone will check the nematode population.

2. ROSE

Varieties: There are number of species, sub-species and clones in roses. Some are cultivated and others are found in a wild state. While many of the cultivated varieties of different growing habits (standard, bush, semi-standard and climbers) and a vast range of flower colours are grown mostly in home gardens merely for beautification. The Edward Rose and Andhra Red Rose are chiefly cultivated as economic flower crops.

Soil and climate: Well drained sandy loam or loam is the most suitable soil for rose cultivation. The above two clones can be grown in the plains of Tamil Nadu.

Propagation: The Edward and Red rose are propagated by layering or through cuttings. The other varieties of roses which are grown in home gardens are propagated by shield budding on Briar rose root stocks.

Preparation of field: Two or three ploughings are given. Pits of 45 × 45 × 45 cm size are dug and 20 kg of well rotten Farm Yard Manure is added per pit before planting. The pit is filled with top soil and irrigated before planting.

Spacing: A spacing of 2.0 × 1.0 m is adopted which can accommodate about 5000 plants per hectare. Planting should be done at the beginning of the rainy season, generally in July.

Irrigation: Watering is done once in 2 days till the plants establish themselves and put forth new growth and thereafter irrigation is given once a week.

Pruning: Pruning is an important practice to get regular flowering. This is done in winter season between October and December. The general principle is to see that the past season shoots are induced to putforth new growth which inturn will produce flowers. Before pruning, little water supply is gradually cut off to force the plants to enter into rest. Pruning is done by heading back the vigorous past-season shoots to half of its growth. All the weak, diseased, criss-crossing and unproductive shoots are to be removed. Pruning should be done always to a bud pointing outward and not inward. The cut ends should be protected with Bordeaux paste or copper oxy chloride paste to control the die-back.

Manuring: After pruning in October, basin is formed around the plants to a depth of 10-15 cm and the following doses of fertilizers are given:

Farm Yard manure – 10 kg
Urea – 66 g
Super phosphate – 25 g
Muriate of potash – 50 g

The above doses of fertilizers are to be applied again in June–July.

Harvest and Yield: The plants will start flowering in the first year itself. The crop will give economic yields from the second year and upto 10th year. For markets, flowers should be cut when the flower buds are in the half open stage. At this stage, the flowers keep for a longer time and will be fresh when they reach the market. The flowers should be cut with long stalks. On an average, one hectare of rose yields 7500 kg of flowers.

Plant Protection

1. **Rose Chaffer beetle:** To control leaf eating pest, monocrotophos @ 0.1% should be sprayed.
2. **Red scale:** Pale red encrustations are often seen on the main stem and branches. The scales suck the cell sap and reduce the yield. To control this scale, rub off the affected area with cotton soaked in kerosene or diesel and remove the affected branches and burn them. At the time of pruning, parathion (2 ml/litre of water) is sprayed and also once during March-April.
3. **Thrips, aphids and jassids:** These sucking pests can be controlled by spraying methyl parathion (2 ml/litre of water.)
4. **Powdery mildew:** This affects the leaves and flower buds. Carbendazim, at the rate of one g/litre of water or wettable sulphur at the rate of 3 g/litre of water may be sprayed to control this disease.

3. CROSSANDRA

It is an important commercial flower of India, which has its origin from India, Madagascar and Africa. In India, area of production is 1716 hectare with an annual production of 4,300 tonnes. In Tamil Nadu area is 988 hectare and production is 2,500 tonnes. Unlike rose, jasmine and tuberose which are highly fragrant, the crossandra is scentless, light weighted and remain fresh for long time. The year round production except the rainy months gives this flower a high demand in the market even when other flowers are not in production. It has occupied an important status in the life of ancient Tamilians. Crossandra *(Crossandra infundibuliformis)*, belonging to the family Acanthaceae is having cultivars like orange, Red, Yellow and Delhi. The cultivar Delhi is a triploid (2 n = 30) and produces more attractive flowers of bright deep orange colour and rest of the varieties are tetraploids (2 n = 40).

Soil and Culture

A tropical plant and grows well in places where the temperature is around 30–32°C. It requires well drained sandy loam soil.

Propagation

Except Delhi crossandra, rest of the varieties is propagated by seeds. Seed rate is 5 kg/ha. Fresh seeds are sown during May–June in raised beds,15 cm apart in lines. The seedling will be ready for transplanting in 60 days or when they have 4-5 pairs of leaves. Delhi Crossandra, being sterile, is propagated through cuttings under mist.

Propagation of Field

The land is ploughed thrice and FYM at 25 t/ha is incorporated at the last ploughing. Ridges are formed 60 cm apart and the seedlings or rooted cuttings are planted on the sides of the ridges at 30 cm spacing. Delhi crossandra is planted at a spacing of 60 × 40 cm.

After Cultivation

High fertility in the soil is essential for good yield. In addition to the basal dressing of FYM, NPK at 75, 50 and 125 kg/ha are to be top dressed three months after planting and the same dose is to be repeated at 5–6 monthly interval for two more years. If Azospirillum @ 2 kg /ha is applied, N level can be reduced to 60 kg/ha. Irrigation is given normally at weekly intervals.

Crossandra commences flowering 2 to 3 months after planting and continues to bear flowers throughout the year with a drop in production during the rainy season. The flowers open in sequence from the base of the spike. The two flowers, which are diagonally opposite in spikes, open at the same time. It takes about 2 days for complete opening of flowers. Therefore, picking of the flowers has to be done on alternate days. Depending upon the length of spike, it takes nearly 15-25 days to complete flowering on a spike. After the flowering is over, the spent spikes are to be removed. The fully opened flowers remain fresh for 1-2 days if picked and in the case of Delhi crossandra, it keeps 3-4 days. Flowers should be picked early in the morning by pulling the corolla out of the calyx, and are packed in polybags or cloth. About 15,000 flowers make one kilogram and 2000 kg of flowers can be obtained per ha per year. Delhi crossandra yields about 2800 kg of flowers per ha/year. The crop can be retained for 2 years normally and if ratoon is allowed, it can be maintained for 3 years.

Plant protection:

1. **Wilt:** Caused by *Rhizoctonia* and *Fusarium solani* is a serious disease. Its symptoms are drooping of leaves, marginal yellowing and rotting of roots and rootlets. Control measure includes drenching the soil around the plant with Carbendazim 0.05%.
2. **Nematodes:** Plants affected by nematodes exhibit stunted growth, curled leaves with purple discolouration, chlorotic symptoms and discolouration of the flowers. Planting in nematode infected area may be avoided. Application of phorate or carbofuran 3 G at 1 kg a.i/ha a week after planting and again 6 months after planting at 3 g and 9 g respectively per metre length is recommended.

4. TUBEROSE

Tuberose (*Polianthes tuberosa L.*) is one of the important economic flower crops whose flowers remain fresh for pretty long time, stands long distance transportation and are used for garland making, bouquet, buttonholes etc. It is a hardy, perennial bulbous plant, bearing rosette leaves and the single or double flowers in a spike.

Soil and climate

It requires well drained loamy soil with irrigation facilities. It prefers a mild climate without extremes of high or low temperatures.

Varieties

There are three types of tuberose: single with one row of corolla pigments, semi-double bearing flowers with two or three rows of segments and double having more than three rows of corolla segments. Culcutta Single and Kolkata Double are the promising types for Tamil Nadu. Other single flower varieties are Mexico, Pune and Bangalore. Recently, two improved varieties have been evolved at NBRI, Lucknow:

Rajat Rekha: Single flowered type with silvery white streak along the middle of the leaf table. It is a mutant evolved by irradiating bulbs of single flowered cultivar. Concrete content has been found to be 0.089 per cent.

Swarna Rekha: Double flowered type with golden yellow steaks along the margins of leaf. It is a gamma ray induced mutant, in which mutation occurred in chlorophyll synthesis resulting in change in leaf colour. Concrete content has been found to be 0.062 per cent.

IIHR, Bangalore has also evolved four new varieties.

Srinagar: A cross between 'Single × Double', bearing single type of flowers on sturdy spikes. The flower bud is slightly pinkish tinged. Florets are bigger and appealing than 'Kolkata Single'. Resistant to *Meloidogyne incognita* nematode, loose flowers are ideal for making garland, while spikes can be used as cut flower. Yield of loose flowers is about 15,000 kg/ha per year, which is 40% higher than 'Kolkata or Mexican Single' and the concrete content of the Hybrid is at par with Mexican Single. It is preferred by farmers and perfumery industries.

Suvasini: A multi whorled variety developed from the cross between 'Single × Double'. Pure white flowers are bold and big, borne on a long spike. Spikes are best suited as cut flower, recording 25% more yield than cv. Double.

Prajwal: This hybrid between 'Srinagar' × 'Mexican Single' bearing single type flowers on tall stiff spikes. The flower buds are slightly pinkish in colour white the flowers are white. The individual flowers are large in size, compared to 'Local Single'. It yields twenty per cent more loose flowers than 'Shrinagar', recommended both for loose flower and cut flower purpose.

Vaibhav: This hybrid ('Mexican Single' × IIHR-2) bears semi-double flowers on medium spikes. The flower buds are greenish in colour in contrast to pinkish buds in 'Suvasini' and 'Local Double'. Flowers are white. Spike yield is 50 per cent higher compared to 'Suvasini'. Hence, it is recommended for cut flower purpose.

Propagation and Planting

Tuberose is commercially propagated vegetatively by means of bulb. As the size of bulb plays an important role on growth and flowering, corms weighing 25–30 g should be selected and planted on the sides of ridges at 45 × 20 cm spacing at 2.5 cm depths during June–July. Usually, corms are planted after 30 days of harvest. Before planting, the corms are dipped in 5000 ppm of cycocel (CCC) which help to increase the yield.

After cultivation

Tuberose responds to manures and fertilizer application. During the field preparation, a basal dose of 25 t of FYM and 25 kg N and 60 kg each of P and K are applied. 25 kg N is applied as top dressing at 2nd, 5th and 8th month after planting in the first year and the same dose is repeated in the second year also. The tuberose field can be irrigated at 10–15 days intervals during the dry months.

Tuberose normally begins to flower in 80–95 days after sprouting and continues to flower throughout the year wherever mild climate prevails. Tuberose is harvested by cutting the spikes from the base for table decoration or the individual flower is picked on daily for making garlands and other floral ornaments. An average of 6000 kg of flowers can be harvested per hectare per year.

Nematodes such as *Meloidogyne incognita* and *M. javanica* are often causing serious problem to tuberose. Affected plants show yellowing, drying up of leaves and retarded growth. Application of aldicarb or carbofuran 3 G g/plant near the root zones followed by immediate irrigation is recommended.

5. HIBISCUS

Hibiscus (*Hibiscus rosa-sinensis*) lacks fragrance, still it is being extensively grown in the tropical areas for its beautiful flowers. Single whorled flowers have petals of colours like red, white, yellow, orange, rose, pink, blood red and blue and multi whorled hibiscus flowers have petals of yellow, red, pink, deep red, rose, orange, reddish yellow, red and white colours mixed. Hibiscus is being cultivated mainly for its flowers and foliage which are also used in many of the religious rituals.

Varieties

Tamil Nadu Agricultural University has released Thilagam, Punnagai and Co-3.

Thilagam

This variety has been developed by inter generic hybridisation between *Hibiscus rosasinensis* and *Malvaviscus arboreus*. The flowers are multi whorled with uniformly red coloured petals from the neck to the tip. Each flower has 30–36 petals in 3 whorls. Since they flower all through the year and the flowers look like roses from a distance they are grown in pots and home gardens for beautification.

Punnagai

This variety is an improved variety of Chandrika. The flowers appear singly at apices or leaf axils of young branches. Flowers are apricot yellow in colour and 14.3 cm in diameter. The petals are wavy and veins from the neck of the flower spread upto the tip of the petals. They are highly suitable for home garden lawns.

Co-3

This variety is a hybrid of yellow sport and Shanti. Flowers are yellow, 19.2 cm in diameter grown in apices or leaf axils. The flower petals are attractive yellow with red lines at the neck regions. Stigmatic lobes are red in colour.

Indian Institute of Horticultural Research has released the following varieties: Aktha, Anuradha, Ashirwal, Bharat Sundari, Chitralekha, Dilruba, Geethanjali, Ratna, Red Chetan, Phulkari, Srimathi, Indira Gandhi, Tribal Queen, Arunodhaya, Basant, Benazir, Jogan, Nartaki, Rasneen, Nilofar, Pakeeza, Tiria, Kaserghatta, Rani, Red Wold, Shanti, Smt. Kamala Nehru. Other than these varieties there are many single whorled and double whorled local varieties in cultivation.

Single Whorled Varieties

Alps, Akkinis, Avrora, Borealis, Australian Single, Flamming Sunset, Glowing Sunset, Gold Mine, Gold Glowlakshmi, Lipstick, Lunasav, Meteor, Mychoice, My Beauty, Nethaji, Pucant Pink, Beauty Persy Lancoster, Bright of Lanka, Rajendra Prasad, Snow White, Sulphuria Sunset, Superb, Superba, Siriyacus, Viceory.

Double Whorled Varieties

Alipur Beauty, Alex Sylvia, Avrora, Baskar, Birliantissima, Centenary, Chitra, Daffodil, Cream, Cagwar, Ambarode, Gulabi, Golden Spring, Golden Giant, Golden Cling, John Walker Geno, Lady Marjyotyaskin, Lord Winlithco, Mahatma, Maywalkar, Trince of Orange, Roy Rastrawang, Goriankiss, Rukhmini, Ruby, Calmon Globe, Siriyacus Double.

Cultivation Hints

Generally hibiscus is propagated by rooted cuttings. Cuttings should be semi-hard wood type. It can be grown in all types of soil especially red loam with high humus and sandy soil are best suited. They come up well in well drained sandy areas. At the beginning of the monsoon the rooted cuttings should be planted during the evening hours and irrigation should be done. The plants produce flowers in young branches during summer and spring season. So the spent branches should be pruned during the winter season to induce new flowering branches. Criss- branches and dried branches should be also pruned to give form to the plant. A pruning height of 1-1½ m should be maintained to facilitate easy harvesting of flowers and leaves.

(B) CUT FLOWERS

'Cut flowers' are fresh flowers harvested in clusters, spikes or in singles along with their stem. Their production is influenced by the demand, customs and fashions of a country. In India cut flowers are grown only in open but in countries like England, U.S.A. and Japan, it is a specialized technique and cut flowers are usually forced in temperature controlled green houses. As most of the cut flowers are grown in open in India, the flowers produced are low in quality and fetch low returns. Protected cultivation of flowers is proposed as one of the essential inputs to boost floriculture in India. Low cost poly green house may be utilized in the flower growing areas to basically prevent the damage from weather factors and other natural agents. Cut flower trade is also well established one from the point of production till they reach the final users. In our country, flower crops like roses, chrysanthemum, orchids, tuberoses, carnations, dahlias, gladioli and gerberas are grown for cut flower production.

Significance of Cut Flower Industry in India

1. Cut flowers contribute nearly 60 percent of the world consumption of the floriculture products which is estimated to be worth of 40 billion.
2. The global floriculture industry is growing at an annual growth rate of 20-30 per cent while Indian market is growing at an annual rate of 2-3 per cent.
3. The per capita consumption of floriculture products is also steadily increasing every year.
4. India's varied agro climate and soil types offer abundant opportunities for cultivation of all kinds of cut flowers almost all through the year.
5. The availability of relatively cheaper labour enables us to produce these crops and supply them to International markets at competitive prices.

The flowers of some species and varieties remain fresh only for a few hours or a day and therefore they are of no value as cut flowers. Longevity of flowers largely depends on the species, season of blooming, prevailing temperature, stage and time of harvesting the bloom etc. Roses, chrysanthemum, tuberoses, carnations, gladiolus, aster, gerbera, snapdragon, tulip and hydrangeas have better shelf-life. Late after-noon or early morning is the ideal time for harvesting of flowers. Sharp knives should always be used for cutting them so that the water conducting system in flower stem is not damaged. Stage of harvest is also an important factor and it varies from species to species or even cultivars to cultivars. Generally, flowers before full opening are harvested. Recently methods have been developed so that some flowers are harvested as 'buds' which facilitate opening them off the plant in specific chemical solutions. This technique is known as "bud opening'. The chemical solution basically contains sucrose only. But opening has many advantages over the cutting of the flowers at the normal stage *viz.,*

1. It reduces the sensitiveness of the flowers to the environmental conditions.
2. It saves space during transport and
3. Vase life is comparatively extended.

This stage of harvesting for bud opening is also specific for each kind. Immediately after harvesting the flowers, the cut end of the flower stalks should be dipped in water. Sometimes, in case of coarse textured flowers the lower 4 to 5 cm of stem is pressed with a piece of wood or hammer in order to facilitate the absorption of water.

Before packing, they are given a treatment called 'pulsing' to extend their shelf-life. It refers to a pre-shipment treatment given to the flowers for a short period by the growers or shippers with high concentrations of certain chemicals, the effect of which last for the entire post-harvest period even when the flowers are held outside. The main ingredient of various pulsing solutions is sucrose, which is used in higher concentrations than in preservative formulations. Specific formulations have been developed for different flowers and sometime, even for different cultivars. Sucrose undoubtedly serves as a respiratory substrate and to certain extent prevents desiccation and probably replaces the depleted natural carbohydrates and eliminates the break down of the other organic compounds, the colour and size of petals especially in rose. For 'Raktagandha' cut roses, 3% Pulsing treatment apart from prolonging the shelf life, promotes opening and improves sucrose for 24 hours was found to be the best. Apart from sugars, pulsing is done with silver nitrate ($AgNO_3$), calcium chloride ($CaCl_2$) calcium nitrate ($CaNO_3$), Sodium thio sulphate (S.T.S.) etc.

Stem blockage is the major cause of water deficit and wilting of cut flowers. Water flow through the stem is blocked by the microbial growth at the base. Certain salts such a Quinoline salts, Zinc ions, nickel are recognized as bactericide as they suppress the microbial growth and control stomata opening.

Packages for Export of Cut-flowers

Packaging for export of cut-flowers is as important as the quality of the flower. Only Corrugated Fibre Board (CFB) packages are accepted in the trade. There are different types of CFB boxes for packaging cut flowers like slotted boxes, telescopic boxes, rigid boxes, ready glazed boxes and folded type boxes. It is better to use telescopic boxes which are made up of two parts *viz.*, a bottom and a lid. An ideal package for export should be small in volume, air-tight, non-water absorbent of the required strength, should have minimal effect on quality and shelf/vase life of flowers.

The size of the packages especially the length of box depends upon the stem length of cut-flowers required for in the trade. For instance, the length of CFB for rose is 100 cm since the stem length ranges from 40 to 60 cm. The strength of the containers should be enough to withstand a stack height of 3 to 5 times to protect the loads from damage due to shock and vibrations. Ventilation holes are not necessary if the flowers are packed and transported in ideal conditions.

The flowers are to be immobilized and cushioned to avoid abrasion bruises due to vibration. This is achieved by lining the inside of the boxes with thin polyethylene films. Fine moist tissue paper shreddings are also spread at each of the box to provide cushioning to the blooms.

The cultivation hints for the important cut flowers are given below:

1. Roses

Rose is one of the important cut flowers in India and recently India has started exporting roses to Europe. Cut rose varieties are of three groups:

(a) **Long stemmed roses:** These generally produce stems of 50–120 cm in length and yield 100–150 stems per year per square metre. They are harvested 40–60 days apart in pronounced flushes depending on temperature and variety. Ex Vivaldi, First red, Grand Gala, Konfetti, Tineke, Skyline

(b) **Medium stemmed roses:** These generally produce stems of 50–70 cm in length and yield upto 220 blooms per square metre per year. Ex. Jaguar, Golden Times, Baronesse, Lambada, Gabriella.

(c) **Small flowered roses:** The yields in this group range from 250–350 stems/sq.m/year with stem length of 30–70 cm. However, these varieties need more labour hours for harvesting. Ex Motrea, Coronette, Calypso, carona. High yielding varieties with longer vase life (upto 16 days) with less of bent neck problem must be selected apart from color (preferably red) and being pest and disease resistant.

The cultivar preference in export market keeps on changing. The varieties like Super Star, Queen Elizabeth and Happiness are the earlier preferred export varieties.

Soil medium and environment: Deep (up to 50 cm) loam soils rich in organic matter with a pH of 5.5-6.5 are most ideally suited for cultivation. Growing medium may consist of soil, sand and FYM in the ratio of 2:1:1. In place of sand, cocopeat or sawdust can also be used. Soil less culture i.e. growing on rock wool is advocated only when severe soil problems exist. The ideal climate for rose growing should have temperature with a minimum of 15°C and maximum of 28°C. Light is important factor which decides the growth. The growth is slowed by day length, i.e. < 12 hours and heavy overcast, cloudy/ mist conditions. High relative humidity exposes the plant to serious fungal diseases.

Planting

Planting is taken up in well prepared raised beds of 1.0–1.6 m wide and 30–40 m in length. About 4–5 beds of 1 m wide with a 60 cm path can be accommodated in a bay of 6.4 m or 8 m wide. There will be two rows in each bed with 14–18 cm between the plants in row. This spacing accommodates 60000–70000 plants per hectare. Normally planting is taken up in April–May or in August–September depending on plant size so that the first harvest comes before Christmas and second flush at the time of Valentine's Day.

After care

Once planting is over, the greenhouse should be kept warm and humid for some period so that the plants start to grow vigorously. During strong sunshine hours, the young plants must frequently be given an overhead spray with water to assist establishment and reduce post planting losses.

Building up of the rose plants

Four weeks after planting when the flower buds have reached the size of a pea, the flower bud is removed without removing any leaves. If lateral shoots are formed in the leaf axil below the removed bud, remove them carefully. The maximum leaf area is required to build up a strong root system. Three to four weeks after removing the main bud, the plant is established and it is possible to cut the shoots down and start production of flowers. It is very important not to remove too much with this cutting. On each shoot, three to four true leaves must be left. Sometimes, instead of cutting down the shoots bending of the shoots can be done. In this case, removal of centre bud and laterals may not be necessary. Instead of cutting the shoots down many growers bend down the shoots and thereby reduce the apical dominance. In this case removal of the center bud and laterals may not be necessary. The ideal is to have the branch to form a bow.

Resting the plants through a dry period

Resting of plants is done during off season (June–August). Irrigation and nutrient application is withheld when the last flower is harvested. The dry period lasts from 4 to 8 weeks. Most of the leaves will drop. Then plants are pruned down to 30–60 cm above the soil. After pruning, start irrigation and feeding slowly. New shoots require shading from very strong sunlight. This method may give few bottom breaks and too weak stems. Another method is harvesting is done through under cutting which also decrease the height of plants. Some growers do not harvest the flowers but remove the flowers and buds when they have reached a certain stage. Before the export season starts, hard pruning is done which is said to give good bottom breaks and stronger stems. There are some operations like disbudding i.e., removal of side buds below the centre bud, under cutting, maintenance of raised beds which have to be carried out on day to day basis.

Nutrition

Regular analysis of soil and plant is necessary from the beginning to give precise amount of fertiliser required by the plants. Organic manure at the rate of 15 kg/m^2 has to be added and mixed with the soil at the time of bed preparation. Before planting, a basic application of fertiliser at the rate of 2 kg superphosphate, 1 kg calcium ammonium nitrate and 1/2 kg of muriate of potash per 10 m^2 is to be incorporated in the soil. As plants grow and establish, nutrition is to be given frequently through irrigation system. Normally 200 ppm each of nitrogen and potash are applied through fertilization. Tissue nutrient content (on dry weight basis) of nitrogen 3.0 to 5.0%, Phosphorus 0.2 to 0.3%, Potassium 1.8 to 3.0 %, Calcium 1.0 to 1.5%, Magnesium 0.25 to 0.35%, Zinc 15–50 mg/kg, manganese 30–250 mg/kg, iron 50–150 mg/kg, copper 5–15 mg/kg and boron 30–60 mg/kg is essential for production of good quality roses.

Irrigation

Irrigation is done through drips and nozzles (mist). The amount of water required depends on temperature, RH and incoming light.

Harvesting

Stage of harvesting varies with varieties and distance of the market. For export, flowers are harvested at tight bud stage and when all sepals have unfurled. Harvesting is done either in early morning or in the evening hours and are placed immediately in buckets containing cleaning water. They are precooled before and after grading. Flowers are packed after grading in cardboard boxes and cold stored (2–10°C) till they are loaded into refrigerated vans and transported to airport.

Post harvest quality requirements

Cut flowers are to be graded according to cultivars, stem length, size and form, stage of maturity, condition flowers and foliage. The flowers with different stem length should not be mixed together and uniform stem length is to be maintained. The size, colour and form of the flower should be the representative of cultivar. The flower stem should be strong and capable of holding the flower in upright condition. The flowers should be free of disease and pests, petal discolouration and brushing injury. According to the market demand and the type of flowers, stem length varies from 15 to 120 cm.

Storage of cut flowers

Rose cut flowers can be stored dry at 0 to 1°C for 15 days. The flowers for dry storage, after harvest, are placed in moisture proof boxes and precooled before sealing in the containers. Cut flowers can be stored wet at 2 to 5°C for 5 days. Flowers can be stored at low pressure storage (40–60 mm Hg) for 8 weeks, and at modified atmospheric storage (0°C) for 3 to 4 weeks.

Pulsing

Pulsing with 3 per cent sucrose for 18 hours at 20°C or 0.5 mm of STS for 45 minutes improves post harvest life and quality of cut flowers.

Holding solution for lengthening the vase life of cut flowers

Vase life of cut roses can be lengthened by keeping in holding solution of 3% D-fructose + 25 ppm Kinetin or 3% D-fructose + $AgNO_3$ or 3% D-fructose + 300 ppm $NiCl_2$ or 3% sucrose + 200 ppm 8-HQC or 3% surcose + 50 ppm DICA or 3% sucrose + 50 ppm DDMH.

Bud opening solution

Tight cut closed buds can be developed to commercial maturity in vase solution containing 1 per cent D-fructose, 100 ppm acetylsalicylic and 250 ppm 8-HQC. Pulsing with 0.2 mm STS for 15 minutes followed by keeping in the holding solution of 300 ppm 8-HQC plus 2 per cent sucrose is effective for bud opening.

Premature harvesting and excessive water loss causes bent neck. Roses are relatively insensitive to ethylene. Fore reviving limp rose, entire flower is to be immersed in water for two or three hours.

2. Chrysanthemum

It is also one of the important cut flowers throughout the world. There are several attractive flower types, differing in colour, form, arrangement and shape of the petals. Most growers allow the flowers to open too far before cutting. Cutting the stem with the flowers slightly on the green side is recommended as it offers better quality to the consumers. The 'spray' type should be cut when the central flower is open and the surrounding bud developed. In the case of standard type they should be cut before the center florets are fully expanded. The varieties which shed pollen should be cut early.

The cultivation of Chrysanthemums depends on what the quality of the end product should be. Chrysanthemum can be grown in open field conditions but since there are limited possibilities of controlling climate, disease and pests, this type of cultivation does not give any guarantee of quality. It goes without saying that the cultivation under a polyhouse with insect proof netting has the highest potential of giving a good quality and exportable product. In all three cases one has to make provisions for lighting and black closing.

Planting

The cuttings are planted on beds of about 12.5 cm wide. Iron nets of 12.5 cm mesh is used for easy distribution of the cuttings. The plants should be planted with their roots fully in the soil and well pressed to make good contact with the soil.

Weeding

Weeds can not be avoided in the green house. They deplete moisture and nourishment from plants. Shortly after cuttings gets established the weed growths are scratched from the ground to uproot them while they are small

Long day and short day periods

The chrysanthemum starts to develop flower buds when the light is switched off during the normal day length i.e., shorter than 13 hours. In tropical areas until about 10° latitude the day length does not exceed 13 hours so blacking-out is not necessary. In areas where the day length exceeds the 13 hours one should interfere by the help of black cloth or plastic.

Length of the crop at starting of short day period

Only when chrysanthemums grow at a day length longer than 16 hours, the plants stay vegetative. The crop has to stay vegetative until it has developed enough levels to reach enough length and sufficient flower buds. Depending on the growing conditions (summer or winter) the length of the plants should be between 30 and 40 cm to start short day conditions.

The number of days it takes to have enough vegetative growth ranges a lot and can be between 18 and 35 days.

The generative period (short day period)

During this period the flower buds are developed and grow out into flowers. Temperatures influence the flowering time (response time) and the form of the umbel. Extreme high or low temperatures have a negative effect on the uniformity and quality of the umbel and can also delay the response time.

Growth regulators

In many countries stem weight is an important quality aspect. In those countries both stem length and weight are being manipulated by using growth regulators. Regulators are used to reduce the growth and increase weights are ALLURE and B-Nine (active ingredient: daminozide). The regulators used to increase growth are called Berelex (gibberellin).

Pinching out the apical bud

As soon as the bud stems begin to elongate it is needed to pinch out the center bud, which in turn improves the spray shape. If pinching is done too early, damage is caused to the side buds. As a rule this treatments is carried out between 2nd and 3rd weeks before harvesting.

Harvest

When the flower buds are big enough the central bud (which is bigger than the others in the spray) has to be removed to get a well formed and uniform flowering spray.

The stems are harvested by pulling them out and breaking of the root system leaving it behind in the bed/field be ploughed into the soil when bed/field is prepared for next crop.

Post harvest

Depending on the variety, plant start yielding flowers after 3-4 months of transplanting. For cut flower purpose stem is cut about 10 cm above the soil to avoid cutting into woody tissues. The lower 1/3rd of stems are stripped off leaves at harvest and as soon as possible

After harvest, the stems have to be cut at equal length (90 cm is the standard), bunched in five putting a rubber band at the base and sliding them into a plastic sleeve and putting the bunches in plastic buckets filled with 20 cm of water. On the day of shipment the bunches can be packed in boxes.

Grading

Chrysanthemums are graded based on the stem length, flower appearance, number of flowers, stem straightness, colour and freshness of flowers. Standard chrysanthemums are graded into Blue, Red, Green and Yellow, whereas spray types are graded into Gold, Silver and Bronze based on the quality parameters. In Dutch market, spray chrysanthemums are graded into extra grade and shorter grade. The lower leaves are stripped off upto 15–20 cm and bundled in units of 5 stems and secured with a rubber band.

Export Standards for Chrysanthemum

Parameters	Standard	Spray	Dwarf
Stem length	88–100 cm.	77–88 cm.	25–38 cm.
Weight	30 g/stem of 90 cm.	30 g/stem of 85 cm.	30 g/stem of 30 cm.
No. of flowers	Only 1flower with 5 buds.	10 flowers.	10–12 flowers.
Diameter	60-80 mm.	35 mm for half bloom. 45 mm for full bloom.	30 mm

Packaging

Most of the Standard chrysanthemums are placed in sleeves and packed in display boxes measuring 91 × 43 × 15 cm. They are placed in the boxes according to the grades. For bulk packing of the spray chrysanthemums, 10, 15 or 20 stems are placed in sleeves according to the grades. Six sleeves, three at each end, are generally packed in each box, measuring 80 × 50 × 23 cm.

Cold storage

The stems in the buckets (after grading) are given a cut using sharp blade and precooled at 1°C minimum of 2 hours before packing. Chrysanthemum can be stored for 3–6 weeks period at 0–3°C.

6. CARNATIONS

Carnation *(Dianthus caryophyllus)* belonging to the family caryophyllaceae, is one of the important cut flower crops in the international flower market. Its large array of colours and excellent keeping quality has made it one of the most demanded flowers. Carnations in general are grown only under protected cultivation of carnation.

Cut carnations are of two types, the Standard Type and Spray Type. The standard varieties (e.g. Scania, Candy white) have one large flower on an individual stem while in spray type (e.g. Etna, Excel) terminal flower bud is removed at an early stage to encourage more even development of the lateral flowers, which then produce a multiple flowered stem.

Bed Preparation

Red sandy loam which is porous is good for carnations. If the soil is very clayey, it can be made more porous by adding paddy husk or similar material. The optimum pH of the soil for growing carnation is 6.5 with an EC of 1.2 at the start and 1.5 at the generative period is ideal.

Soil should have the combination of 1:1:1:1 of red soil, FYM, sand and Paddy husk or similar material, but this can vary according to the condition of the soil. Besides, organic manures like neem cake, pongamia cake are to be applied @ 500 g per sq.mt. After adding the organic matter, the soil should be fumigated by either Methyl Bromide, Metam sodium, Basamid or Formaldehyde (formalin). After fumigation the soil should be thoroughly turned and the fumigation chemical should be removed by leaching with plain water 3 to 4 times with turning of soil each time.

Generally before bed preparation, the following basal dose has to be added. This can vary as per the soil analysis report.

    (a) Single super phosphate   – 200 g/sq.mt.
    (b) Potassium sulphate   – 150 g/sq.mt.
    (c) Magnesium sulphate   – 50 g/sq.mt.
    (d) Boron   – 2 g/sq.mt.

The ideal size of bed is 75 cm wide, 25 m long and 0.30 m high with a path of 40 cm wide between any two beds.

Support Netting

Before planting, the netting should be ready (at least the first layer). The main frames on either ends can be fixed before bed making and the beds can be cut as per the frames. The frames can be either made from 'L' angles or pipes. The main frame 'L' angle should be 40 mm with 5 mm thickness. The height of the frame above the ground should not be less than 4 ft and 2 ft should go below the ground with concrete. The main frame should be supported by 2 'L' angles of similar gauge welded from $1/3^{rd}$ height from the top. This will prevent the frame from collapsing inside. Four horizontal angles should be welded or bolted to the Vertical angle at distance of 20 cm except the bottom, which should not be more than 15 cm from the bed level. The bottom angle can be made flexible.

The intermediate stands should be at a distance of 3 m. each, if this distance increases the net tends to collapse. The intermediate stand can be of 25 mm 'L' angle with 3 mm thickness with corresponding horizontal angles.

Netting can be prefabricated or done on the spot. The netting can be with GI wire of 16 gauge for length wise fixing and nylon thread for width. The nylon wire should not be of very bright colors like red, yellow or orange, as they will attract insects/pests.

As the plants grow the II, III and IV layer can be put. Care should be taken to train the plants into the net so that the stem is straight and the stem length will not be reduced because of bent stems.

Planting

Shade net; misting lines; drip lines and support netting should be in place before going for planting. Bed should be moderately wet. The beds should be drenched lightly with a fungicide (captan 2 g/litre) before planting. Preferably the planting should be done in the evening so that the plant has time for recovering overnight. The planting distance is 15 cms × 15 cms. This will come automatically if the above procedure of netting is done. Small shallow holes should be made in the beds with the help of a stick and a little amount of coco peat to be put in the hole. The plants should be removed from the packet carefully so as not to damage the existing roots. The plants should be planted with part of the root zone being exposed, without pressing the root zone. Deep planting will lead to rotting. The thumb rule is after planting and misting if about 25% plants lodge to one side then the planting depth is correct. The lodged plants should be straightened the next day.

Regular misting has to be carried out for atleast 14 days or till the roots penetrate the media and are able to take the water through the roots. Care should be taken to see that a thin layer of moisture always remain on the leaves. The last spray for the day should be done 1 hour before sunset because moisture on the leaves through out the night may lead to fungal infection. No irrigation is given until the establishment period of 14 days end. Next one week plain water irrigation has to be given judicially depending on the weather condition and the soil moisture level. Shade net can be removed or used as per requirements after the establishment.

Pinching

There are 3 methods of pinching 'Single pinch', 'One half pinch' and 'Double pinch'. For general Indian condition 'single pinch' is preferred. In this method, when the plants attain about 6 nodes in about 21 days from the date of planting, the plant is pinched at the 5th node. Once the pinching is done the plants start giving breaks at the nodes. Care should be taken to remove if there are more than six shoots emerging. By doing this 4 to 6 good flowering shoots will be got.

Irrigation

Over watering and poor drainage cause root death and stunted growth. Water logging would cause deprival of oxygen to plants. The growing medium should be evenly moist.

Fertigation

Fertigation has to be done through drip. Growers adopt different combination of fertilizers and schedule. One basic program/schedule is given below. Changes must/should be made as per the soil and water test results.

Nutrient	Till bud formation	From Bud Formation To Harvest
	g per square metre per week*	g per square metre per week*
Ammonium nitrate	3.0	2.0
Potassium nitrate	5.0	7.5
Monoammonium phosphate	2.0	2.0
Magnesium sulphate	2.5	2.5
Boron	0.1	0.1
Trace elements/micronutrients	1.0	1.0

* The above dose is given on Monday and Thursday while on Tuesday and Fridays, Potassium nitrate @ 5 g and Calcium nitrate @ 8 g alone are given. On Wednesday and Sunday, plain water alone is given while on Saturday, fungicide is drenched.

Training

Training is a very important and continuous operation in growing of carnations. This operation is making the plants go through the net openings for it to grow straight with out bending at the bottom. This has to be done on a daily basis, so that there is no loss of stem length while harvesting.

Disbudding

Disbudding of standard carnations is done as lateral buds can be easily removed and should be done on a continuous basis; it is essential to maintain flower size and quality.

Disbudding refers to removal of side buds so that the central/terminal bud receives maximum food for the full development. In standard carnations, side buds should be removed where as in spray carnations, the terminal bud has to be removed and it should be done on a continuous basis; it is essential to maintain flower size and quality.

Harvesting

The flowers will be ready for harvest in most varieties between 105 to 120 days. Very sharp cutter or secateur is used so that minimum damage is done to the main stem. Both hands are to be used while harvesting the flowers, or else there is a chance of damaging other shoots and the plant itself. The harvested flowers should be put into a bucket containing water as soon as possible.

There are different stages of harvest done as per the market demand.

- **Tight bud stage:** This is not advisable as chances are that it may not open at all after harvest.
- **Paint brush stage:** Ideal stage for harvest for long distance travel or for use after a couple of days.
- **Semi-open stage:** Some market demands this stage as the flowers can be used in a day or two after harvest. Ideal for short distance travel.
- **Open stage:** This is ready to use harvest stage and not suitable for travel.

Standard carnation flowers are cut when they are still in tight bud stage with the petals emerging. Spray type carnations are cut when two flowers are open and the remaining buds are showing colour. Harvested flowers are bunched together based on their physical measurements like length of stem, diameter of flower etc. For a good post harvest life, flower stems have to be trimmed at the base and should be immediately placed in a bucket of preservative solution (pH 4.5) for 2 to 4 hours with 2–5 per cent sucrose and a biocide not phytotoxic to carnations after keeping in preservative solution.

Post Harvest Handling

When the bucket in the field is full, it is to be transported to the grading hall. The grading specification varies from market to market. However, commonly followed grades based on stem length of cut carnation are:

Code	Stem length (cm)
40	40–50
50	50–60
60	60–70

The carnation flower is very susceptible to the gas ethylene, when flowers get exposed to the gas, they become sleepy and are useless. Pulsing of cut stem with 1 mm silver thiosulphate + 10 per cent sucrose for 8 hours improve vase life. Other recommended floral preservatives on carnation are given below:

1. AgNO$_3$ 30 mg/l + Sucrose 4%
2. 8-HQC 300–400 mg/l + Sucrose 4%
3. Copper nitrate 30–300 mg/l + Sucrose 3%
4. Benzalkone 50 mg/l + Sucrose 10%
5. STS (Pulsing 20 hours) + Sucrose 10%

Commercial flower preservative containing STS are: Chrysal, AVB, Florissant 100 and Forever. 2 ml of either Chrysal or AVB or Florissant 100 in 1 lit of water is used. Flowers of carnation cut at paint brush stage can be stored under refrigeration for 3 weeks at 4 ±1°C, both under wet and dry conditions. The vase life of cut flowers can be extended by improving both pre- and post-harvest conditions. After storage, keeping the cut stems in 100 ppm STS + 2% sucrose in holding solution proves to be the best.

Packing

After grading and treatment with flower preservative, the flowers are packed in bunches and sleeved in plastic sheets or newspaper according to the demand of the customer. The sleeved bunches are packed in cardboard boxes (30 cm height × 50 cm width × 122 cm length) that are brought into the cold store and cooled down.

Flowers are also packed in cartons lined with polyethylene and pre-cooled without lid. The plastic is then loosely folded on top of the stems and the lid is closed. These cartons are stored in cool chambers designed to maintain 0°C with good air circulation and a constant relative humidity of 90–95 per cent.

Storage

The carnation flowers can be stored between 4°C to 6°C to keep the flowers in good condition, but not lower. Carnations are sensitive to ethylene, hence it should not be stored with fruit, vegetables, cuttings or cut flowers that release ethylene gas. Carnation flowers can be stored for two to four weeks before marketing.

Disorders

Calyx Splitting

Cultivars with too many petals are susceptible to calyx splitting. Varying temperature and environmental conditions also influence calyx splitting. Selection of cultivars that are less prone to splitting, regulation of temperature and maintenance of optimal fertilizer level can minimize this disorders. This can also be reduced by placing a rubber band or 6mm wide clear plastic tape is used around the calyx of the flowers which have just started opening. The following varieties are less prone to splitting, cv. Espana, Cabaret, Red Corso, Pamir, Raggio-di-Sole.

Slabside

This is another malformation during cool periods. The bud does not open evenly so that petals protrude on one side only, giving a lopsided shape to the flower. This problem can be avoided in heated greenhouses but not when temperatures are too cool.

Plant protection

Aphids

Aphids suck the sap from leaves of growing plants and can be most serious insect pest of carnation. They also transmit carnation ring spot and carnation mosaic virus diseases. They are controlled by spraying of metasystox (2 ml/l), malathion (2 ml/l), endosulfan (2 ml/l) or rogor (2 ml/l).

Thrips

They suck sap from leaves causing distortion. Spraying of rogor (1 ml/l) or sumithion (3.5 ml/l) or malathion (2 ml/l) controls thrips.

Red spider mite

Spraying with Kelthane (2.5 ml/l) or Wettable sulphur 3 g/lit controls mite effectively.

Nematodes

Nematodes can be eliminated by growing plants in fumigated soil. Application of furadan, aldicarb or nemaphox controls nematode infestation.

7. ANTHURIUM

Anthurium, a tropical plant, is cultivated for its colourful spathe and attractive foliage, belonging to family Araceae. Since the mid 1980's, Anthurium's popularity as a flowering pot plant has increased dramaticaly and has become a popular addition to many foliage growers' product lines. Currently numerous cultivars with different flower sizes, shapes, colors and some with delicate fragrances are available for the consumer.

Anthuriums can be divided into four basic groups; *A. andreanum* cultivars, inter-specific hybrids between *A. andreanum* cultivars and dwarf species currently referred to as 'Andreacola' types, *A. scherzerianum* hybrids and foliage Anthuriums.

Anthurium andreanum, a generally large one, produces flower of white, pink, red, red-orange and green colours. 'Andreacola' cultivars are small to intermediate in overall size, fuller; more compact and generally produce smaller but more numerous flowers than andreanum cultivars. 'Andreacola' cultivars tend to have thicker, dark green leaves with white, pink, red and lavender colored flowers. *A. scherzerianum*, the first widely cultivated Anthurium pot plant, is a small, compact plant with primary flower colors of white (sometimes with polka-dots), pink and red.

Foliage Anthuriums come in numerous shapes and sizes and represent a minor portion of the total Anthurium pot market. However, it should be noted that most foliage Anthuriums are durable plants with distinct forms.

Growing Environment

For a luxuriant growth, the cultivated Anthurium requires 18°C during night whereas a higher day temperature of 24°C is essential for initiation of flowering. Anthuriums do not prefer temperatures above 35°C. If the temperature exceeds 35°C coupled with low relative humidity leaf scorching is common.

Anthuriums have a moderate to low light requirement and require 60 to 80% shade. To ensure the highest possible production during winter the shading should be removed at the onset of winter. In commercial practice it is advisable to have 50% shade, net on the top and a 25% shade net below it, so that the light levels at the plant growth can be modified depending upon the ambient light conditions. Anthurium can be shaded with saran or with UV stabilized agro shade nets for providing uniform shading. The shade nets can be operated manually by closely watching the external light intensity.

Anthurium thrives well in areas having high relative humidity levels, which are common in coastal areas of Kerala, Tamil Nadu, Andhra Pradesh, Maharashtra, high rainfall areas of Western, Eastern Ghats and North Eastern hilly regions. The optimum relative humidity levels range from 50 to 85% and at humidity levels less than 40%, the vegetative growth is slower and the flower development is poor. The relative humidity levels can be maintained by either overhead sprinkling/misting/fogging or ground level sprinkling. In areas where the water has lot of salt concentrations overhead sprinkling/fogging/ misting is not advisable.

Growing Structures

Depending upon the area of cultivation, the growing structures can be modified to suit the local climatic conditions. Commercial scale Anthurium cultivation can be taken up in low cost poly houses or shade houses.

Low Cost Poly Houses

In areas where the relative humidity levels are low and the temperature levels are high, low cost polyhouses are more suitable for commercial cultivation of Anthuriums. By creating a protected climate by cladding UV stabilized polythene, the relative humidity levels can be easily maintained and by providing top or side ventilation, the hot air can be easily expelled. Inside the polyhouse, the top portion can be provided with layers of two shade nets (50% and 25%). If the light intensity is very high a thin film of calcium chloride or lime can be sprayed on the external surface of the polyhouse to reduce the incoming radiation.

Shade Houses

In many of the commercial Anthurium growing areas, shade houses are the most popular structures because of the simplicity of construction and low cost . The shade houses can be constructed by taking the support of existing tree trunks if the Anthuriums are grown in multistoried plantation cropping. In open areas wooden, granite or bamboo pillars can be erected at regular intervals connected by GI wires as a mesh on top. Shade net can be covered on the top to provide the required shade (75%, 25%, shade nets). Provision can be made for installing overhead sprinkler/misters/fogging systems by running GI wires from one end to another end. Ground level misting/fogging is also possible by providing suitable foggers or misters at the plant level. The sides are also covered with a single layer of shade net. Such structures are more suitable in areas where the ambient relative humidity levels are already high and the frequent air exchange through the porous side nets do not result in drastic reduction in the relative humidity levels.

Growing Media

Anthurium requires a highly organic, well aerated medium with good water retention. However, good drainage in the medium is a must. An ideal medium for pots or ground potting should have good water holding capacity, high porosity, good aeration, low salt concentration (especially Na^{2+}, Cl^- and Ca^{2+} ions), must provide good anchorage, should have optimum pH (5.0) and EC (0.6 m mhos/cm^2) and should have good structure and texture. Various natural derivatives which satisfy the above conditions can be used to grow Anthuriums commercially. The range of media that are used include Sugarcane bagasse, Coffee leaf mould, spent ground coffee, coffee husk, cured coffee pulp, coffee parchment, rock wool, saw dust, tree bark, chicken manure, ground nut shells, peat, wood shavings, brick, gravel, rubble etc. In many Anthurium growing countries, coconut husk and cocopeat have become the most popular media for Anthurium cultivation.

India has a rich wealth of coconut plantations throughout southern India, the byproducts of which can be exploited for Anthurium cultivation. The coconut waste generated from the industry and in the trade can be collected and washed repeatedly to wash of the excess salts especially sodium chloride. The coconut waste is allowed to decompose by using microbial (*Pleurotis* sp.) inoculation to the stock of coconut husk/cococoir/coir dust.

A. CULTIVATION IN POTS

Anthurium especially the miniature type, *A. scherzerianum* and its hybrids have to be grown in pots to use them as potted plants. When young the plants can be planted in smaller pots (4") and as they grow they can be shifted to bigger pots (6"–10" dia. pots). Depending on the plant size, the pots have to be arranged in such a way that the foliage does not overlap with the leaves of adjacent pots.

B. CULTIVATION IN BEDS

When planting is done in soil, the plants have to be grown in 1.00 meter wide raised beds with gentle slope which facilitates good drainage. The beds have to be raised approximately 30 cm in height which get compressed due to frequent irrigation and other cultural practices. The beds are prepared by digging about 2 feet deep and filling the pits with the chosen growing medium (cocopeat, coffee husk etc). The dug out soil is thoroughly mixed with chosen growing medium and is used as top layer to create the raised beds. The mixing of soil with the medium which is otherwise loose and friable, prevents runoff of the medium.

Planting

The optimum spacing for commercial cultivation of Anthurium is 45 × 45 cm, which accommodates 5 plants/m^2.

Anthurium should be irrigated at least twice a day in summer months. In field grown plantations besides ground level irrigation, the plants can also be watered using overhead sprinklers. Irrigation can be accomplished by flooding beds or by installing microsprinklers at the plant level. Rainwater is considered to be the best quality water for anthurium cultivation. Water from underground source varies in its salt concentration hence; it needs to be tested before using for irrigation. By providing good drainage excessive salt accumulation can be overcome.

Nutrition

Anthuriums prefer smaller doses of fertilizers at frequent intervals rather than larger doses in longer intervals. Fertigation or root zone application of fertilizers is preferred. For pot cultivation it is advisable to apply 10 g of complex fertilizer (17:17:17 NPK) in 1000 ml of water (1%) once in a month. Anthuriums can also efficiently take up nutrients through foliage. A foliar spray of 30:10:10 NPK at 0.2% in combination with 100 ppm GA_3 produces higher number of quality flowers and highest number of suckers per plant. Foliar application of 0.1% urea at monthly intervals is found to be beneficial. Adequate levels of calcium and magnesium are also necessary for optimum yield. Deficiency of calcium often results in color break in the spathes. Application of lime or $CaNO_3$ @ 5 g per plant of at monthly intervals stabilizes the spathe color. Organic manure is also beneficial. Cow dung or neem cake mixed with 10-15 times with water and kept for 3-4 days. The slurry is then filtered and applied to the media.

Propagation

Anthurium can be very easily grown from seeds, but it is an extremely slow process. The berries (fruit of anthurium) pop out extremely slow process. The seed (some times two) is enveloped by juicy, mucilaginous pulp which depending needs to be removed completely before sowing. Depending upon the type of medium used the seeds will germinate in 10 to 40 days. The seedlings can be transferred to individual pots when they attain 3–4 leaf stages.

The highly condensed underground stem often produces shoots at the plant base. Such multiple shoots may be with or without roots and they can be separated from mother plant and planted in individual pots. The ability to produce suckers is variety dependent and suckering capacity can be improved by exogenous application of growth regulators like BAP (75 ppm) at monthly interval. Anthurium starts producing suckers once they attain an age of 12–16 months.

Pruning

Pruning helps to improve air circulation and subsequently the health of the plant. On an average only 3–5 leaves are to be retained on the plant. An excessive amount of leaves may cause the flower stem to curve. Therefore the leaves must be cut regularly. It is advisable to remove the small offshoots, maintaining only the large offshoots since they give a reasonable amount of flowers of a decent size.

Harvesting

Anthurium flowers are harvested when the spathes completely unfurl and the spadix is well developed. Development of true flowers on the spadix is also used as a criterion for harvesting the blooms. When one third of the flowers on the spadix mature, change of color can be observed that moves from base to tip of spadix. At that stage the flowers are harvested. Harvesting has to be done during cooler parts of the day i.e., early morning or late evening.

Grading

The flowers having any sort of blemishes, black spots, discoloured spathe, short stalk as well as deformed flowers should be removed before grading. The flowers should be spotlessly clean and shining.

Post harvest life

One important factor causing senescence is the increase in loss of water from flower parts coupled with the reduced rate of water uptake during storage. Various holding solutions, preservatives and growth regulators are used for the long-term storage of Anthurium cut flowers. BA at 25 ppm and 8 HQ at 30 ppm resulted in delayed spadix necrosis, spathe blueing and longest vase life (27 days) in Hawaiian Red. Reduced water loss from the flower by wax coating also increases post-harvest life up to 30%. The best wax of those tested was a carnauba base wax (FMC-819). This also imparts a sheen to the flowers which is an added advantage.

8. ORCHIDS

The time of picking varies from species to species. The flowers of *Cattleya* should be cut from 3 to 5 days after the bud dehisces or splits depending upon the weather conditions. The flower stem is then placed in water at 15° to 20°C. Orchid flowers are never kept in temperature below 15°C because transparent water spots may appear on the petal or flowers may wilt. They normally keep fresh for 10–12 weeks. The flowers of *Phalaenopsis* are cut when they are fully open either individually or as spray is cut above the bracts on the lone stems, a new flower shoot will develop. *Cymbidium* flowers can be cut individually but more often the entire spikes should be removed and individual flowers are cut at the time of grading or packing. For shipping the flower, stems are inserted through gum rubber caps into small tubes filled with water.

9. GLADIOLUS

It is grown in some parts of eastern, northern and western regions of India. Beauty Spot, Cherry Blossoms, American Beauty and Friendship are some of the varieties suitable as cut flower production. These flowers keep fresh for 7 to 9 days. When the first florets of the spikes open it should be cut immediately in the morning by a sharp knife and placed in water. The cutting of the spike for export market should be done at the stage when the first three buds of the spike are showing colour and are still firm. After detaching the spikes from the parent plant, they are tied in bundles of about 10 dozens, rolled tightly in a brown paper or a cardboard leaving about 30cm of the cut ends of the spikes uncovered. They are first graded according to their colour, size of spike and number of florets per spike. Then a bunch of 12 spikes are held together in 2 rubber bands one below the first florets and the other near the end of the stalk and then tied into bundles for consignment. For local market, when the first floret opens, the spike should be removed, leaving 4 to 6 leaves intact. These spikes will keep for 15 days and all florets open simultaneously.

10. GERBERAS

They last long and should be cut at the proper time, when the first outer row of staminate flowers release pollen. Premature cutting will cause wilting. They do not require refrigeration.

26

Flower Arrangement and Dry Decorations

FLOWER ARRANGEMENT AND DRY DECORATIONS

Flower arranging is one of the arts in horticulture. There are basically two styles in flower arrangements *viz.*, occidental and oriental styles. The occidental or western style or British style involves mainly massing of flowers and the beauty in this type of flower arrangements is mainly from massing. The oriental or eastern style involves mainly line arrangement with foliage and flowers. Historically speaking it involved religious symbolism and zen Buddhism as is practiced in Japan is the spirit behind this kind of art. The origin of this style is traceable to 9th or 10th century. This style is called as 'Ika-bana', a Japanese word. Irrespective of the styles each flower arrangement contains the following three components:

1. **Containers:** The containers or vases for flower arrangement differ in size, shape and material. It may be made up of porcelain, ceramic, brass, bell metals, bamboo, drift wood etc. Glass containers are not generally preferred as vases. The size of the containers varies from tall vases to flat saucers. Symmetrically or asymmetrically shaped containers are preferred depending upon the type of flower arrangement. The color of the containers should not be bright; as such containers will detract the beauty of the blossoms. Besides the above containers,

split bamboos or pieces of bamboo with two or three internodes having holes can be also employed as natural containers especially for line arrangement. Sometimes drift wood (pieces of any lignified woody stem) with a depression on its top portion may be also used as containers for line arrangement.

2. **Mechanism:** It refers to the physical mechanism by which flowers are held in position in containers. Pin-holders, clay balls or a crushed wire mesh are all commonly used in containers to hold the flowers in position. Deep containers generally do not require a mechanism. However, it is a must for flat and shallow containers. In some cases, the flowers have long and weak stalks. In such cases the mechanism alone will not suffice to keep the flowers in position where a thin G.I. wire can be pushed inside the stalks to retain in position. Mechanism of flower arrangement should be such that all the stalks of the cut flowers must originate from a point. The exposed stalk of the cut flowers as far as possible must be 1½ times the height of the containers.

3. **Flowers:** The term 'flowers' in flower arrangement include real flowers, foliage, dried twig, fruits (fresh or dry). Dry and fresh flowers may be combined or arranged separately. Color combination adopted in a flower arrangement is highly based on individual taste and preferences.

In massing or grouping flowers, care must be taken to have proper balance and proportion between the container and the flowers. When different flowers are combined, it is advisable to combine orange colored flowers with red flowers or blue flowers with violet coloured flowers. It is always attractive to combine all light colors or all bright colors in a vase. In line arrangement one can have opposing line, parallel line or straight line arrangements. Three flowers are usually kept in line arrangements at different heights to signify heaven, man and earth.

Shapes of flower arrangement: Flower arrangements are generally more pleasing to the eye when their outline creates a geometric pattern. The outline of the arrangement is the frame work from which the floral composition is constructed. The most commonly formed shapes are circular, triangular, radiating, crescent, horizontal, hogarithan curve (Fig. 26.1).

Prolonging the Vase Life of Flowers

How long the flowers kept in the containers remain fresh is referred to as vase life of flowers. This is dependent upon many factors like temperature, relative humidity, light, air velocity, composition of the holding water etc. The vase life can be extended by the following methods:

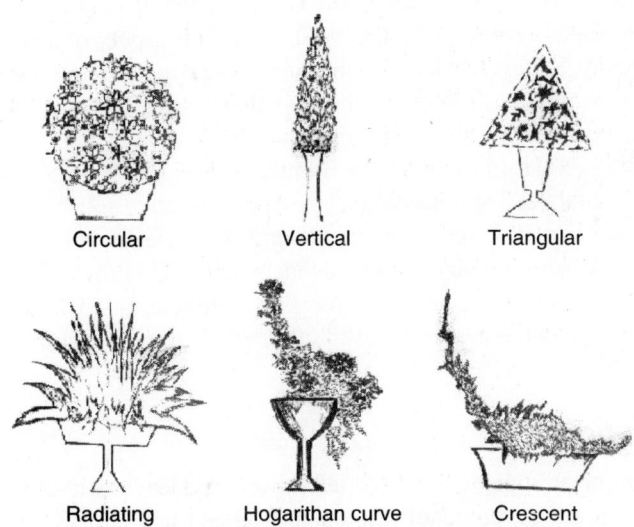

Circular   Vertical   Triangular

Radiating   Hogarithan curve   Crescent

Fig 26.1 Commonly formed shapes during flower arrangement

1. **Prevention of pollination**: Flowers begin to fade after fertilization. Therefore main purpose is to delay this natural process. This explains why double flowers keep longer than single flowers. To prevent pollination the easiest way.is to remove the stamens from the flowers when they open. By this method, tne vase life of Lily flowers could be extended by 10 days.

2. **Prevention of blocking of the conducting vessels of the flower stalks**: Air pockets sometimes block the conducting vessels, thereby causing the flowers to fade quickly. Blocking can be prevented by the following methods:

   (a) By dipping the stems of cut flowers in boiling water for 20–30 seconds (e.g. Rose, Iceland poppy, Dahlia, Zinnia, Chrysanthemum etc.).

   (b) By burning the stems of the cut flowers for 15 seconds (e.g. poppy).

   (c) By dipping in dilute acids like $N/500$ nitric acid.

   (d) Splitting the stems in woody flowers like Hydrangea, Rhododendron before placing them in cold water.

   (e) Sometimes blocking is also due to bacterial growth at the cut end. This can be prevented by any copper fungicide treatment or by dropping a copper coin into the vessel or treating with 0.1% boric acid.

3. **Addition of sugar or salt to the water:** Maintenance of improved water status seems to be the most important aspect of extension of longevity. Sugars when added to water in the container, are translocated and accumulated in flowers, increase their osmotic concentration and improve their ability to absorb more and more water to maintain turgidity. Salts like potassium nitrate, potassium chloride also delay the gradual decline in turgidity. Use of preservative solutions has been known for many years in lengthening the vase life of cut flowers. Preservatives normally include sugars, biocide, anti-ethylene compound and hydrated compound. Silver thio sulphate is an anti-ethylene compound in the preservative solution while quinoline compound is a biocide.

Dry Decoration

The art of making arrangements with dried flowers and leaves was a flourishing craft as early as the seventeenth century in England and America. But now it has passed from the hobby stage to big business in many countries. In recent years, there has developed a trend to fragrance a mixture of flowers for keeping in rooms, kitchens, toilets etc. This is known as 'Potpourri', wherein our living environment is made fragrant with natural items rather than the ozone depleting 'aerosols' as the case before.

When compared to the fresh flower this dried arrangement has many advantages.

1. It is not dependent on season or weather if once good materials are collected.
2. It lasts almost indefinitely.
3. It helps the individuals to improve their skill in design by affording time to study composition and to correct mistakes.
4. Dried arrangements are certainly time savers as they can be made by conveniently in advance and then enjoyed as needed.

Materials for Dried Arrangement

The materials usually employed for dried arrangement can be grouped into three categories as follows:

1. Cultivated flowers and vegetables including seed, pods, vines and grains.
2. Naturally available materials like flowers, seed pods, fern leaves, certain weeds and grasses.
3. Pods, cones, capsules, fruits, branches, leaves, berries and flowers of broadleaved evergreen trees and shrubs.

Preservation of Plant Materials:

There are a few methods of preserving flowers and foliage.

1. **Hanging upside-down:** Many flowers dry best by this method. It is the easiest method and is preferable to those materials which keep their shape and do not fall apart while drying. Materials intended for drying are tied and allowed to hang upside down from string without touching each other.

2. **Burying in sand or borax:** Fine and dry sand are poured to a depth of 10 cm into a box or pan. The flowers which should be dry are stripped of all foliage and may be kept upside down on the sand and then covered with a layer of sand over them. After two weeks time. the sand may be poured off carefully and the flowers may be taken out after gently wiping them free of sand with a soft brush. They are then ready for arrangement.

   Borax may be used in almost the same way as sand and is a good medium for preserving materials when space is limited. Alum and silica gel are also used in the place of borax. The form of flowers is also well preserved in borax method. Candytuft, daisies, marigold, narcissus, chrysanthemum, snapdragons, sunflowers, tithonia, all roses, eunonymes and coleus leaves dry well in sand or borax.

3. **Glycerin method:** This method is used for foliage only. The clean foliage materials are kept inside a jar containing a solution made up of 2/3 of water and 1/3 glycerin. The material has to be kept there till full absorption has taken place. It takes at least two weeks for most materials to do so. Materials so preserved keep well for 5 to 10 years.

4. **Pressing:** Leaves and flowers can also be preserved by pressing between several thickness of newspapers and some heavy object is kept over them. It takes about three weeks for complete drying.

Design for Dry Arrangement

The same basic principles of design apply both to dried as well as fresh arrangement. Here beauty is lacking as against in some fresh flowers, design is given much importance in dried arrangement. Before deciding on a design, one has to consider the composition to be a line or mass arrangement. In line arrangement, the main emphasis is a definite outline with restrained use of interesting forms of leaves, stems, buds or seed pods. On the other hand, in mass arrangement abundant use of materials is there to have a mass effect in colour and form.

The pattern most frequently used are the crescent, S-curve, triangle, fan, Round, Oval, Pyramid, L-Vertical: and horizontal. The other considerations *viz.*, balance harmony, scale, repetition, focal point, rhythm and unities are also important in dried arrangement. The focal point or center of interest, where all important lines meet, is the spot where the eyes come to rest. This is a natural place for the largest flowers, the key note of the whole composition. The focal point should be in proper scale to the rest of the composition. Hydrangeas, Celosia, Magnolia blossoms make good focal points.

Colour in Dry Arrangement

Flowers and foliage properly dried retain much of their colour and variety in colour may be brought out by choosing correct materials. The commonly used colour blends are monochromatic, analogous; complementary and triad. When one hue with its light and dark values or tones are brought in an arrangement, it is 'monochromatic'. When neighbouring and closely related colours, those, adjacent on the colour wheel say 'blue and green, yellow and orange are combined it is called 'analogous'. On the other hand, contrasted opposite on colour wheel, as green and red are combined, it is known as 'complementary'. Three colours equidistant on the colour wheel, as yellow, blue and red are when combined, it is a 'triad'. Containers made up of dull metals like copper, brass or earthen pots and wooden vessels are commonly preferred. Besides, bottles, jugs, wrought iron vessels, old lamps and tree caddies also make excellent containers.

Bases

Dry arrangements are made 'distinctive' by keeping them over a proper base. A commercial base of round, oval or rectangular blocks of any good wood is preferred. Bases give formality and dignity. A well-proportioned base is essential for any dried arrangement. The surfaces of the base blocks may be polished to make them shining. The kinds of mechanism to hold the materials in fresh arrangements may also be used for dried arrangement. The dust noticed on the arrangements may be cleaned gently with a small water colour brush or wiped with a slightly moistened cloth or brush.

27

Horticultural Shows

HORTICULTURAL SHOWS

Rapid progress has been made in horticulture in our country due to the work of some horticultural societies. These organizations conduct annual horticultural shows (often popularly referred to as flower shows) and garden competitions to stimulate interest of the public in gardening and encourage people to form and maintain gardens. These shows have the advantage of exhibiting at one place the best of a very large number or kinds of plants cultivated in an area: Besides, this serves as a place of proud display of new plant introductions and acquisitions. The Horticultural Societies at Chennai, Kolkata, Mumbai, Bangalore, Trivandrum and Ootacamund are more than a century old and still in active function. Similar societies have been formed in most of the district headquarters in recent years. The exhibitions and other activities of these societies are very popular with the garden enthusiasts. In a show one can exhibit fruits, vegetables, spices, fruit products, cut flowers, various types of flower arrangements, dry decorations, floral carpets, vegetable carving and house plants. Gardens owned by institutions, industrial establishments and house gardening are also adjudged for their excellence in layout and maintenance.

Merits of a Garden Entering in a Competition

For participation in a garden competition, preparation for atleast 2–3 years is necessary. A good garden should have a combination of annuals, perennials, shrubs, climbers, pot plants, besides a lawn, a compost pit and a working shed. Sowing of the annuals is adjusted in such a way that they are in bloom during the expected date of the show. 'Pinching off' and 'disbudding' are done to stimulate branching and production of large sized flowers. While judging the garden, general maintenance, proportion of various garden components and collection of different plants are also taken into account. A kitchen garden for competition should be at their best with a combination of different groups of vegetables *viz.*, roots, beans, pods, fruit vegetables, cucurbits, leaf vegetables together with herbs.

Exhibiting Fruits

It is important that one must satisfy all conditions strictly before exhibiting fruits. If six fruits of each variety are required, it means only six and not five or seven. In any show, the number of fruits to be kept is specified. For instance, six numbers are to be kept in fruits like mango, oranges, lime, lemon, guava, pomegranate sapota and Annonaceous fruits. Pine apple and papaya are kept in three in each kind. Another important factor is the maturity of the fruits. It is imperative that the fruit is properly ripe at the time of judging. The time of harvest is adjusted in such a way that fruits should be ripe at the time of the show. Harvesting and handling must be proper to avoid any damage to the fruits. The size of the stem to be retained is also important. In mango a 5 cm long 'button' should be retained while for citrus fruits length should be the minimum possible.

Good specimens free from blemishes and diseases, with proper maturity and uniform appearance and size are to be selected. Uniform fruits of medium size avoiding those which are too big or small are alone to be exhibited.

Exhibiting Vegetables

Vegetables should be uniform in size, colour, form, clean, tender and fresh without any blemishes. Tenderness is an important factor as while judging the quality the tender ones fetch additional scores.

Exhibiting Cut Flowers

General qualities which, are to be checked are uniformity, freshness, colour, size, form and stem length etc. Uniformity is very important as large flowers arranged with small flowers or *vice-versa* will affect the quality. The cut flowers must be fresh and bright in colour. In exhibiting a spike of flowers like Gladiolus, the first flower should be fresh and the maximum number of flowers should be open. A large stem is the general qualification but must be proportionate to the size of the vase. All precautions must be taken to extend the vase-life of these cut flowers.

Exhibiting Pot Plants

The plants should have been grown in the pot itself and a plant lifted recently from the ground and sown in a pot is liable for disqualification. Healthy plants alone should be exhibited. The pots may be given a 'terracota' colour wash. The foliage plants in the pot should be syringed with clean water once at least every other day and fed with liquid manure prepared from oil cake once in 20 days. While grouping these pot plants, they should harmonize with each other in colour and size and also with the back ground.

The flower arrangements based on Western and Japanese styles should be true to the design. As far as possible, principles governing the arrangements have to be strictly followed. For dry decoration, fresh materials are not used or if the situation compels, it is limited to a very small share.

Rangoli (floral carpets) with flowers, the art depicting a carpet design in the ground with the help of floral materials, is a time consuming one. Therefore, arrangement must be started early so that one can complete it in time. No coloured powder should be used. The flower colours selected for the arrangements must be appealing and catching the eye of the visitors. Vegetable carving, the art of developing some mimicking forms with the help of any kind of vegetables may be also included as an entry in some shows and for these work vegetables alone should be used. Food products intended for display should have been a home made one and accompanied by a written script on the recipe.

Part 5

OLERICULTURE

28

Olericulture

CLASSIFICATION OF VEGETABLES

There are more than fifty different kinds of vegetables. If the method of growing each is dealt in detail, it is likely to make things complicated and cause a lot of repetition. To eliminate this repetition, it is desirable to classify the vegetables into some groups. An arrangement of the kinds based on alphabets is though best for reference; it does not contribute to an understanding of relationship or similarity of cultural requirements. There are four general methods of classification namely (1) botanical classification (2) classification based on hardiness (3) classification based on parts used and (4) classification based on essential methods of culture.

1. Botanical Classification

Plant kingdoms are divided into four great groups and among them 'Spermatophyta' representing seed plants are again divided into two division's *viz.*, Gymnospermae and Angiospermae. No vegetable belongs to the former division and we are concerned only with the latter division namely Angiospermae. They are again divided into Monocotyledoneae and Dicotyledoneae. These are further divided into family, genus, species, sub species and botanical varieties. The groupings of the vegetables are as follows:

The botanical classification though brings out the relationship that exists between groups of different vegetables, in many cases, the purpose of the classification namely grouping of those vegetables requiring similar cultural requirements is not met. This does not give the final solution of avoiding repetition. The cultural operations of the vegetables belonging the same family are not always similar. Potato and tomato belong to the same family but their requirements are very different. The requirements of carrot are quite different from those of celery. Similarly most of the vegetables belonging to the family cucurbitaceae have similar requirements excepting chowchow which is perennial in habit demanding some special cultural requirements.

2. Classification Based on Hardiness

Based on their ability to withstand the frost injury, vegetables are classified as hardy and tender. The hardy group of vegetables will endure ordinary frost without injury while tender group will be killed. Aspagaraus is a typical example for hardy vegetables. The vegetables are often grouped as 'cool season' and 'warm season' crops according to their temperature requirement. The cool season vegetables or winter season vegetables often comprise of edible parts such as root, stem, leaves and buds or immature flowers. There are two exceptions to this rule which are warm season crops viz., sweet potato and Newzealand Spinach. Similarly warm season crops or summer season vegetables comprise of edible parts mainly of mature and immature fruits, exception being sweet potato and tapioca. Though this classification is simple, the purpose of classification is not fulfilled as all the cool season vegetables do not require similar type of cultivation. Peas and cabbage are cool season vegetables but their cultural requirements are entirely different. Besides with the improvement in the plant genetics, the distinction between cool season and summer season is slowly diminishing. This is true because cauliflower once considered as a cool season vegetable is grown throughout the year in the plains of certain parts of Tamil Nadu.

3. Classification Based on Parts used as Vegetable

(a) **Leaves or stems:** Cabbage, asparagus, amaranthus, knolkhol.
(b) **Fruits:** Melon, tomato, egg-plant, beans and peas.
(c) **Flower parts:** Cauliflower and broccoli
(d) **Underground:** Sweet potato, beet, radish, carrot, turnip, onion (roots, tubers, bulbs and corms) and garlic.

This method of classification is also not satisfactory because each group covers a number of crops differing widely in their cultural requirements.

4. Classification Based on Method of Culture

This system of classification is based on essential methods of culture and those vegetable crops which require similar cultural requirements are grouped together. This enables it possible to give general cultural practices for the group without the necessity of repetition in discussing individual crops. This is the most satisfactory method of classification and is being widely adopted. The grouping based on this method of classification is as follows:

Group 1: Perennial vegetables, Drumstick, Chowchow, Asparagus, Coccinea, Sesbania (Agathi) etc.

Group 2: Pot herbs or greens. Amaranthus, Spinach

Group 3: Salad crop, Celery, Lettuce, Endive, Chicory

Group 4: Cole crop, Cabbage, Cauliflower, Broccoli, Brussels sprout, Knolkhol, Chinese cabbage.

Group 5: Root crops, Beet root, Radish, Carrot, Turnip

Group 6: Bulb crops, Onion, Leek, Garlic, Shallot

Group 7: Potato

Group 8: Tuber crops, Tapioca, Sweet potato, Dioscorea, Elephant yam, Colocasia, Coleus

Group 9: Peas and Beans, Pea, French bean, Broad bean, Field bean, Cowpea, Cluster bean

Group 10: Solanaceous Vegetables, Tomato, Brinjal and Chillies,

Group 11: Cucurbits, Cucumber, Musk melon, Ash gourd, Pumpkin, Bitter gourd, Snake gourd, Ribbed gourd, Bottle gourd,

Group 12: Okra

TYPES OF VEGETABLE GROWING

Vegetable growing can be classified into the following types according to the purpose for which they have been developed.

1. Kitchen Garden

A home garden or kitchen garden refers to the raising of vegetable crops in the back yard of a house. But with the growing demand for the valuable land in urban areas the city dwellers may not have any land adjoining his apartment, yet the enthusiasts have interest in growing vegetables over the roof or in the verandah. Special arrangements are made on the roof to grow vegetables in soils placed on floors after using water proof cementing. Other vegetables are grown in boxes, pots and such other structures. All these refer only to the kitchen gardening. Growing vegetables in one's own garden is not only an art but also a continuous training for the person and the family. It supplies fresh vegetables. It constitutes a healthy hobby and the spare time of the family is well utilized. The fresh vegetables not only ensure a better balanced diet but there is also a delicate psychology behind the taste as every one will definitely appreciate the produce obtained by their own effort.

Selection of Site

There will be a limited choice for the selection of a site for the kitchen garden. It is usually the backyard of the house. This is itself convenient as the members of the family can give a constant care to the vegetables during the leisure time and the waste water from the bath rooms, kitchen etc, can easily be diverted to the vegetable beds. The size of a kitchen garden depends upon the availability of land and number of persons for whom vegetables are to be provided. No choice in the shape of the kitchen garden wherever possible rectangular garden is preferred to a square one. By close attention to succession cropping and intercropping, five cents of land may be adequate to supply vegetables for an average family of husband, wife and three children.

The main aim in a kitchen garden is the maximum output and a continuous supply of vegetables for the table throughout the year. By following certain principles in the layout of kitchen garden, the above objective can easily be fulfilled:

1. The perennial plants should be located on one side of the garden, usually on the rear end of the garden so that they may not shade other crops, compete for nutrition with the culture of other vegetable crops.
2. Adjacent to the foot path all around the garden and the central foot path may be utilized for growing different short duration green vegetables like coriander, Ceylon Spinach, fenugreek, alternanthera, mint and amaranthus. Each type of this green can be grown along each side of foot path and these crops can be rotated in different seasons.
3. The fence or trellises around the home garden may be utilized for growing light creepers like basella, coccinea, sponge gourd and bitter gourd. These may be also rotated in different seasons,
4. The compost pits are placed in two corners of the garden. They are meant for garden wastes and kitchen wastes. Pandals may be erected over the compost pits and trained with the creeper vegetables like lablab, ribbed gourd and snake gourd. This will hide off the compost pits from view.
5. Pandals may also be erected over the central foot bath, grapes varieties like Anab-e-Shahi or Black Prince may be trained over it.
6. Both the sides of the central foot path may be utilized to train tomato plants on single stemmed with the support of stakes.
7. The bunds separating the beds may be used for growing root crops or onion.
8. The conveniently divided small plots may be utilized to produce as much as possible by following a very intensive method of cultivation. This is possible by following continuous crop pattern in the form of succession and companion cropping.

Model Kitchen Garden and the Cropping Arrangements

Our dietician recommends under our condition to consume 300 g of vegetables daily by an adult and based on this, a kitchen garden should supply 1.5 kg of fresh vegetables to an average family size of two adults and three children. This quantity of fresh vegetables can be assured from a kitchen garden, laid out in an extent of 5 cents (200 sq. m) following the above principles (Fig. 28.1).

It is advisable to make a plan before undertaking the planting of a garden. The location of plots, crops to be grown, the probable date of planting, spacing between the plants, inter cropping and succession planting should be clearly indicated in the plan. This will guide the gardener to get constant supply of all kinds of vegetables avoiding a glut of anyone crop.

A cropping pattern which may prove helpful for kitchen garden under Tamil Nadu condition (excepting the hill station) is suggested below:

It may be observed from the above crop arrangements that throughout the year some crop is grown in each plot without a gap between any two crops (succession cropping) and wherever possible two crops (one long duration and the other a short duration one) are grown together in the same plot (companion cropping).

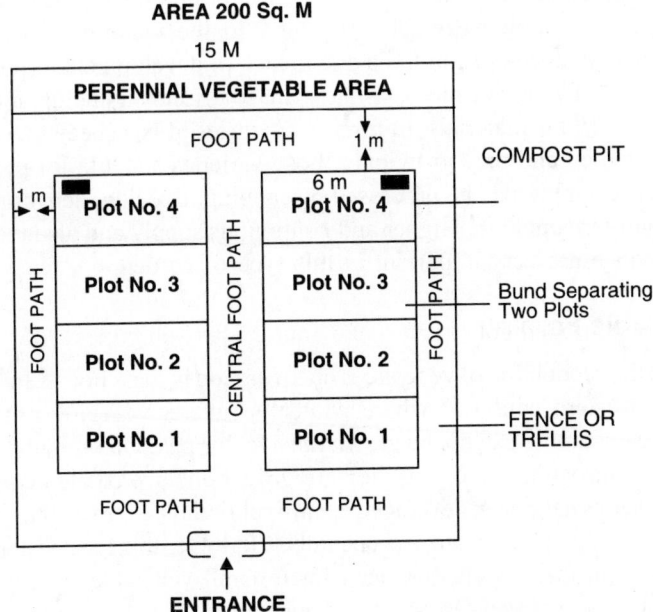

Figure 28.1 Layout of model kitchen garden.

2. Market Gardening

This type of vegetable garden is intended to supply vegetables for the local market. Therefore they are confined to the immediate vicinities of cities and towns. Such gardens are usually located within a radius of about 15 to 30 km from large towns and cities. Since these types of gardens are close to urban lands are generally costly. Hence intensive method of cultivation is followed. The gardener must be a versatile person as he is to grow a number of vegetables throughout the year to supply all kinds of vegetables to the market.

3. Truck Garden

This refers to producing some special crops in relatively larger quantities for distant market. The location of this type of gardens is determined by the soil and climatic factors suitable for raising those particular crops. Cultivation is more extensive. The commodities are usually marketed through middle-men. In Tamil Nadu, growing of cabbage and cauliflower in Nilgiris and tomatoes, in Periyakulam tract and sending them to distant places like Chennai are good examples of truck gardens.

4. Growing of Vegetables for Processing

This is a type of garden which comes up around processing factories. They are intended to supply vegetables regularly to the factories. This type of garden is not yet developed in India due to lack of development of processing industries. Few factories which exist now are solely depending upon the local market for their raw materials. In foreign countries, it is a specialized type of vegetable garden and they grow only those varieties suitable for processing only. The vegetables may be processed by canning, freezing, dehydration and pickling and fermentation. Higher and continuous supply and also the quality of the produce are very important in this type of garden.

5. Vegetable Forcing

It refers to the production of vegetables out of season in glass house. In England and other western countries where winter is severe, vegetables especially tomato and cucumber can not be grown outside during cooler months. As they are required throughout the year they are grown under protected conditions. Special varieties have been developed exclusively suitable for vegetable forcing in vegetable crops like tomato and cucumber. In India, this type of garden has little scope and there is also no need for it as all vegetables can be grown throughout the year in one part or other of our country. Yet certain big companies have recently ventured in this field in order to capitalize the off-season marketing of important vegetables like tomato, capsicum etc.

6. Vegetable Seed Industry

Growing of vegetables to produce pure, true-to-type seed of good quality is another specialized activity. Soil, climate and disease free conditions are factors influencing location of seed producing areas. A thorough knowledge of the crop, its growth habit, mode of pollination, proper isolation distances are all of prime importance in seed industry. The handling of the seed crop, curing, thrashing, cleaning, grading, packing and storage are the specialized techniques and a thorough knowledge on these aspects must be possessed by the grower. In India this is an expanding industry having vast scope. At present, certified seeds are produced for sale in government forms and authorized private nurseries. Seed law is in force enforcing standards, compulsory seed testing and certification in India. In and around Coimbatore and Madurai some leading vegetable seed producing gardens have come into existence recently.

CULTURAL ASPECTS OF VEGETABLES

This chapter deals with the general cultural aspects of vegetable crops.

Soil

A sandy loam or salty loam rich in organic matter with good drainage is the best suited type of soil for growing most of the vegetable crops. In sandy soils, early crop can be obtained but yield will be low. The tuber and root crops prefer a red loamy soil and in clayey and clay loam soil the roots and tubers may not develop properly. In peaty soils, cabbage, potato and carrot will grow luxuriously. Highly acidic or highly alkaline soils should be generally avoided.

Climate

Vegetable crops can be grouped as winter and summer vegetables according to their temperature requirement. Summer vegetables generally prefer a tropical climate. These crops cannot withstand frost. Successful production of these vegetables is favoured at a temperature range of 13° to 27°C. The plains of Tamil Nadu offer a good scope for the cultivation of most of the warm season vegetable groups throughout the year. However the crops like tomato and chillies do not set fruits during summer months in Tamil Nadu if the day temperature exceeds 35°C. The summer season vegetables can be grown from sea level upto an altitude of 1000 m in Tamil Nadu and above 1000 m, these crops can be grown during the summer months alone i.e. from March to June. The winter vegetables are generally grown in Tamil Nadu at an elevation above 1500 m where cool most climate prevails. These cool season vegetables can be raised in plains during the winter months from September to February. However, certain crops like peas, butter beans are so sensitive that they will not grow in plains at all. In recent years, certain varieties of cabbage, cauliflower and beetroot have been acclimatized to the level that one can grow them throughout the year in the plains of Tamil Nadu.

Certain vegetable crop like tomato, chilli, tapioca and sweet potato are grown as rainfed crops in some parts of Tamil Nadu and the success of these crops depend upon the timely receipt of the monsoon. On the other hand heavy rain during the fruiting season is harmful to tomato and in certain crops during seed maturation cause damage.

Season of Cropping

In general the warm season vegetables can be grown in two seasons *viz.,* June–July to October (Adipattam) and December–January to May–June (Thaipattam). In some parts of Tamil Nadu there is a third season *viz.,* March–April to June–July, however, the high temperature prevailing during these months pose poor fruit setting in crops like tomato and chilli. The vegetable crops like tapioca, bhendi, amaranthus and field lablab do not exact in their season, requirement as such they can be grown throughout the year. The winter season vegetables are generally grown in three seasons under South Indian hills as follows:

1. Autumn season: June–July to November–December
2. Summer season: April to August
3. Irrigated: January–February to June–July

The crop during the irrigated season is possible wherever facilities for irrigation are available in the hills and also the area is frost-free. The winter season vegetables can be grown in the plains between August-February.

Nursery

The nursery for vegetables is generally selected near a water source, partially shaded and not exposed to wind. The soil should be thoroughly prepared to make it fine tilth and free from any clods or stones. Equal parts of sand, red earth and well rotten farm yard manure or compost may be mixed with the soil and raised beds of 10 to 15 cm height and one metre width of any convenient length may be formed. The seed before sowing may be treated with fungicide *viz.,* Thiram or Captan or dry Cerason @ 2 g/kg of seed to prevent the seed borne diseases. The seeds are sown in lines spaced at 10 cm. Care must be taken to sow the seeds sparsely as thick sowing may enhance the incidence of a disease *viz.,* damping-off in the seedlings. In some parts of Tamil Nadu instead of line sowing the seeds are broadcast in the beds. However, the later method is not desirable for the seedlings.

After sowing, the seeds must be covered with fine sand and soil. The nursery must be watered twice a day in the initial stages until germination is completed, thereafter, once in a day till 15 to 20 days after sowing. It is desirable to harden the seedlings before transplanting by withholding watering or restricting the supply of water for the last two or three days. The seedlings will be ready for transplanting in about 25 to 40 days depending on the kind of the vegetable.

Plug Seedlings

Generally plastic trays or protrays having different sizes on cells are used for raising seedlings. Mainly two kinds of plastic protrays are used in raising the seedlings. One tray which may have cavities of 3.75 cm (1.5") in size, whereas the other tray may have cavities of 2.5 cm (1.0') in size. These trays help in proper germination, provide independent area for each seed to germinate, reduce the mortality rate, maintain uniform and healthy growth of seedlings, are easy in handling and storing, reliable and economical in transportation. These plastic trays may be fixed in thermocool base trays having the same number and size of cavities before filling the media. In case, thermocool base is not available, only trays placed on floor or firm base may be used.

Mostly artificial soil-less media is used for raising healthy and vigorous seedlings of vegetable in plastic protrays. Mainly three ingredients *viz.*, coco-peat, vermiculite and perlite are used as root medium for raising the nursery. These ingredients are mixed in 3:1:1 (V/V) ratio before filling in the required containers or plastic protrays.

Sowing of Seeds and Fertigation

Seeds selected for sowing in plastic protrays must be generally true to type and of good quality (high germination > 95 % and do not contain of other cultivars or weed seeds etc). Seeds should be purchased fresh for each growing season and should be typically sold in sealed packages that indicate the crop, variety, its germination percentage and status of seed treatment, if any. Seeds are usually planted/sown at a shallow depth after pressing the media with finger in a gentle way into the potting plugs or cells which are filled with artificial media. The actual depth of sowing depends on the crop and the size of the seeds. After sowing of seeds a thick layer of vermiculite is given to cover the seeds for better germination as this media is having water holding capacity. After seed sowing the pro-trays are kept in the germination room at the optimum required temperature for early and better germination. Usually one to two seed is sown per cell and after the seeds germinate during the seedling production process, the seedlings are thinned to one per cell the thinned can be gap filled in those cells where seed germination could not take place.

The size of the cavities depends upon the kind of crops to be sown in the nursery trays. For cucumber, muskmelon, tomato, brinjal, we use the trays having 187 cavities of 3.75 cm (1.5") size and for lettuce, cabbage, cauliflower and capsicum, we use the trays having 345 cavities of 2.5 cm (1.0") size are prepared. For raising good and healthy seedlings, the optimum Electrical Conductivity of irrigation water should be 1.6 to 1.7 with a pH of 6.6 to 6.7 and the optimum temperature should be 20°C during the winter. The optimum temperature for raising seedlings during summer season is 30°C. Water is applied after sowing of the seed in each tray through fine sprinkler boom uniformly in the trays. Hundred percent humidity is maintained and when level comes down irrigation is applied. Special treatment of growth regulators @ 140 ppm is applied just after germination (emergence of seedlings) of seeds in winter season and 70 ppm special treatment is applied for summer season nursery. Nutrients are applied in the form of N:P:K (1:1:1) @ 140 ppm once a week through the fine sprinkler to maintain the uniformity in application of nutrients. When temperature goes below 20°C in winter, application of special treatment is not recommended. For maintaining the optimum temperature we can use the germination room as per requirements of the crops.

Advantages

1. Seedlings can be raised under adverse climatic conditions where it is not possible under open field conditions.
2. Healthy seedlings can be raised in short period as compared to the time taken under open field nursery.
3. There is no chance of soil borne fungus or virus infection to the seedlings as the nursery is grown in soil-less sterilized media and insects can not enter under the protected conditions.
4. Drastic reduction in the mortality in transplanting of the seedlings as compared to the traditional system of nursery raising. No mortality, no transplanting shock and a quick establishment of the seedlings due to perfect development of root system.
5. Early planting is accomplished by raising such nursery.
6. It is suitable for raising the nursery of sexually and asexually propagated vegetable and ornamental crops.
7. Management of insect/pests and diseases under greenhouse/protected conditions is quite easy particularly the infection of viruses.

8. Looking to the very high cost of hybrid seed of vegetables we can reduce the seed rate to 30–40 per cent in comparison with the traditional nursery raising system, as individual seeds are sown in each cell which produces a very healthy seedling.
9. Easy for transportation alter packing for long distances.
10. Farmers can get the nursery ready from such nursery greenhouse any time as per the requirement.

Preparation of the field

The land must be ploughed 3–4 times to get fine tilth. Afterwards the ridges and furrows or beds can be formed at desired spacing depending upon the crop to be grown.

Planting

Some vegetable seeds can be directly sown in the main field and others have to be transplanted as seedlings after raising them in the nursery. The following are the vegetable crops which will not stand transplanting. Therefore, they have to be sown directly in the field.

1. Amaranthus 2. Fenugreek 3. All cucurbitaceous vegetables 4. Carrot 5. Radish 6. All beans group including peas and bhendi.

Transplanting

The seedlings are removed from the nursery and are immediately transplanted in the main field. Transplanting is preferably done in the evening hours on sunny days and throughout the day if weather is cloudy and drizzling.

Spacing

Spacing between rows and between plants within a row varies from crop to crop and even between variety to variety in a crop. Spreading type of varieties requires wider spacing while dwarf bushy types require closer spacing. Proper spacing has to be adopted for maximum growth and productivity. Hybrid varieties require wider spacing than ordinary varieties (e.g. tomato).

Irrigation

First irrigation is normally given just before transplanting in the case of transplanted vegetable crops and soon after sowing in direct sown crops. Third day after sowing or planting the second irrigation is given which is highly essential for germination of the seeds or establishment of the seedlings. Therefore, second irrigation is known as 'life irrigation'. Subsequent irrigation may be given as and when required depending upon the moisture conditions. Generally, irrigation is given at an interval of 7–10 days.

Manures and manuring

Farm yard manure or well rotten compost is recommended at the rate of 25 to 30 tonnes per hectare. They are incorporated in the field during the last ploughing while preparing the land.

To maximise the yield, fertilizers containing N, P and K are recommended for almost all the vegetables. Phosphorus and Potash and half the quantity of Nitrogen are applied as basal dose before sowing or transplanting. The remaining quantity of nitrogenous fertilizer is applied as top dressing at 25 to 45 days after sowing or transplanting. For vegetables like Brinjal and Bhendi foliar spraying of urea (1% concentration) is effective in increasing the fruit yield. Hybrid varieties require more quantity of N & K than the ordinary varieties (e.g. Tomato).

After cultivation

10–15 days after sowing or transplanting thinning of the excess seedlings should be done. In general two seedlings are retained per hill in most of the vegetable crops.

Weeding

First weeding and hoeing are normally done 10–15 days after transplanting and second weeding cum earthing up is taken up 25–30 days after sowing or transplanting. Subsequent weedings are attempted as and when necessary. Recently chemical weedicides have come into usage to control weeds. Pre-emergence application of pendementhalin (stomp) or Fluchloralin (Basilin) @ 1.0 litre/ha is recommended within 3 to 4 days of transplanting or sowing i.e. before the emergence of weeds to kill most of the grasses and broad leaved weeds. Post-emergence application of 2, 4-D (Fernoxone) @ 1 kg/ha is to be followed 20 to 25 days after the pre emergence application to kill the broad leaved weeds.

Harvesting

In most of the vegetables, harvesting time is a very important factor as it affects the edibility of the produce. Correct stage of maturity is assessed in each kind of vegetable and the produce is then harvested. The harvested materials are not kept in open as in certain cases like tomato it is susceptible to an injury known as sunscald. The cultural hints of some important vegetable grown in Tamil Nadu are given in Table 28.1.

Table 28.1 Cultural hints for important vegetables.

S. No.	Name of the Vegetables	Important varieties	Method of planting	Seed rate/ha	Spacing in cm	Season of sowing	Manures and manuring (Kg/ha)	Duration of the crops (in days)	Yield (t/ha)	Other particulars
1.	Tomato	Co.1, Co.2, Co.3, COTH2, PKM.1, Pusa Ruby	Sown in nursery and transplanted 25 days after sowing	375 g	60 × 60	June–Nov, Nov–Apr, Feb–May	N-150 P-100 K-50 Hybrid Varieties require 250:250:250	135–150	25–50	Harvest the half ripe fruits at 3 days interval
2.	Chillies	Co.1, Co.2, K-1, K-2, MDU-1, Pusa Jwala, PKM.1	Sown in nursery and transplanted 40-45 days after sowing	1.0 to 1.5 kg	30 × 30 or 45 × 30	June-July to Dec-Jan, Sep.–Oct. to Mar–Apr., Mar.–Apr. to Sep.–Oct.	N-70 P-35 K-35	210–240	Dry pods: 2–2.5 Green chillies 7.5-10.3	
3.	Brinjal	Co.1, MDU-1, PKM.1, Pusa Purple Long, Arka Navneet (F_1) Palur-1	Sown in nursery and transplanted 45 days after sowing	375-500 g	75 × 60	May-Oct., Dec.-May	N-100 P-75 K-50	150–165	20-25	–
4.	Bhendi	Co.1, Co.2, Co.Bh.H.1, MDU-1, Pusa Sawani, Punjab-Padmini	Seeds dibbled in rows	7.5 kg	60 × 30	July–Nov., March-April	N-40 P-50 K-30	90–110	7–10	Immature fruits are edible and should be harvested at 3 days interval

(Contd.)

Table 28.1 Cultural hints for important vegetables.

S. No.	Name of the Vegetables	Important varieties	Method of planting	Seed rate/ha	Spacing in cm	Season of sowing	Manures and manuring (Kg/ha)	Duration of the crops crops (in days)	Yield (tons/ha)	Other particulars
5.	Pumpkin	Co.1, Co-2,	Seeds dibbled in pits formed in long channels	1.0 kg	250-200	June-July to Oct. Jan-Feb. to May-June	N-12 P-24 K-24	135-180	20-25	Spray ethrel 250 ppm four times at weekly intervals commencing from 10-15 days after sowing
6.	Ash gourd	Co.1, Co.2	Seeds dibbled in pits formed in long channels	2.5 kg	300 × 300	June-July to Oct.-Jan.-Feb. to May-June	N-12 P-24 K-24	140-150	20	Spray ethrel 250 ppm
7.	Bitter gourd	Co.1, MDU.1	Seeds dibbled in pits formed in long channels	2.5 kg	250 × 200	June-July to Oct., Jan.-Feb. to May-June	N-20 P-30 K-60	140-150	14	Allowed to climb over a pandal, immature fruits are edible
8.	Snake gourd	Co.1, Co.2, PKM.1	Seeds dibbled in pits formed in long channels	1.5 kg	250 × 200	June-July to Oct., Jan.- Feb. to May-June	N-12 P-24 K-24	135-145	15-18	Allowed to climb over a pandal, immature fruits are edible
9.	Peas	Bonnevilla, Telephone, Gloriosa	Seeds sown in raised beds (Hills) and in ridges (plains)	80-100 kg	40 × 100 in hills, 10 in plains	Hills: Feb-Mar to May-June plains Oct.-Jan	N-180 P-125 K-100	80 to 100	5-7.5	In Tamil Nadu, peas grown in plain

(Contd.)

Table 28.1 Cultural hints for important vegetables.

S. No.	Name of the Vegetables	Important varieties	Method of planting	Seed rate/ha	Spacing in cm	Season of sowing	Manures and manuring (Kg/ha)	Duration of the crops (in days)	Yield (tons/ha)	Other particulars
10.	French beans	Watex, Premier, Contender, TKD.1	Sown in lines Also grown in the beds (hills), sown on the side of the ridges (plains)	Hills 125 kg, Plains 62.5 kg	Hills: 50 × 20, Plains: Oct-Jan.	Hills: Mar-May Plains: Oct-Jan.	N-50 P-75 K-75	90-100	4-5	Picking should be done at correct maturity
11.	Bellary onion	Bellary Red, N-53, Arka Niketan, Arka Pragati, Arka Kalyan	Sown in nursery and transplanted in the main field after 45 days	8-12 kg	30-45 × 8 × 10, transplanted on both the side of the ridges	May-June	N-75 P-150 K-75	145-150	15-18	Harvest the bulbs when 75% of the tops fall.
12.	Tapioca	Co.1, Co.2 H.2304 H.1687 Mulluvadi-1	Medium stem cuttings of 15 cm long with 8 to 10 nodes planted on ridges	12,000 cutting	Irrigated 75 × 75, Rainfed 60 × 60	Irrigated Throughout year Rainfed: Sept. – Oct.	N-60 P-60 K-120	8½ to 10 months	Irrigated 30-40 Rainfed -15-20	50% of N and K are applied as basal and remaining as to dressing 3 months after planting. Tubers can be lifted when cracking of soil at the root zone is seen

(Contd.)

Table 28.1 Cultural hints for important vegetables.

S. No.	Name of the Vegetables	Important varieties	Method of planting	Seed rate/ha	Spacing in cm	Season of sowing	Manures and manuring (Kg/ha)	Duration of the crops (in days)	Yield (tons/ha)	Other particulars
13.	Potato	Kufri Jothi, Kufri Muthu, Kufri Swarna, President	Healthy sprouted tubers of 40-50 g planted in furrows	3000 to 3750 kg	45 × 30	Summer: Mar-Apr. to July-Aug. Autumn: Aug.-Dec. Irrigated Feb.-June	N-100 P-125 K-50	100 to 140	20-25	Fresh tubers exhibit dormancy broken by treating carbon-disulphate @ 30 g/100 kg of seeds. Harvest the crop when the foliage turns yellow and starts withering.
14.	Carrot	Zino, Pusa Kesar, Half Long Danvers, Ooty-1	Seeds dibbled in ridges	3-4 kg	Hills: 25 × 5, Plains: 30 × 15	Hills: Throughout the year, Plains: Aug.–Sep.	N-180 P-125 K-100	Hills: 120 Plains: 100	25 to 30	Thinning is essential
15.	Cabbage	September, Golden-Acre. Pusa Drum Head, Hero	Sown in nursery and transplanted 30-35 days after sowing	375 g	Hills: 45 × 45, Plains: 60 × 45	Hills: Jan-Feb, April, June-July. Plains: Aug.-Sep.	N-180 P-125 K-100	105-200	30 to 40	Harvest when the head is firm but tender.

(Contd.)

Table 28.1 Cultural hints for important vegetables.

S. No.	Name of the Vegetables	Important varieties	Method of planting	Seed rate/ha	Spacing in cm	Season of sowing	Manures and manuring (Kg/ha)	Duration of the crops (in days)	Yield (tons/ha)	Other particulars
16.	Cauliflower	Snow Early-Synthetic, Serrana, Giant Ball	Sown in nursery and transplanted 40-45 days after sowing	400-500 g	Hills: 45 × 45 Plains: 60 × 45	Hills: Jan-Feb, April, June – July. Plains: Aug. – Sep.	N-180 P-125 K-100	90-105	20 to 30	The young developing inflorescence (curd) may be covered with the outer leaves to exclude light to obtain a pure white curd.
17.	Drum stick	Jaffna (perennial) K.M.1, PKM.1 (annual) PKM.2	Seedlings or limb cutting are planted in pits		300 to 500 cm on either side for perennial and 200 cm on either side for annual	June-July	Ammonium sulphate 40 gm/tree	Start bearing 6 months after planting and may be retained for 10 to 15 years	600 to 700 fruits per tree	Annual moringa can be pruned every year to a height of a 60 cm from ground level.

Index